Instrument Procedures Handbook

U.S. Department of Transportation
FEDERAL AVIATION ADMINISTRATION

Skyhorse Publishing

First published in 2015
First Skyhorse Publishing edition 2017

All rights to any and all materials in copyright owned by the publisher are strictly reserved by the publisher. All inquiries should be addressed to Skyhorse Publishing, 307 West 36th Street, 11th Floor, New York, NY 10018.

Skyhorse Publishing books may be purchased in bulk at special discounts for sales promotion, corporate gifts, fund-raising, or educational purposes. Special editions can also be created to specifications. For details, contact the Special Sales Department, Skyhorse Publishing, 307 West 36th Street, 11th Floor, New York, NY 10018 or info@skyhorsepublishing.com.

Visit our website at www.skyhorsepublishing.com.

10 9 8 7 6 5 4 3 2 1

Library of Congress Cataloging-in-Publication Data is available on file.

Cover design by Federal Aviation Administration

Print ISBN: 978-1-5107-2548-5
Ebook ISBN: 978-1-5107-2549-2

Printed in China

Preface

This handbook supersedes FAA-H-8261-16, Instrument Procedures Handbook, dated 2014. It is designed as a technical reference for all pilots who operate under instrument flight rules (IFR) in the National Airspace System (NAS). It expands and updates information contained in the FAA-H-8083-15B, Instrument Flying Handbook, and introduces advanced information for IFR operations. Instrument flight instructors, instrument pilots, and instrument students will also find this handbook a valuable resource since it is used as a reference for the Airline Transport Pilot and Instrument Knowledge Tests and for the Practical Test Standards. It also provides detailed coverage of instrument charts and procedures including IFR takeoff, departure, en route, arrival, approach, and landing. Safety information covering relevant subjects such as runway incursion, land and hold short operations, controlled flight into terrain, and human factors issues also are included.

This handbook conforms to pilot training and certification concepts established by the FAA. The discussion and explanations reflect the most commonly used instrument procedures. Occasionally, the word "must" or similar language is used where the desired action is deemed critical. The use of such language is not intended to add to, interpret, or relieve pilots of their responsibility imposed by Title 14 of the Code of Federal Regulations (14 CFR).

It is essential for persons using this handbook to also become familiar with and apply the pertinent parts of 14 CFR and the Aeronautical Information Manual (AIM). The CFR, AIM, this handbook, AC 00-2.15, Advisory Circular Checklist, which transmits the current status of FAA advisory circulars, and other FAA technical references are available via the internet at the FAA Home Page http://www.faa.gov. Information regarding the purchase of FAA subscription aeronautical navigation products, such as charts, Airport/Facility Directory, and other publications can be accessed at http://faacharts.faa.gov.

This handbook is available for download, in PDF format, from the FAA's Regulations and Policies website at:

http://www.faa.gov/regulations_policies/handbooks_manuals/aviation/instrument_procedures_handbook/

This handbook is published by the United States Department of Transportation, Federal Aviation Administration, Flight Technologies and Procedures Division, Flight Procedures Standards Branch, AFS-420, 6500 S. MacArthur Blvd, Ste 104, Oklahoma City, OK 73169.

Comments regarding this publication should be sent, in email form, to the following address: 9-AMC-AFS420-IPH@faa.gov

John Boyello

for John S. Duncan
Director, Flight Standards Service

09/21/2015

Acknowledgments

The following individuals and their organizations are gratefully acknowledged for their valuable contribution and commitment to the publication of this handbook:

FAA: Project Managers: Maj Brian Strack (USAF); Assistant Project Managers: Gilbert Baker, Dustin Davidson; Editor: Tara Savage; Technical Illustrator: Melissa Spears; Subject Matter Experts: Dean Alexander, John Bickerstaff, Barry Billmann, John Blair, John Bordy, Larry Buehler, Dan Burdette, Kel Christianson, Wes Combs, Jack Corman, Rick Dunham, Dave Eckles, Gary Harkness, Hooper Harris, Harry Hodges, John Holman, Bob Hlubin, Gerry Holtorf, Steve Jackson, Scott Jerdan, Alan Jones, Alex Krause, Norm Le Fevre, Bill McWhirter, Barry Miller, John Moore, T.J. Nichols, Jim Nixon, Dave Olsen, Jerel Pawley, Gary Petty, Gary Powell, Phil Prasse, Jeff Rawdon, Mark Reisweber, Christopher Rice, Dave Reuter, Jim Rose, Jim Seabright, Eric Secretan, Ralph Sexton, Tom Schneider, Lou Volchansky, Dan Wacker, and Mike Webb.

The Instrument Procedures Handbook was produced by the Federal Aviation Administration (FAA) with the assistance of Safety Research Corporation of America (SRCA).

The FAA also wishes to acknowledge the following contributors:

Aircraft Owners and Pilots Association (www.aopa.org) for images used in chapter 5.

Volpe National Transportation Systems Center (www.volpe.dot.gov) for images used in chapter 5.

Gary and Cecil Tweets at ASAP, Inc. (www.asapinc.net) for images used in chapter 1.

The staff of the U.S. Air Force Flight Standards Agency and Advanced Instrument School (HQ AFFSA/AIS) for various inputs.

Additional appreciation is extended to the Airman Certification System Working Group for its technical recommendations, support, and inputs.

Notice

The United States Government does not endorse products or manufacturers. Trade or manufacturers' names appear herein solely because they are considered essential to the objective of this handbook.

FAA-H-8083-16A, Instrument Procedures Handbook, 2015

Summary of Changes

This handbook supersedes FAA-H-8083-16, Instrument Procedures Handbook dated 2014, and contains substantial changes, updates, and reorganization. It must be thoroughly reviewed.

Chapter 1

- This Chapter contains updated information and reorganization of important concepts and principles related to obstacle avoidance and departure planning. The presentation retains the same logical order as earlier versions, and includes updated graphics for clarity.

- The section related to Surface Movement Guidance and Control System contains significant revisions to better reflect advancements in the way the system operates, as well as the Advisory Circulars published related to the subject.

- The section related to Diverse Vector Areas (DVAs) contains significant revision reflecting policy changes.

- Several subject matter areas and graphics discussed in this Chapter contain changes made in order to better align with updates and changes made to the Airman's Information Manual (AIM).

- Various editorial and graphics issues were addressed revised as appropriate.

Chapter 2

- This Chapter contains various updates to Sectors and Altitudes, as well as various editorial changes throughout.

- Several graphics were updated or changed as appropriate.

Chapter 3

- This Chapter was updated with various editorial and graphics changes as appropriate.

Chapter 4

- This Chapter contains a significant number of changes and updates specific to the subject of Approaches:

- Changed internet references related to on-line flight planning and filing.

- Updated verbiage and information regarding Vertical Descent Angles (VDAs) and Visual Descent Points (VDPs).

- Revised verbiage and illustration related to GLS approaches and associated minimums.

- Added discussion regarding alerting functions that are part of the Performance-Based Navigation concept and associated systems.

- Addressed changes to RNP approach naming convention issues.

- Hot and cold weather altimetry limitations and their associated FAA-directed procedure implementation changes were addressed and discussed.

- New information related to Terminal Arrival Areas (TAAs) was presented and discussed. This information now aligns with advances in the subject matter presented in the AIM.

- Several updates were made regarding RNAV and GPS-based approaches in general, under several sub-sections. Associated graphics and illustrations updated.

- Multiple changes were made to the section discussing ILS and parallel ILS approaches.

- Several editorial and graphics issues were addressed as appropriate. While significant information was updated for this version, there are multiple policy changes pending that will be further changed or discussed in subsequent versions of this Handbook. .

Chapter 5
- This Chapter contains several editorial and graphics updates as deemed appropriate (fixed browser links, etc).

Chapter 6
- This Chapter contains several editorial and graphics updates as deemed appropriate (fixed browser links, etc).

Chapter 7
- This Chapter remains "Helicopter Instrument Procedures" and contains updated illustrations and graphics pertinent to information discussed within the Chapter.

Appendices
- The Appendices in remain intact in this version. There are information and policy changes pending that, due to time constraints, will be addressed in subsequent versions of this Handbook

Table of Contents

Departure Procedures

Introduction

Thousands of instrument flight rules (IFR) takeoffs and departures occur daily in the National Airspace System (NAS). In order to accommodate this volume of IFR traffic, air traffic control (ATC) must rely on pilots to use charted airport sketches and diagrams, as well as departure procedures (DPs) that include both standard instrument departures (SIDs) and obstacle departure procedures (ODPs). While many charted (and uncharted) departures are based on radar vectors, the bulk of IFR departures in the NAS require pilots to navigate out of the terminal environment to the en route phase.

IFR takeoffs and departures are fast-paced phases of flight, and pilots often are overloaded with critical flight information. While preparing for takeoff, pilots are busy requesting and receiving clearances, preparing their aircraft for departure, and taxiing to the active runway. During IFR conditions, they are doing this with minimal visibility, and they may be without constant radio communication if flying out of a non-towered airport. Historically, takeoff minimums for commercial operations have been successively reduced through a combination of improved signage, runway markings and lighting aids, and concentrated pilot training and qualifications. Today at major terminals, some commercial operators with appropriate equipment, pilot qualifications, and approved Operations Specifications (OpSpecs) may takeoff with visibility as low as 300 feet runway visual range (RVR). One of the consequences of takeoffs with reduced visibility is that pilots are challenged in maintaining situational awareness during taxi operations.

Surface Movement Safety

One of the biggest safety concerns in aviation is the surface movement accident. As a direct result, the Federal Aviation Administration (FAA) has rapidly expanded the information available to pilots, including the addition of taxiway and runway information in FAA publications, particularly the IFR U.S. Terminal Procedures Publication (TPP) booklets and Airport/Facility Directory (A/FD) volumes. The FAA has also implemented new procedures and created educational and awareness programs for pilots, ATC, and ground operators. By focusing resources to attack this problem head on, the FAA hopes to reduce and eventually eliminate surface movement accidents.

Airport Sketches and Diagrams

Airport sketches and diagrams provide pilots of all levels with graphical depictions of the airport layout. National, Aeronautical Information Systems (AIS), formerly known as Aeronautical Products (AeroNav), provide an airport sketch on the lower left or right portion of every instrument approach chart. [Figure 1-1] This sketch depicts the runways, their length, width and slope, the touchdown zone elevation, the lighting system installed on the end of the runway, and taxiways. Graphical depictions of NOTAMS are also available for selected airports as well as for temporary flight restriction (TFRs) areas on the defense internet NOTAM service (DINS) website.

For select airports, typically those with heavy traffic or complex runway layouts, AIS also prints an airport diagram. The diagram is located in the IFR TPP booklet following the instrument approach chart for a particular airport. It is a full page depiction of the airport that includes the same features of the airport sketch plus additional details, such as taxiway

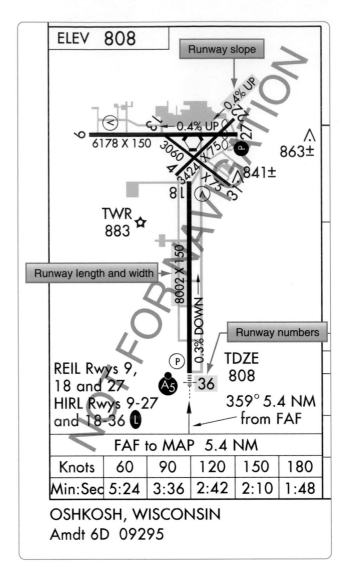

Figure 1-1. Airport diagram included on the Oshkosh, Wisconsin VOR RWY 9 Approach Chart as depicted in the IFR TPP.

identifiers, airport latitude and longitude, and building identification. The airport diagrams are also available in the A/FD and on the AIS website, located at www.aeronav.faa. gov. [Figure 1-2]

Airport Facility Directory (A/FD)

A/FD, published in regional booklets by AIS, provides textual information about all airports, both visual flight rules (VFR) and IFR. The A/FD includes runway length and width, runway surface, load bearing capacity, runway slope, runway declared distances, airport services, and hazards, such as birds and reduced visibility. [Figure 1-3] Sketches of airports also are being added to aid VFR pilots in surface movement activities. In support of the FAA Runway Incursion Program, full page airport diagrams and "Hot Spot" locations are included in the A/FD. These charts are the same as those published in the IFR TPP and are printed for airports with complex runway or taxiway layouts.

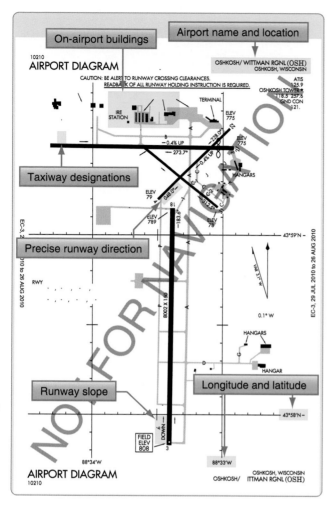

Figure 1-2. Airport diagram of Oshkosh, Wisconsin as depicted in the A/FD.

Surface Movement Guidance Control System (SMGCS)

The Surface Movement Guidance Control System (SMGCS) was developed to facilitate the safe movement of aircraft and vehicles at airports where scheduled air carriers were conducting authorized operations. Advisory Circular 120-57 was developed in 1992 and followed by AC 120-57A in 1996. In 2012, FAA Order 8000.94 was published to provide procedures for establishing Airport Low-Visibility Operations (LVO) and Surface Movement Guidance and Control Systems. It established the necessary FAA headquarters and operating services, roles, responsibilities, and activities for operations at 14 CFR Part 139 airports using RVRs of less than 1,200 feet for each runway. The Order applies to all users of the system at all levels who are formally listed. The FAA requires the commissioning of an "FAA approved LVO/SMGCS Operation" for all new Category III ILS supported runways. Since there are no regulatory takeoff minimums for 14 CFR Part 91 operations, the information provided by this AC and Order must

be understood so that the general aviation pilot can understand LVO and SMGCS during day or night.

The SMGCS low visibility taxi plan includes the enhancement of taxiway and runway signs, markings, and lighting, as well as the creation of SMGCS visual aid diagrams. [Figure 1-4] The plan also clearly identifies taxi routes and their supporting facilities and equipment. Airport enhancements that are part of the SMGCS program include, but are not limited to:

- Controllable Stop bars lights—these consist of a row of red, unidirectional, in-pavement lights that can be controlled by ATC. They provide interactions with and aircraft that prevent runway incursions during takeoff operations. These are required for operations at less than 500 ft RVR

- Non-Controllable Stop bars lights—these are red, unidirectinoal lights place at intersections where a restriction to movement is required. They must be in continuous operation at less than 500 ft RVR.

- Taxiway centerline lead-on lights—guide ground traffic under low visibility conditions and at night. These lights consist of alternating green/yellow in-pavement lights.

- Runway guard lights—either elevated or in-pavement, may be installed at all taxiways that provide access to an active runway. They consist of alternately flashing yellow lights. These lights are used to denote both the presence of an active runway and identify the location of a runway holding position marking.

- Geographic position markings—ATC verifies the position of aircraft and vehicles using geographic position markings. The markings can be used either as hold points or for position reporting. These checkpoints or "pink spots" are outlined with a black and white circle and designated with a number or a number and a letter.

- Clearance bar lights—three yellow in-pavement clearance bar lights used to denote holding positions for aircraft and vehicles. When used for hold points, they are co-located with geographic position markings.

Both flight and ground crews, Part 121 and 135 operators, are required to comply with SMGCS plans when implemented at their specific airport. All airport tenants are responsible for disseminating information to their employees and conducting training in low visibility operating procedures. Anyone operating in conjunction with the SMGCS plan must have a copy of the low visibility taxi route chart for their given airport as these charts outline the taxi routes

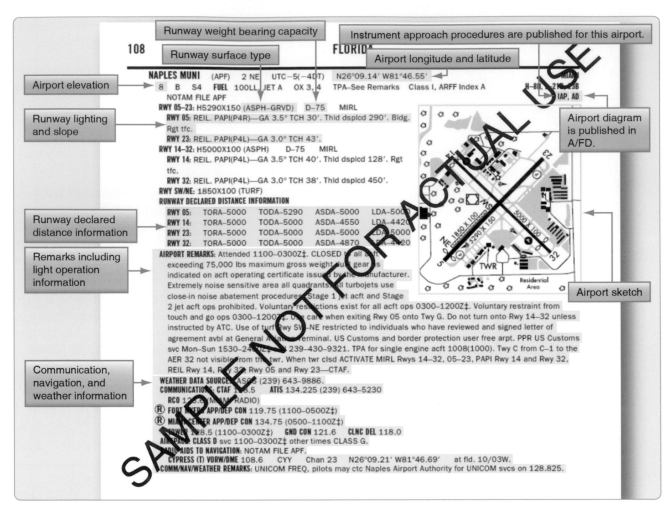

Figure 1-3. Excerpts from the Airport Facility Directory of Naples Muni, Naples, Florida.

and other detailed information concerning low visibility operations. These charts are available from private sources outside of the FAA. Government sources for SMGCS charts may be available in the future. Part 91 operators are expected to comply with the guidelines listed in AC 120-57, and should expect "Follow Me" service (when available) when low visibility operations are in use. Any SMGCS outage that would adversely affect operations at the airport is issued as a Notice to Airmen (NOTAM).

Advanced Surface Movement Guidance Control System (A-SMGCS)

With the increasing demand for airports to accommodate higher levels of aircraft movements, it is becoming more and more difficult for the existing infrastructure to safely handle greater capacities of traffic in all weather conditions. As a result, the FAA is implementing runway safety systems, such as Airport Surface Detection Equipment-Model X (ASDE-X) and Advanced Surface Movement Guidance and Control System (A-SMGCS) at various airports. The data that these systems use comes from surface movement radar and aircraft transponders. The combination of these data

sources allows the systems to determine the position and identification of aircraft on the airport movement area and decreases the potential of collisions on airport runways and taxiways.

Additional information concerning airport lighting, markings, and signs can be found in the Aeronautical Information Manual (AIM) and the Pilot's Handbook of Aeronautical Knowledge, Appendix 1, as well as on the FAA's website at http://www.faa.gov/airports/runway_safety/.

Airport Signs, Lighting, and Markings

Flight crews use airport lighting, markings, and signs to help maintain situational awareness. These visual aids provide information concerning the aircraft's location on the airport, the taxiway in use, and the runway entrance being used. Overlooking this information can lead to ground accidents that are entirely preventable. If you encounter unfamiliar markings or lighting, contact ATC for clarification and, if necessary, request progressive taxi instructions. Pilots are encouraged to notify the appropriate

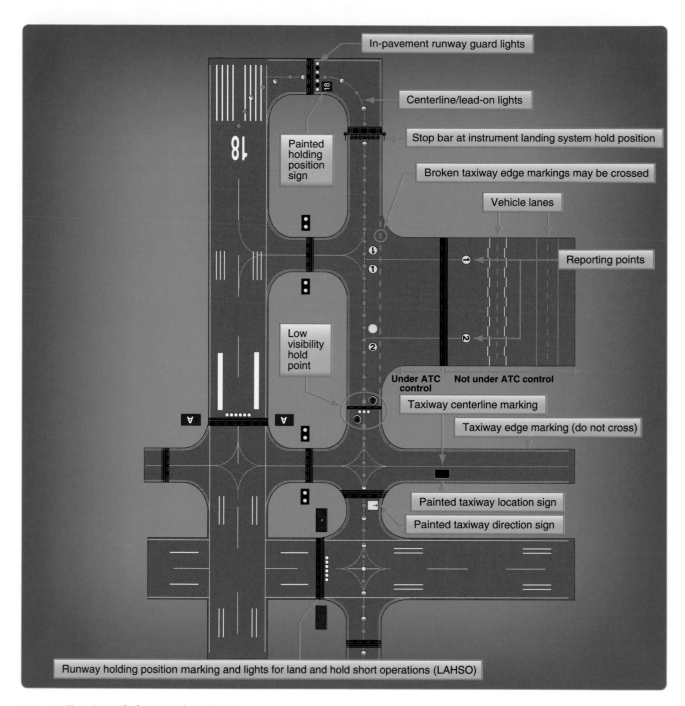

Figure 1-4. Key airport lighting and markings.

authorities of erroneous, misleading, or decaying signs or lighting that would contribute to the failure of safe ground operations.

Runway Incursions

On any given day, the NAS may handle almost 200,000 takeoffs and landings. Due to the complex nature of the airport environment and the intricacies of the network of people that make it operate efficiently, the FAA is constantly looking to maintain the high standard of safety that exists at airports today. Runway safety is one of its top priorities.

The FAA defines a runway incursion as:

> "Any occurrence at an aerodrome involving the incorrect presence of an aircraft, vehicle, or person on the protected area of a surface designated for the landing and takeoff of aircraft."

The four categories of runway incursions are listed below:

- Category A—a serious incident in which a collision was narrowly avoided.

- Category B—an incident in which separation decreases and there is a significant potential for collision that may result in a time critical corrective/evasive response to avoid a collision.

- Category C—an incident characterized by ample time and/or distance to avoid a collision.

- Category D—an incident that meets the definition of runway incursion, such as incorrect presence of a single vehicle/person/aircraft on the protected area of a surface designated for the landing and takeoff of aircraft but with no immediate safety consequences.

Figure 1-5 highlights several steps that reduce the chances of being involved in a runway incursion.

The FAA recommends that you:
• Receive and understand all NOTAMs, particularly those concerning airport construction and lighting.
• Read back, in full, all clearances involving holding short, line up and wait, and crossing runways to ensure proper understanding.
• Abide by the sterile cockpit rule.
• Develop operational procedures that minimize distractions during taxiing.
• Ask ATC for directions if you are lost or unsure of your position on the airport.
• Adhere to takeoff and runway crossing clearances in a timely manner.
• Position your aircraft so landing traffic can see you.
• Monitor radio communications to maintain a situational awareness of other aircraft.
• Remain on frequency until instructed to change.
• Make sure you know the reduced runway distances and whether or not you can comply before accepting a land and hold short clearance or clearance for shortened runway.
• Report confusing airport diagrams to the proper authorities.
• Use exterior taxi and landing lights when practical.

NOTE:
The sterile cockpit rule refers to a concept outlined in 14 CFR Part 121, sections 121.542 and 135.100 that requires flight crews to refrain from engaging in activities that could distract them from the performance of their duties during critical phases of flight.

Figure 1-5. FAA recommendations for reducing runway incursions.

In addition to the SMGCS program, the FAA has implemented additional programs to reduce runway incursions and other surface movement issues. They identified runway hotspots, designed standardized taxi routes, and instituted the Runway Safety Program.

Runway Hotspots

ICAO defines runway hotspots as a location on an aerodrome movement area with a history or potential risk of collision or runway incursion and where heightened attention by pilots and drivers is necessary. Hotspots alert pilots to complex or potentially confusing taxiway geometry that could make surface navigation challenging. Whatever the reason, pilots need to be aware that these hazardous intersections exist, and they should be increasingly vigilant when approaching and taxiing through these intersections. These hotspots are depicted on some airport charts as circled areas. [Figure 1-6] The FAA Office of Runway Safety has links to the FAA regions that maintain a complete list of airports with runway hotspots at http://www.faa.gov/airports/runway_safety/.

Standardized Taxi Routes

Standard taxi routes improve ground management at high-density airports, namely those that have airline service. At these airports, typical taxiway traffic patterns used to move aircraft between gate and runway are laid out and coded. The ATC specialist (ATCS) can reduce radio communication time and eliminate taxi instruction misinterpretation by simply clearing the pilot to taxi via a specific, named route. An example of this would be Los Angeles International Airport (KLAX), where North Route is used to transition to Runway 24L. [Figure 1-7] These routes are issued by ground control, and if unable to comply, pilots must advise ground control on initial contact. If for any reason the pilot becomes uncertain as to the correct taxi route, a request should be made for progressive taxi instructions. These step-by-step routing directions are also issued if the controller deems it necessary due to traffic, closed taxiways, airport construction, etc. It is the pilot's responsibility to know if a particular airport has preplanned taxi routes, to be familiar with them, and to have the taxi descriptions in their possession. Specific information about airports that use coded taxiway routes is included in the Notices to Airmen Publication (NTAP).

Taxi and Movement Operations Change

As of June 30, 2010, controllers are required to issue explicit instructions to cross or hold short of each runway that intersects a taxi route. Following is a summary of these procedural changes:

- "Taxi to" is no longer used when issuing taxi instructions to an assigned takeoff runway.

- Instructions to cross a runway are issued one at a time. Instructions to cross multiple runways are not issued. An aircraft or vehicle must have crossed the previous runway before another runway crossing

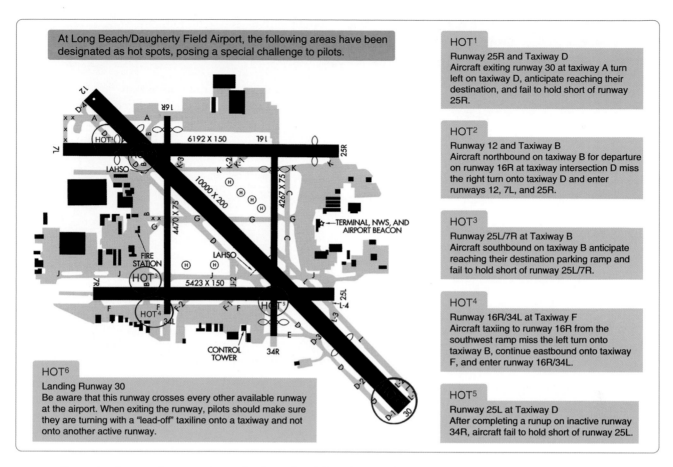

At Long Beach/Daugherty Field Airport, the following areas have been designated as hot spots, posing a special challenge to pilots.

HOT¹
Runway 25R and Taxiway D
Aircraft exiting runway 30 at taxiway A turn left on taxiway D, anticipate reaching their destination, and fail to hold short of runway 25R.

HOT²
Runway 12 and Taxiway B
Aircraft northbound on taxiway B for departure on runway 16R at taxiway intersection D miss the right turn onto taxiway D and enter runways 12, 7L, and 25R.

HOT³
Runway 25L/7R at Taxiway B
Aircraft southbound on taxiway B anticipate reaching their destination parking ramp and fail to hold short of runway 25L/7R.

HOT⁴
Runway 16R/34L at Taxiway F
Aircraft taxiing to runway 16R from the southwest ramp miss the left turn onto taxiway B, continue eastbound onto taxiway F, and enter runway 16R/34L.

HOT⁵
Runway 25L at Taxiway D
After completing a runup on inactive runway 34R, aircraft fail to hold short of runway 25L.

HOT⁶
Landing Runway 30
Be aware that this runway crosses every other available runway at the airport. When exiting the runway, pilots should make sure they are turning with a "lead-off" taxiline onto a taxiway and not onto another active runway.

Figure 1-6. Example of runway hot spots located at Long Beach/Daugherty Field Airport (KLGB).

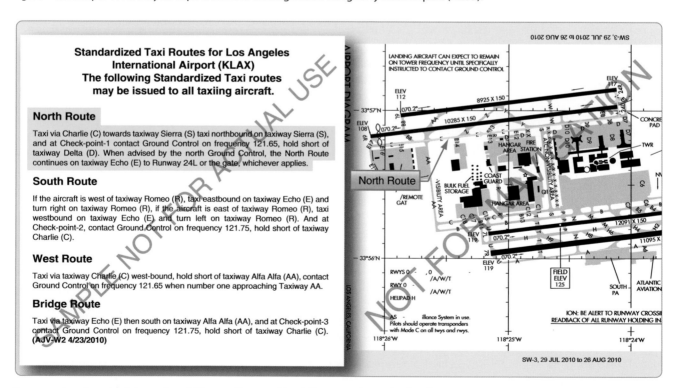

Standardized Taxi Routes for Los Angeles International Airport (KLAX)
The following Standardized Taxi routes may be issued to all taxiing aircraft.

North Route
Taxi via Charlie (C) towards taxiway Sierra (S) taxi northbound on taxiway Sierra (S), and at Check-point-1 contact Ground Control on frequency 121.65, hold short of taxiway Delta (D). When advised by the north Ground Control, the North Route continues on taxiway Echo (E) to Runway 24L or the gate, whichever applies.

South Route
If the aircraft is west of taxiway Romeo (R), taxi eastbound on taxiway Echo (E) and turn right on taxiway Romeo (R), if the aircraft is east of taxiway Romeo (R), taxi westbound on taxiway Echo (E) and turn left on taxiway Romeo (R). And at Check-point-2, contact Ground Control on frequency 121.75, hold short of taxiway Charlie (C).

West Route
Taxi via taxiway Charlie (C) west-bound, hold short of taxiway Alfa Alfa (AA), contact Ground Control on frequency 121.65 when number one approaching Taxiway AA.

Bridge Route
Taxi via taxiway Echo (E) then south on taxiway Alfa Alfa (AA), and at Check-point-3 contact Ground Control on frequency 121.75, hold short of taxiway Charlie (C). **(AJV-W2 4/23/2010)**

Figure 1-7. Los Angeles International Airport diagram, North Route, and standardized taxi route.

is issued. This applies to any runway, including inactive or closed runways.

- Never cross a runway hold marking without explicit ATC instructions. If in doubt, ask!

Reminder: You may not enter a runway unless you have been:

1. Instructed to cross or taxi onto that specific runway;

2. Cleared to take off from that runway; or

3. Instructed to line up and wait on that specific runway.

For more information on the change, refer to FAA Order JO 7110.65, which can be found at www.faa.gov.

Weather and the Departure Environment

Takeoff Minimums

While mechanical failure is potentially hazardous during any phase of flight, a failure during takeoff under instrument conditions is extremely critical. In the event of an emergency, a decision must be made to either return to the departure airport or fly directly to a takeoff alternate. If the departure weather were below the landing minimums for the departure airport, the flight would be unable to return for landing, leaving few options and little time to reach a takeoff alternate.

In the early years of air transportation, landing minimums for commercial operators were usually lower than takeoff minimums. Therefore, it was possible that minimums allowed pilots to land at an airport but not depart from that airport. Additionally, all takeoff minimums once included ceiling, as well as visibility requirements. Today, takeoff minimums are typically lower than published landing minimums, and ceiling requirements are only included if it is necessary to see and avoid obstacles in the departure area.

The FAA establishes takeoff minimums for every airport that has published Standard Instrument Approaches. These minimums are used by commercially operated aircraft, namely Part 121 and Part 135 operators. At airports where minimums are not established, these same carriers are required to use FAA designated standard minimums: 1 statute mile (SM) visibility for single- and twin-engine aircraft, and 1/2 SM for helicopters and aircraft with more than two engines.

Aircraft operating under 14 CFR Part 91 are not required to comply with established takeoff minimums. Legally, a zero/zero departure may be made, but it is never advisable. If commercial pilots who fly passengers on a daily basis must comply with takeoff minimums, then good judgment and common sense would tell all instrument pilots to follow the established minimums as well.

AIS charts list takeoff minimums only for the runways at airports that have other than standard minimums. These takeoff minimums are listed by airport in alphabetical order in the front of the TPP booklet. If an airport has non-standard takeoff minimums, a ▼(referred to by some as either the "triangle T" or "trouble T") is placed in the notes sections of the instrument procedure chart. In the front of the TPP booklet, takeoff minimums are listed before the obstacle departure procedure. Some departure procedures allow a departure with standard minimums provided specific aircraft performance requirements are met. [Figure 1-8]

Takeoff Minimums for Commercial Operators

While Part 121 and Part 135 operators are the primary users of takeoff minimums, they may be able to use alternative takeoff minimums based on their individual OpSpecs. Through these OpSpecs, operators are authorized to depart with lower-than-standard minimums provided they have the necessary equipment and crew training.

Operations Specifications (OpSpecs)

Within the air transportation industry, there is a need to establish and administer safety standards to accommodate many variables. These variables include a wide range of aircraft, varied operator capabilities, the various situations requiring different types of air transportation, and the continual, rapid changes in aviation technology. It is impractical to address these variables through the promulgation of safety regulations for each and every type of air transport situation and the varying degrees of operator capabilities. Also, it is impractical to address the rapidly changing aviation technology and environment through the regulatory process. Safety regulations would be extremely complex and unwieldy if all possible variations and situations were addressed by regulation. Instead, the safety standards established by regulation should usually have a broad application that allows varying acceptable methods of compliance. The OpSpecs provide an effective method for establishing safety standards that address a wide range of variables. In addition, OpSpecs can be adapted to a specific certificate holder or operator's class and size of aircraft and type and kinds of operations. OpSpecs can be tailored to suit an individual certificate holder or operator's needs.

Part 121 and Part 135 certificate holders have the ability, through the use of approved OpSpecs, to use lower-than-standard takeoff minimums. Depending on the equipment

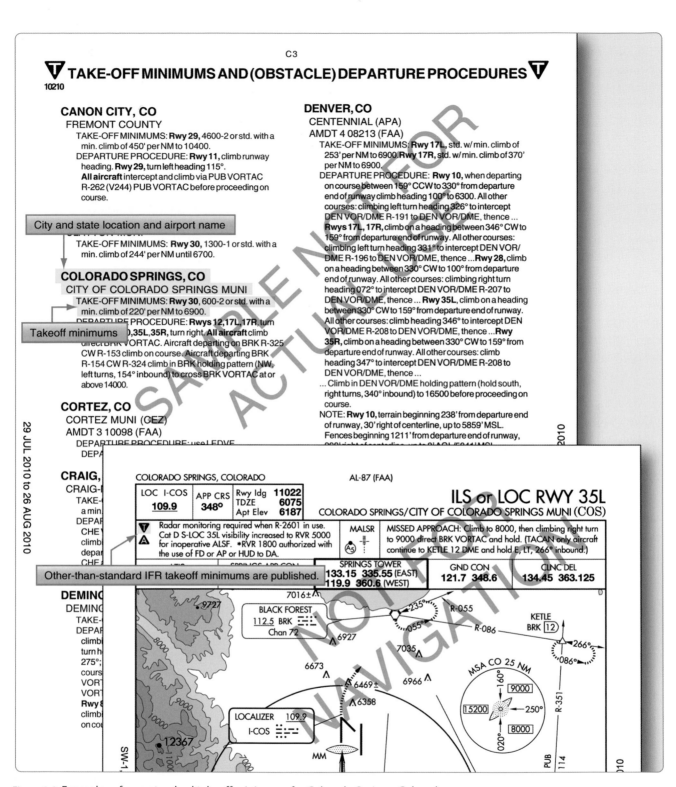

Figure 1-8. Examples of non-standard takeoff minimums for Colorado Springs, Colorado.

installed in a specific type of aircraft, the crew training, and the type of equipment installed at a particular airport, these operators can depart from appropriately equipped runways with as little as 300 feet RVR. Additionally, OpSpecs outline provisions for approach minimums, alternate airports, and weather services in Volume III of FAA Order 8900.1, Flight Standards Information Management System (FSIMS).

Ceiling and Visibility Requirements

All takeoffs and departures have visibility minimums (some may have minimum ceiling requirements) incorporated into the procedure. There are a number of methods to report visibility and a variety of ways to distribute these reports, including automated weather observations. Flight crews should always check the weather, including ceiling and visibility information, prior to departure. Never launch an IFR flight without obtaining current visibility information immediately prior to departure. Further, when ceiling and visibility minimums are specified for IFR departure, both are applicable.

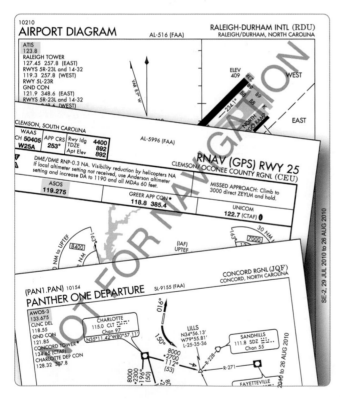

Figure 1-9. Examples of weather information of various flight information publications (FLIP).

Weather reporting stations for specific airports across the country can be located by reviewing the A/FD. Weather sources along with their respective phone numbers and frequencies are listed by airport. Frequencies for weather

sources, such as Automatic Terminal Information Service (ATIS), Digital Automatic Terminal Information Service (D-ATIS), Automated Weather Observing System (AWOS), Automated Surface Observing System (ASOS), and FAA Automated Flight Service Station (AFSS) are published on approach charts as well. [Figure 1-9]

Visibility

Visibility is the ability, as determined by atmospheric conditions and expressed in units of distance, to see and identify prominent unlighted objects by day and prominent lighted objects by night. Visibility is reported as statute miles, hundreds of feet, or meters.

Prevailing Visibility

Prevailing visibility is the greatest horizontal visibility equaled or exceeded throughout at least half the horizon circle, which need not necessarily be continuous. Prevailing visibility is reported in statute miles or fractions of miles.

Runway Visibility Value (RVV)

Runway visibility value is the visibility determined for a particular runway by a transmissometer. A meter provides continuous indication of the visibility (reported in statute miles or fractions of miles) for the runway. RVV is used in lieu of prevailing visibility in determining minimums for a particular runway.

Tower Visibility

Tower visibility is the prevailing visibility determined from the airport traffic control tower at locations that also report the surface visibility.

Runway Visual Range (RVR)

Runway visual range is an instrumentally derived value, based on standard calibrations, that represents the horizontal distance a pilot sees down the runway from the approach end. It is based on the sighting of either high intensity runway lights or on the visual contrast of other

Conversion	
RVR (feet)	Visibility (sm)
1,600	1/4
2,400	1/2
3,200	5/8
4,000	3/4
4,500	7/8
5,000	1
6,000	11/4

Figure 1-10. RVR conversion table.

targets, whichever yields the greater visual range. RVR, in contrast to prevailing or runway visibility, is based on what a pilot in a moving aircraft should see looking down the runway. RVR is horizontal visual range, not slant visual range. RVR is reported in hundreds of feet, so the values must be converted to SM if the visibility in SM is not reported. [Figure 1-10] It is based on the measurement of a transmissometer made near the touchdown point of the instrument runway and is reported in hundreds of feet. RVR is used in lieu of RVV and/or prevailing visibility in determining minimums for a particular runway.

Types of RVR

The following are types of RVR that may be used:

- Touchdown RVR—the RVR visibility readout values obtained from RVR equipment serving the runway touchdown zone.

- Mid-RVR—the RVR readout values obtained from RVR equipment located near the runway midpoint .

- Rollout RVR—the RVR readout values obtained from RVR equipment located nearest the rollout end of the runway.

- Far End RVR—when four RVR visibility sensors (VS) are installed, the far end RVR VS is the touchdown RVR VS on the reciprocal runway. The far end sensor will serve as additional information.

RVR is the primary visibility measurement used by Part 121 and Part 135 operators with specific visibility reports and controlling values outlined in their respective OpSpecs. Under their OpSpecs agreements, the operator must have specific, current RVR reports, if available, to proceed with an instrument departure. OpSpecs also outline which visibility report is controlling in various departure scenarios.

Adequate Visual Reference

Another set of lower-than-standard takeoff minimums is available to Part 121 and Part 135 operations as outlined in their respective OpSpecs document. When certain types of visibility reports are unavailable or specific equipment is out of service, the flight can still depart the airport if the pilot can maintain adequate visual reference. An appropriate visual aid must be available to ensure the takeoff surface can be continuously identified, and directional control can be maintained throughout the takeoff run. Appropriate visual aids include high intensity runway lights, runway centerline lights, runway centerline markings, or other runway lighting and markings. With adequate visual references and appropriate OpSpec approval, commercial operators may take off with a visibility of 1600 RVR or ¼ SM.

Figure 1-11. AWSS installation at Driggs-Reed, Idaho.

Ceilings

Ceiling is the height above the earth's surface of the lowest layer of clouds or obscuring phenomena that is reported as broken, overcast, or obscuration and not classified as thin or partial.

Automated Weather Systems

An automated weather system consists of any of the automated weather sensor platforms that collect weather data at airports and disseminate the weather information via radio and/or landline. The systems consist of the ASOS/Automated Weather Sensor System (AWSS) and the AWOS. These systems are installed and maintained at airports across the United States by both government (FAA and National Weather Service (NWS)) and private entities. They are relatively inexpensive to operate because they require no outside observer, and they provide invaluable weather information for airports without operating control towers. [Figure 1-11]

AWOS and ASOS/AWSS offer a wide variety of capabilities and progressively broader weather reports. Automated systems typically transmit weather every one to two minutes so the most up-to-date weather information is constantly broadcast. Basic AWOS includes only altimeter setting, wind speed, wind direction, temperature, and dew point information. More advanced systems, such as the ASOS/AWSS and AWOS-3, are able to provide additional information, such as wind speed, wind gust, wind direction, variable wind direction, temperature, dew point, altimeter setting, and density altitude. ASOS/AWSS stations providing service levels A or B also report RVR. The specific type of equipment found at a given facility is listed in the A/FD. [Figure 1-12]

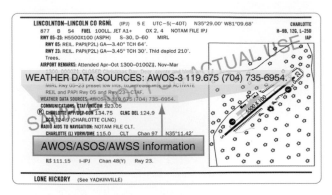

Figure 1-12. A/FD entry for an AWOS station.

The use of the aforementioned visibility reports and weather services are not limited for Part 91 operators. Part 121 and 135 operators are bound by their individual OpSpecs documents and are required to use weather reports that come from the NWS or other approved sources. While every operator's specifications are individually tailored, most operators are required to use ATIS, RVR reports, and selected reports from automated weather stations. All reports coming from an AWOS-3 station are usable for Part 121 and Part 135 operators. Each type of automated station has different levels of approval as outlined in individual OpSpecs. Ceiling and visibility reports given by the tower with the departure information are always considered official weather, and RVR reports are typically the controlling visibility reference.

Automatic Terminal Information Service (ATIS)

ATIS is another valuable tool for gaining weather information. ATIS is available at most airports that have an operating control tower, which means the reports on the ATIS frequency are only available during the regular hours of tower operation. At some airports that operate part-time towers, ASOS/AWSS information is broadcast over the ATIS frequency when the tower is closed. This service is available only at those airports that have both an ASOS/AWSS on the field and an ATIS-ASOS/AWSS interface switch installed in the tower.

Each ATIS report includes crucial information about runways and instrument approaches in use, specific outages, and current weather conditions including visibility. Visibility is reported in statute miles and may be omitted if the visibility is greater than five miles. ATIS weather information comes from a variety of sources

depending on the particular airport and the equipment installed there. The reported weather may come from a manual weather observer, weather instruments located in the tower, or from automated weather stations. This information, no matter the origin, must be from NWS approved weather sources for it to be used in the ATIS report.

Digital Automatic Terminal Information Service (D-ATIS)

The digital ATIS (D-ATIS) is an alternative method of receiving ATIS reports. The service provides text messages to aircraft, airlines, and other users outside the standard reception range of conventional ATIS via landline and data link communications to the flight deck. Aircraft equipped with data link services are capable of receiving ATIS information over their Aircraft Communications Addressing and Reporting System (ACARS) unit. This allows the pilots to read and print out the ATIS report inside the aircraft, thereby increasing report accuracy and decreasing pilot workload.

Also, the service provides a computer-synthesized voice message that can be transmitted to all aircraft within range of existing transmitters. The Terminal Data Link System (TDLS) D-ATIS application uses weather inputs from local automated weather sources or manually entered meteorological data together with preprogrammed menus to provide standard information to users. Airports with D-ATIS capability are listed in the A/FD.

It is important to remember that ATIS information is updated hourly and anytime a significant change in the weather occurs. As a result, the information is not the most current report available. Prior to departing the airport, you need to get the latest weather information from the tower. ASOS/ AWSS and AWOS also provide a source of current weather, but their information should not be substituted for weather reports from the tower.

IFR Alternate Requirements

On AIS charts, standard alternate minimums are not published. If the airport has other than standard alternate minimums, they are listed in the front of the approach chart booklet. The presence of a triangle with an **A** on the approach chart indicates the listing of alternate minimums should be consulted. Airports that do not qualify for use as an alternate airport are designated with an **A** N/A. [Figure 1-13]

The requirement for an alternate depends on the aircraft category, equipment installed, approach navigational aid (NAVAID), and forecast weather. For example, airports with only a global positioning system (GPS) approach procedure cannot be used as an alternate by

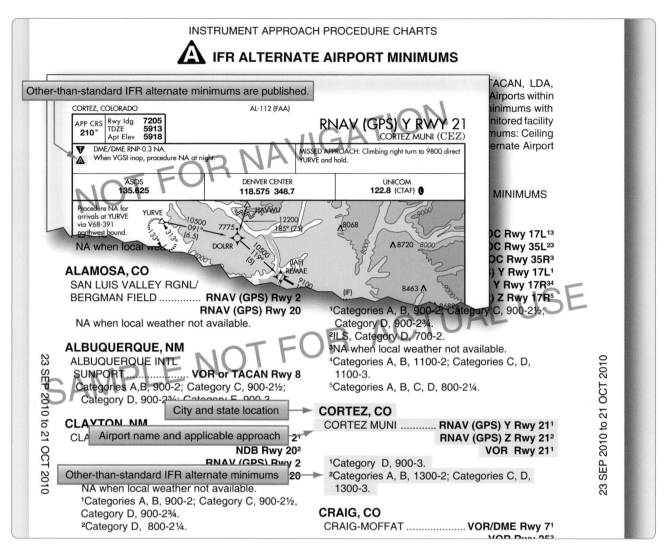

Figure 1-13. Examples of IFR alternate minimums.

TSO-C129 or C196 users unless certain requirements are met (see Aeronautical Information Manual) even though the "N/A" has been removed from the approach chart. For select area navigation (RNAV) GPS and GPS approach procedures, the "N/A" is being removed so they may be used as an alternate by aircraft equipped with an approach-approved Wide Area Augmentation System (WAAS) receiver complaying with (TSO-C145 or C146) or TSO-C129 or C196 meeitng certain requirements (see AIM). Because GPS is not authorized as a substitute means of navigation guidance when conducting a conventional approach at an alternate airport, if the approach procedure requires either distance measuring equipment (DME) or automatic direction finder (ADF), the aircraft must be equipped with the appropriate DME or ADF avionics in order to use the approach as an alternate.

For airplane 14 CFR Part 91 requirements, an alternate airport must be listed on IFR flight plans if the forecast weather at the destination airport, for at least 1 hour before and for 1 hour after the estimated time of arrival (ETA), the ceiling is less than 2,000 feet above the airport elevation, and the visibility is less than 3 SM. A simple way to remember the rules for determining the necessity of filing an alternate for airplanes is the "1, 2, 3 Rule." For helicopter 14 CFR Part 91, similar alternate filing requirements apply. An alternate must be listed on an IFR flight plan if at the ETA and for 1 hour after the ETA, the ceiling is at least 1,000 feet above the airport elevation, or at least 400 feet above the lowest applicable approach minima, whichever is higher, and the visibility is at least 2 SM.

Not all airports can be used as alternate airports. An airport may not be qualified for alternate use if the airport NAVAID is unmonitored, or if it does not have weather reporting capabilities. For an airport to be used as an alternate, the forecast weather at that airport must meet certain qualifications at the ETA. Standard airplane alternate minimums for a precision approach are a 600-foot ceiling and a 2 SM visibility. For a non-precision approach, the minimums are an 800-foot ceiling and a 2 SM visibility. Standard alternate minimums apply unless higher alternate minimums are listed for an airport. For helicopters, alternate weather minimums are a ceiling of 200 feet above the minimum for the approach to be flown, and visibility at least 1 SM but never less than the minimum visibility for the approach to be flown.

Alternate Minimums for Commercial Operators

IFR alternate minimums for Part 121 and Part 135 operators are very specific and have more stringent requirements than Part 91 operators.

Part 121 operators are required by their OpSpecs and 14 CFR Part 121, sections 121.617 and 121.625 to have a takeoff alternate airport for their departure airport in addition to their airport of intended landing if the weather at the departure airport is below the landing minimums in the certificate holder's OpSpecs for that airport. The alternate must be within 2 hours flying time for an aircraft with three or more engines with an engine out in normal cruise in still air. For two engine aircraft, the alternate must be within 1 hour. The airport of intended landing may be used in lieu of an alternate provided that it meets all the requirements. Domestic Part 121 operators must also file for alternate airports when the weather at their destination airport, from 1 hour before to 1 hour after their ETA, is forecast to be below a 2,000-foot ceiling and/or less than 3 miles visibility.

For airports with at least one operational navigational facility that provides a straight-in non-precision approach, a straight-in precision approach, or a circling maneuver from an instrument approach procedure determine the ceiling and visibility by:

- Adding 400 feet to the authorized CAT I height above airport (HAA)/height above touchdown elevation (HAT) for ceiling.

- Adding one mile to the authorized CAT I visibility for visibility minimums.

This is one example of the criteria required for Part 121 operators when calculating minimums. Part 135 operators are also subject to their own specific rules regarding the selection and use of alternate minimums as outlined in their OpSpecs and 14 CFR Part 135, sections 135.219 through 135.225, and are similar to those used by Part 121 operators.

Typically, dispatchers who plan flights for these operators are responsible for planning alternate airports. The dispatcher considers aircraft performance, aircraft equipment and its condition, and route of flight when choosing alternates. In the event changes need to be made to the flight plan en route due to deteriorating weather, the dispatcher maintains contact with the flight crew and reroutes their flight as necessary. Therefore, it is the pilot's responsibility to execute the flight as planned by the dispatcher; this is especially true for Part 121 pilots. To aid in the planning of alternates, dispatchers have a list of airports that are approved as alternates so they can quickly determine which airports should be used for a particular flight. Dispatchers also use flight planning software that plans routes including alternates for the flight. This type of software is tailored for individual operators and includes their normal flight paths and approved airports. Flight planning software and services are provided through private sources.

Though the pilot is the final authority for the flight and ultimately has full responsibility, the dispatcher is responsible for creating flight plans that are accurate and comply with the CFRs. Alternate minimum criteria are only used as planning tools to ensure the pilot in command and dispatcher are thinking ahead to the approach phase of flight. In the event the flight would actually need to divert to an alternate, the published approach minimums or lower-than-standard minimums must be used as addressed in OpSpecs documents.

Departure Procedures

Instrument departure procedures are preplanned IFR procedures that provide obstruction clearance from the terminal area to the appropriate en route structure. Primarily, these procedures are designed to provide obstacle protection for departing aircraft. There are two types of Departure Procedures (DPs): Obstacle Departure Procedures (ODPs) and Standard Instrument Departures (SIDs).

When an instrument approach is initially developed for an airport, the need for an ODP is assessed. If an aircraft may turn in any direction from a runway within the limits of the assessment area and remain clear of obstacles that runway passes what is called a diverse departure assessment, and no ODP is published. A diverse departure assessment ensures that a prescribed, expanding amount of required obstacle clearance (ROC) is achieved during the climb-out until the aircraft can obtain a minimum 1,000 feet ROC in non-mountainous areas or a minimum 2,000 feet ROC in mountainous areas. Unless specified otherwise, required obstacle clearance for all departures, including diverse, is based on the pilot crossing the departure end of the runway (DER) at least 35 feet above the DER elevation, climbing to 400 feet above the DER elevation before making the initial turn, and maintaining a minimum climb gradient of 200

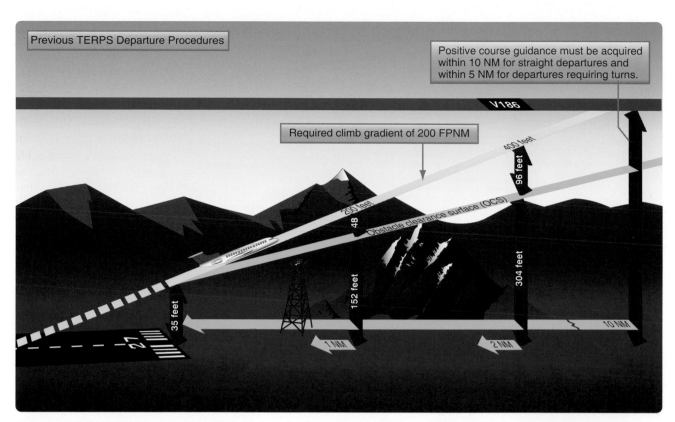

Figure 1-14. Previous TERPS departure procedures.

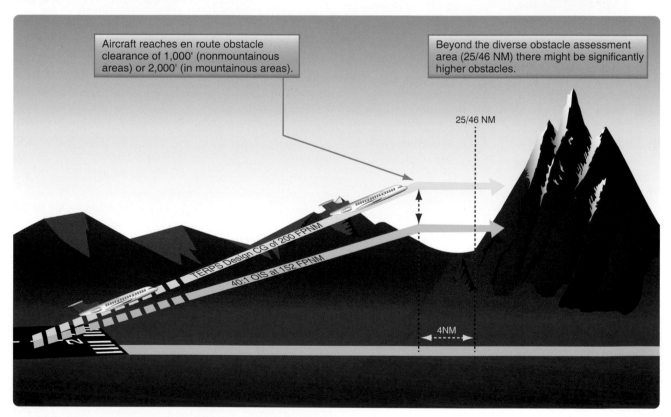

Figure 1-15. Diverse Departure Obstacle Assessment to 25/46 NM.

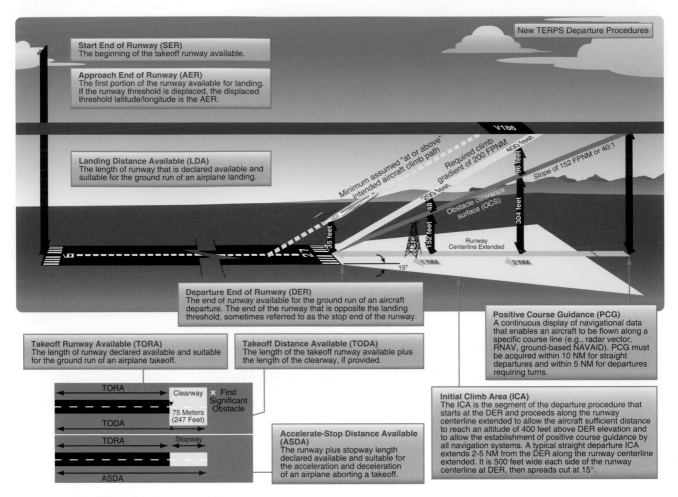

Figure 1-16. New TERPS departure procedures.

feet per nautical mile (FPNM), unless required to level off by a crossing restriction until the minimum IFR altitude is reached. Following ODP assessment, a SID may still be established for the purposes of ATC flow management, system enhancement, or noise abatement.

Design Criteria

The design of a departure procedure is based on terminal instrument procedures (TERPS), which is a living document that is updated frequently (FAA Order 8260.3). Departure design criterion begins with the assumption of an initial climb of 200 FPNM after crossing the DER at a height of at least 35 feet. [Figure 1-14] The aircraft climb path assumption provides a minimum of 35 feet of additional obstacle clearance above the required obstacle clearance (ROC), from the DER outward, to absorb variations ranging from the distance of the static source to the landing gear, to differences in establishing the minimum 200 FPNM climb gradient, etc. The ROC is the planned separation between the obstacle clearance surface (OCS) and the required climb gradient of 200 FPNM. The ROC value is zero at the DER elevation and increases along the departure route until

the ROC value appropriate for en route flight is achieved. The appropriate ROC value for en route operations is typically achieved about 25 NM for 1,000 feet of ROC in non-mountainous areas, and 46 NM for 2,000 feet of ROC in mountainous areas.

If taking off from a runway using a diverse departure (a runway without a published ODP), beyond these distances the pilot is responsible for obstacle clearance if not operating on a published route, and if below the MEA or MOCA of a published route, or below an ATC-assigned altitude. [Figure 1-15]

Recent changes in TERPS criteria make the OCS lower and more restrictive. [Figure 1-16] However, there are many departures today that were evaluated under the old criteria that allowed some obstacle surfaces to be as high as 35 feet at the DER. [Figure 1-14] Since there is no way for the pilot to determine whether the departure was evaluated using the previous or current criteria, and until all departures have been evaluated using the current criteria, pilots need to be very familiar with the departure environment and associated obstacles, especially if crossing the DER at less than 35 feet.

All departure procedures are initially assessed for obstacle clearance based on a 40:1 Obstacle Clearance Surface (OCS). If no obstacles penetrate this 40:1 OCS, the standard 200 FPNM climb gradient provides a minimum of 48 FPNM of clearance above objects that do not penetrate the slope. The departure design must also include the acquisition of positive course guidance (PCG), typically within 5 to 10 NM of the DER for straight departures. Even when aircraft performance greatly exceeds the minimum climb gradient, the published departure routing must always be flown.

Airports publish the declared distances in the A/FD. These include takeoff runway available (TORA), takeoff distance available (TODA), accelerate-stop distance available (ASDA), and landing distance available (LDA). These distances are calculated by adding to the full length of paved runway any applicable clearway or stop-way and subtracting from that sum the sections of the runway unsuitable for satisfying the required takeoff run, takeoff, accelerate/stop, or landing distance as shown in Figure 1-16.

Optimally, the 40 to 1 slope would work for every departure design; however, due to terrain and manmade obstacles, it is often necessary to use alternative requirements to accomplish a safe, obstacle-free departure design. In such cases, the design of the departure may incorporate a climb gradient greater than 200 FPNM, an increase in the standard takeoff minimums to allow the aircraft to "see and avoid" the obstacles, a standard climb of 200 FPNM with a specified reduced takeoff length, or a combination of these options and a specific departure route.

If a departure route is specified, it must be flown in conjunction with the other options.

The obstacle environment may require a climb gradient greater than 200 FPNM. In these cases, the ROC provided above obstacles is equivalent to 24 percent of the published climb gradient. The required climb gradient, for obstacle purposes on ODPs and SIDs, is obtained by using the formulas:

Standard Formula   DoD Option*

$$CG = \frac{O - E}{0.76\,D} \qquad\qquad CG = \frac{(48D + O) - E}{D}$$

O = obstacle mean sea level (MSL) elevation
E = climb gradient starting MSL elevation
D = distance (NM) from DER to the obstacle

Examples:

$$\frac{2049 - 1221}{0.76 \times 3.1} = 351.44 \qquad \frac{(48 \times 3.1 + 2049) - 1221}{3.1} = 315.10$$

Round to 352 FPNM   Round to 316 FPNM
*Military only

These formulas are published in FAAO 8260.3 (TERPS) for calculating the required climb gradient to clear obstacles.

The following formula is used for calculating SID climb gradients for other than obstacles (i.e., ATC requirements):

$$CG = \frac{A - E}{D}$$

A = "climb to" altitude
E = climb gradient starting MSL elevation
D = distance (NM) from the beginning of the climb

Example:

$$\frac{3000 - 1221}{5} = 355.8 \text{ round to } 356 \text{ FPNM}$$

NOTE: The climb gradient must be equal to or greater than the gradient required for obstacles along the route of flight.

The published climb gradient, obstacle or otherwise, is treated as a plane which must not be penetrated from above until reaching the stated height or has reached the en route environment (e.g., above the MEA, MOCA). Departure design, including climb gradients, does not take into consideration the performance of the aircraft; it only considers obstacle protection for all aircraft. TERPS criteria assume the aircraft is operating with all available engines and systems fully functioning. Development of contingency procedures, required to cover the case of an engine failure, engine out procedures (EOPs) or other emergency in flight that may occur after liftoff, is the responsibility of the operator. When a climb gradient is required for a specific departure, it is vital that pilots fully understand the performance of their aircraft and determine if it can comply with the required climb. The standard climb of 200 FPNM is not an issue for most aircraft. When an increased climb gradient is specified due to obstacle issues, it is important to calculate aircraft performance, particularly when flying out of airports at higher altitudes on warm days. To aid in the calculations, the front matter of every TPP booklet contains a rate of climb table that relates specific climb gradients and typical groundspeeds [Figure 1-17].

Low, Close-In Obstacles

Obstacles that are located within 1 NM of the DER and

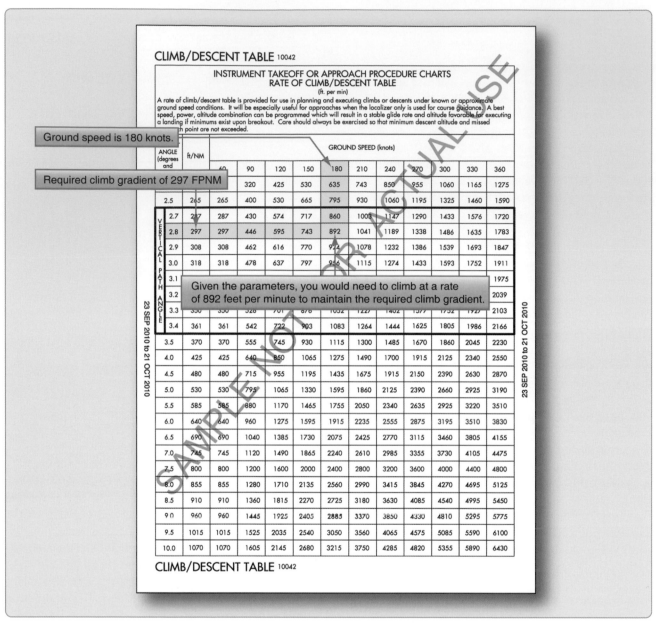

Figure 1-17. Rate of climb table.

penetrate the 40:1 OCS are referred to as "low, close-in obstacles" and are also included in the TPP. These obstacles are less than 200 feet above the DER elevation, within 1 NM of the runway end, and do not require increased takeoff minimums. The standard ROC to clear these obstacles would require a climb gradient greater than 200 FPNM for a very short distance, only until the aircraft was 200 feet above the DER. To eliminate publishing an excessive climb gradient, the obstacle above ground level (AGL)/MSL height and location relative to the DER is noted in the Takeoff Minimums and (Obstacle) Departure Procedures section of a given TPP booklet. The purpose of this note is to identify the obstacle and alert the pilot to the height and location of the obstacle so they can be avoided. This can be accomplished in a variety of ways:

- The pilot may be able to see the obstruction and maneuver around the obstacle(s) if necessary;

- Early liftoff/climb performance may allow the aircraft to cross well above the obstacle(s);

- If the obstacle(s) cannot be visually acquired during departure, preflight planning should take into account what turns or other maneuver(s) may be necessary immediately after takeoff to avoid the obstruction(s).

These obstacles are especially critical to aircraft that do not lift off until close to the DER or which climb at the minimum rate. [Figure 1-18]

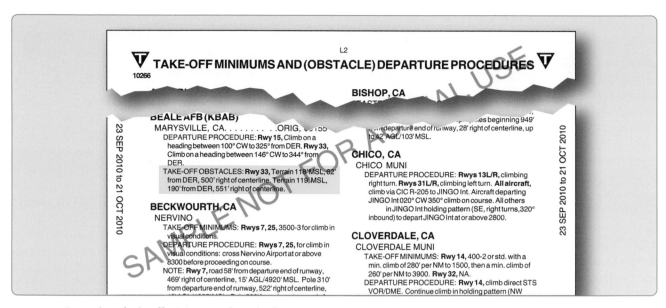

L2

▽ TAKE-OFF MINIMUMS AND (OBSTACLE) DEPARTURE PROCEDURES ▽

10266

23 SEP 2010 to 21 OCT 2010

BEALE AFB (KBAB)
MARYSVILLE, CA. . . . . ORIG, 09155
  DEPARTURE PROCEDURE: **Rwy 15,** Climb on a
  heading between 100° CW to 325° from DER. **Rwy 33,**
  Climb on a heading between 146° CW to 344° from
  DER.
  TAKE-OFF OBSTACLES: **Rwy 33,** Terrain 118' MSL, 62'
  from DER, 500' right of centerline. Terrain 119' MSL,
  190' from DER, 551' right of centerline.

BECKWOURTH, CA
NERVINO
  TAKE-OFF MINIMUMS: **Rwys 7, 25,** 3500-3 for climb in
  visual conditions.
  DEPARTURE PROCEDURE: **Rwys 7, 25,** for climb in
  visual conditions: cross Nervino Airport at or above
  8300 before proceeding on course.
  NOTE: **Rwy 7,** road 58' from departure end of runway,
  469' right of centerline, 15' AGL/4920' MSL. Pole 310'
  from departure end of runway, 522' right of centerline,

BISHOP, CA
EAST...
  ...m departure end of runway, 28' right of centerline, up
  to 42' AGL/103' MSL.

CHICO, CA
CHICO MUNI
  DEPARTURE PROCEDURE: **Rwys 13L/R,** climbing
  right turn. **Rwys 31L/R,** climbing left turn. **All aircraft,**
  climb via CIC R-205 to JINGO Int. Aircraft departing
  JINGO Int 020° CW 350° climb on course. All others
  in JINGO Int holding pattern (SE, right turns, 320°
  inbound) to depart JINGO Int at or above 2800.

CLOVERDALE, CA
CLOVERDALE MUNI
  TAKE-OFF MINIMUMS: **Rwy 14,** 400-2 or std. with a
  min. climb of 280' per NM to 1500, then a min. climb of
  260' per NM to 3900. **Rwy 32,** NA.
  DEPARTURE PROCEDURE: **Rwy 14,** climb direct STS
  VOR/DME. Continue climb in holding pattern (NW

23 SEP 2010 to 21 OCT 2010

Figure 1-18. Examples of takeoff minimums obstacle clearance.

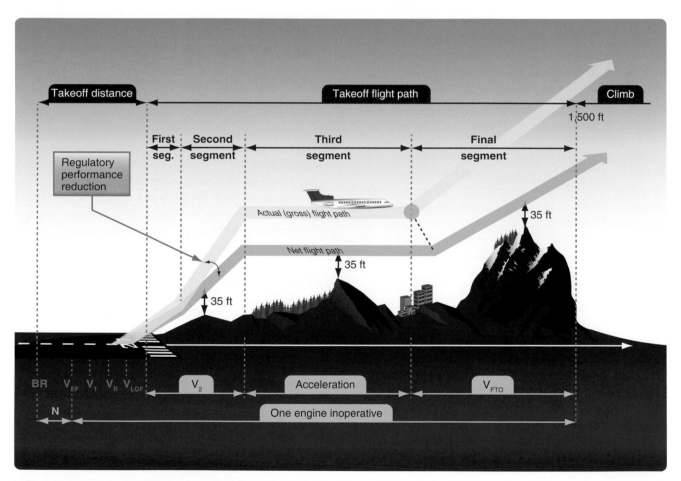

Figure 1-19. Part 25 turbine-powered, transport category airplane OEI actual (gross) takeoff flight path and OEI net takeoff flight path.

One-Engine-Inoperative (OEI) Takeoff Obstacle Clearance Requirements

Large and turbine-powered, multiengine transport category airplanes and commuter category airplanes operated under Part 121 or Part 135 have additional takeoff obstacle clearance requirements beyond the scope of the IFR departure procedure requirements addressed by TERPS.

Part 25 transport category and Part 23 commuter category airplane certification rules define the one-engine-inoperative (OEI) takeoff flight path, which is normally constructed from a series of segments beginning from 35 feet above the runway surface at the end of the OEI takeoff distance and ending at a minimum height of 1,500 feet above the runway elevation. However, the OEI net takeoff flight path assessment may continue above 1,500 feet if necessary to ensure obstacle clearance.

The actual, or gross, OEI flight path represents the vertical OEI climb profile that the airplane has been demonstrated capable of achieving using takeoff procedures developed for line operations based on the airplane's weight, configuration, and environmental conditions at the time of takeoff. The OEI net takeoff flight path represents the actual OEI takeoff flight path that has been degraded by an amount specified by the certification rules to provide a safety margin for expected variations under operational conditions. Subpart I of Part 121 and Part 135 require that the OEI net takeoff flight path be at least 35 feet above obstacles that are located within the prescribed lateral distance either side of the flight path The actual obstacle clearance capability, under optimum conditions after experiencing an engine failure on takeoff, is equal to the difference between gross and net flight path, plus the additional 35 feet. [Figure 1-19]

Advisory Circular (AC) 120-91, Airport Obstacle Analysis, provides guidance and acceptable criteria for use in determining the safe lateral clearance from obstacles, when developing takeoff and initial climb out airport obstacle analyses and engine out obstacle avoidance procedures to comply with the intent of these regulatory requirements. Pilots departing an airport under IFR and operating under Part 121 or 135 are required by 14 CFR 91.175(f)(4) to use an engine-inoperative takeoff obstacle clearance or avoidance procedure that assures compliance with the obstacle clearance requirements (subpart I) of those rules. The assessment of OEI takeoff obstacle clearance is separate and independent of the IFR departure procedure and associated all-engines- operating climb gradient requirements. While the Part 91 operating rules governing large and turbine-powered airplanes and commuter category airplanes do not require the use of an OEI takeoff obstacle clearance or avoidance procedure, such use is encouraged for Part 91 operators of these airplanes.

Unlike TERPS, which assesses obstacle clearance beginning at the DER, the OEI net takeoff flight path obstacle assessment begins at the point where the aircraft reaches 35 feet above the runway at the end of the OEI takeoff distance. Therefore, the OEI net takeoff flight path assessment may begin before the DER allowing for the use of a portion of the runway for the OEI climb. The OEI net takeoff flight path obstacle clearance assessment must also account for clearance of the low, close-in obstacles that are noted on the IFR departure procedure, but are not necessarily cleared when complying with the TERPS-based IFR climb gradient.

The OEI net takeoff flight path is unique for each airplane type and is assessed on each takeoff for the required obstacle clearance directly against those obstacles located beneath the OEI flight track and within the prescribed lateral distance from the flight path centerline. TERPS, on the other hand, provides a required climb gradient that represents a surface that the aircraft's all-engines-operating climb profile must remain above throughout the IFR climb until reaching the en route environment. These two methods of assessing obstacle clearance are necessarily quite different. TERPS is used by the procedure designer to determine a lateral path that is usable by a wide variety of aircraft types, and establishes a clearance plane that aircraft must be able to stay above to fly the procedure. A Part 25 transport category and Part 23 commuter category airplane's OEI takeoff flight path is established by or on behalf of the operator for a particular aircraft type and then limit weights are determined that assure clearance of any obstacles under that flight path (or within the prescribed lateral distance from the flight path centerline).

It may be necessary for pilots and operators of these categories of airplanes to use the services of an aircraft performance engineer or airport/runway analysis service provider as means of compliance with the requirements of Part 121 subpart I, or Part 135 subpart I concerning OEI net takeoff flight obstacle clearance and takeoff field length requirements. [Figure 1-20] Airport/runway analysis involves the complex, usually computerized, computations of aircraft performance, using extensive airport/obstacle databases and terrain information. This yields maximum allowable takeoff and landing weights for particular aircraft/engine configurations for a specific airport, runway, and range of temperatures. The computations also consider

Runway Conditions: Dry				Sample Aircraft Engine Type Flaps 0						KAPA/APA Denver - Centennial
Elevation = 5883 ft				Max Structural Takeoff Weight Limit = 28000						
				Runways - lbs						SEC.
OAT		N1	A/1							SEG.
F	C		ON	10	17L	17R	28	35L	35R	CLIMB
50	10	98.91	97.11	21880 R	29380 C	25250 R	21210 R	26690 R	29460 C	29000
52	11	98.73		21750 R	29340 C	25070 R	21090 R	26500 R	29460 C	29000
54	12	98.56		21620 R	29190 C	24900 R	20950 R	26310 R	29460 C	29000
55	13	98.47		21550 R	29100 C	24820 R	20890 R	26210 R	29460 C	29000
57	14	98.29		21420 R	28950 O	24650 R	20750 R	26020 R	29460 C	29000
59	15	98.11		21290 R	28790 O	24490 R	20620 R	25830 R	29460 C	29000
61	16	97.95		21160 R	28620 O	24320 R	20490 R	25650 R	29360 C	29000
63	17	97.80		21020 R	28450 O	24160 R	20360 R	25460 R	29210 C	29000
64	18	97.72		20960 R	28360 O	24080 R	20300 R	25370 R	29120 C	29000
66	19	97.56		20820 R	28170 O	23930 R	20170 R	25180 R	28870 R	29000
68	20	97.40		20670 R	27980 O	23770 R	20030 R	25000 R	28630 R	28930
70	21	97.20		20530 R	27720 O	23610 R	19880 R	24810 R	28380 R	28630
72	22	97.00		20380 R	27460 O	23420 R	19740 R	24630 R	28110 R	28330
73	23	96.90		20310 R	27370 R	23340 O	19670 R	24530 R	27970 R	28180
75	24	96.70		20170 R	27110 R	23170 O	19530 R	24350 R	27690 R	27870
77	25	96.50		20020 R	26860 R	22990 O	19390 R	24150 R	27390 R	27570
79	26	96.24		19880 R	26590 R	22810 O	19240 R	23960 R	27100 R	27230
81	27	95.97		19710 R	26330 R	22630 O	19070 R	23760 R	26800 R	26890
82	28	95.84		19630 R	26190 R	22540 O	18990 R	23660 R	26650 R	26720
84	29	95.58		19460 R	25920 R	22320 O	18820 R	23460 R	26350 R	26380
86	30	95.31		19290 R	25610 R	22120 O	18650 R	23260 R	26050 C	26040
88	31	95.04		19130 R	25300 R	21930 O	18490 R	23070 R	25740 C	25690
90	32	94.77		18980 R	24990 R	21730 O	0 R	22870 R	25440 C	25350
91	33	94.64		18900 R	24840 R	21630 O	0 R	22770 R	25290 C	25180
93	34	94.37		18740 R	24530 R	21440 O	0 R	22570 R	24990 C	24830
95	35	94.10		18590 R	24220 R	21240 O	0 R	22380 R	24690 C	24490
RUNWAY DIM		Length = ft		4800	10002	7000	4800	7000	10002	---
		Slope = %		−0.62	0.9	0.93	0.62	−0.93	−0.9	---
LVLOFF ALT			ft	7383	7383	7383	7383	7383	7383	---
WIND CORR		lbs/kt hw		67	5	31	62	42	4	---
		lbs/kt tw		N/A	−254	−227	N/A	−223	−131	---
QNH		lbs/.1"Hg > 29.92		75	10	85	0	88	5	0
		lbs/.1"Hg < 29.92		−87	−127	−99	−86	−113	−114	−123
ANTI-ICE			lbs	−1210	−1400	−1480	−1180	−1690	−830	−70
LIMIT CODES				R = RUNWAY LIMIT		O = OBSTACLE LIMIT		B = BRAKE LIMIT		C = CLIMB LIMIT

DATE: 07/01/2011

FOR SAMPLE USE ONLY

Figure 1-20. Airport/runway analysis example.

flap settings, various aircraft characteristics, runway conditions, obstacle clearance, and weather conditions. Obstacle data also is available from these service providers for operators who desire to perform their own analysis using the OEI climb performance and flight path data furnished in the Airplane Flight Manual or when using an aircraft electronic performance program supplied by the manufacturer or other service provider.

Airport/runway analysis is typically based on the assumption that the pilot will fly a straight-out departure following an engine failure on takeoff. However, when a straight-out departure is not practical or recommended, a special OEI turn procedure can be developed for each applicable runway. This OEI turn procedure may follow the path of a published IFR departure procedure or it may follow an independent path designed to avoid otherwise onerous obstacles and thereby maximize the allowable takeoff weight and payload. Graphic depiction of the OEI procedure is often available to give the pilot a pictorial

representation of the special OEI procedure. An engine failure during takeoff is a non-normal condition, and therefore the actions taken by the pilot including the use of an OEI turn procedure takes precedence over noise abatement, air traffic, SIDs, DPs, and other normal operating considerations.

It must be understood that the airport/runway analysis assesses obstacle clearance using the OEI net takeoff flight path data provided in the Airplane Flight Manual and the selected lateral obstacle assessment area. A takeoff weight limit provided on the analysis does not necessarily ensure compliance with the all-engines-operating climb gradient published on an IFR departure procedure even if the track of the OEI special procedure and the IFR departure procedure are identical.

Categories of Departure Procedures

There are two types of DPs: those developed to assist pilots in obstruction avoidance, known as ODPs, printed

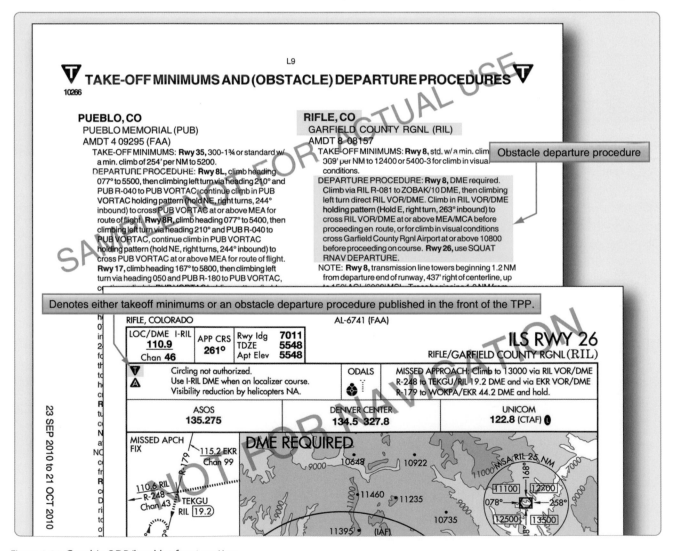

Figure 1-21. Graphic ODP/booklet front matter.

either textually or graphically, and those developed to communicate ATC clearances, SIDs, always printed graphically.

Obstacle Departure Procedures (ODPs)

The term ODP is used to define procedures that simply provide obstacle clearance. ODPs are only used for obstruction clearance and do not include ATC related climb requirements. In fact, the primary emphasis of ODP design is to use the least restrictive route of flight to the en route structure or to facilitate a climb to an altitude that allows random (diverse) IFR flight, while attempting to accommodate typical departure routes.

An ODP must be developed when obstructions penetrate the 40:1 departure OCS, as described in Order 8260.3. Only one ODP will be established for a particular runway. This is considered the default IFR departure procedure for a given runway and is intended for pilot awareness and use in the absence of ATC radar vectors or SID assignment. Text is not published to allow an option to use a SID or alternate maneuver assigned by ATC (e.g., "Climb heading 330 to 1200 before turning or use Manchester Departure" or "Turn right, climb direct ABC very high frequency (VHF) omnidirectional range (VOR) or as assigned by ATC."). ODPs are textual in nature. However, due to the complex nature of some procedures, a visual presentation may be necessary for clarification and understanding. If the ODP is charted graphically, the chart itself includes the word "Obstacle" in parentheses in the title. Additionally, all newly-developed RNAV ODPs are issued in graphical form.

All ODPs are listed in the front of the AIS approach chart booklets under the heading Takeoff Minimums and Obstacle Departure Procedures. Each procedure is listed in alphabetical order by city and state. The ODP listing in the front of the booklet includes a reference to the graphic chart located in the main body of the booklet if one exists. [Figure 1-21]

ODP Flight Planning Considerations

ODPs are not assigned by ATC unless absolutely necessary to achieve aircraft separation. It is the pilot's responsibility to determine if there is an ODP published for that airport. If a Part 91 pilot is not given a clearance containing an ODP, SID, or radar vectors and an ODP exists, compliance with such a procedure is the pilot's choice. A graphic ODP may also be filed in an instrument flight plan by using the computer code included in the procedure title. As a technique, the pilot may enter "will depart (airport) (runway) via textual ODP" in the remarks section of the flight plan. Providing this information to the controller clarifies the intentions of the pilot and helps prevent a potential pilot/controller

misunderstanding. If the ODP is not included in the pilot's clearance, the pilot should inform ATC when an ODP is used for departure from a runway so that ATC can ensure appropriate traffic separation.

During planning, pilots need to determine whether or not the departure airport has an ODP. Remember, an ODP can only be established at an airport that has instrument approach procedures (IAPs). An ODP may drastically affect the initial part of the flight plan. Pilots may have to depart at a higher than normal climb rate, or depart in a direction opposite the intended heading and maintain that for a period of time, any of which would require an alteration in the flight plan and initial headings. Considering the forecast weather, departure runways, and existing ODP, plan the flight route, climb performance, and fuel burn accordingly to compensate for the departure procedure.

Additionally, when close-in obstacles are noted in the Takeoff Minimums and (Obstacle) Departure Procedures section, it may require the pilot to take action to avoid these obstacles. Consideration must be given to decreased climb performance from an inoperative engine or to the amount of runway used for takeoff. Aircraft requiring a short takeoff roll on a long runway may have little concern. On the other hand, airplanes that use most of the available runway for takeoff may not have the standard ROC when climbing at the normal 200 FPNM.

Another factor to consider is the possibility of an engine failure during takeoff and departure. During the preflight planning, use the aircraft performance charts to determine if the aircraft can still maintain the required climb performance. For high performance aircraft, an engine failure may not impact the ability to maintain the prescribed climb gradients. Aircraft that are performance limited may have diminished capability and may be unable to maintain altitude, let alone complete a climb to altitude. Based on the performance expectations for the aircraft, construct an emergency plan of action that includes emergency checklists and the actions to take to ensure safety in this situation.

Standard Instrument Departures (SIDs)

A SID is an ATC-requested and developed departure route, typically used in busy terminal areas. It is designed at the request of ATC in order to increase capacity of terminal airspace, effectively control the flow of traffic with minimal communication, and reduce environmental impact through noise abatement procedures.

While obstacle protection is always considered in SID routing, the primary goal is to reduce ATC/pilot workload

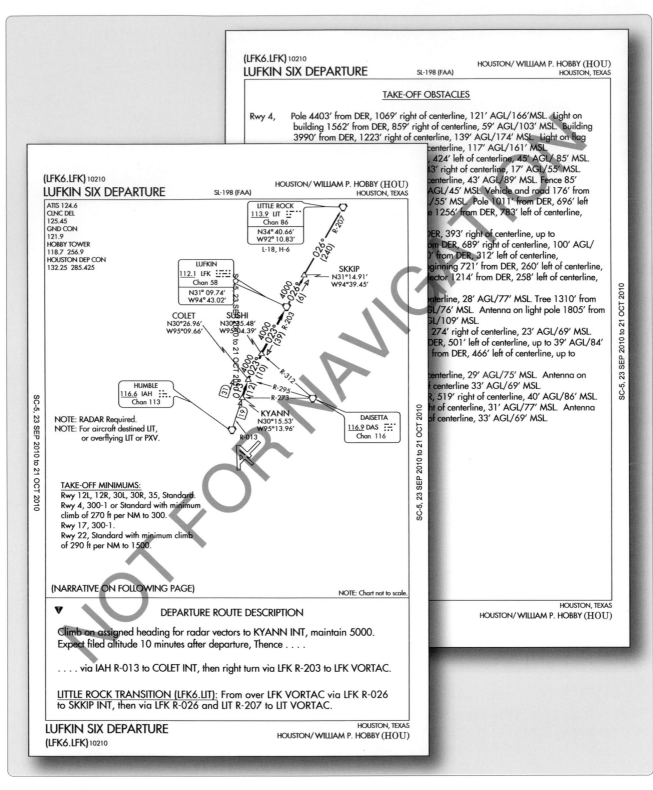

Figure 1-22. SID chart.

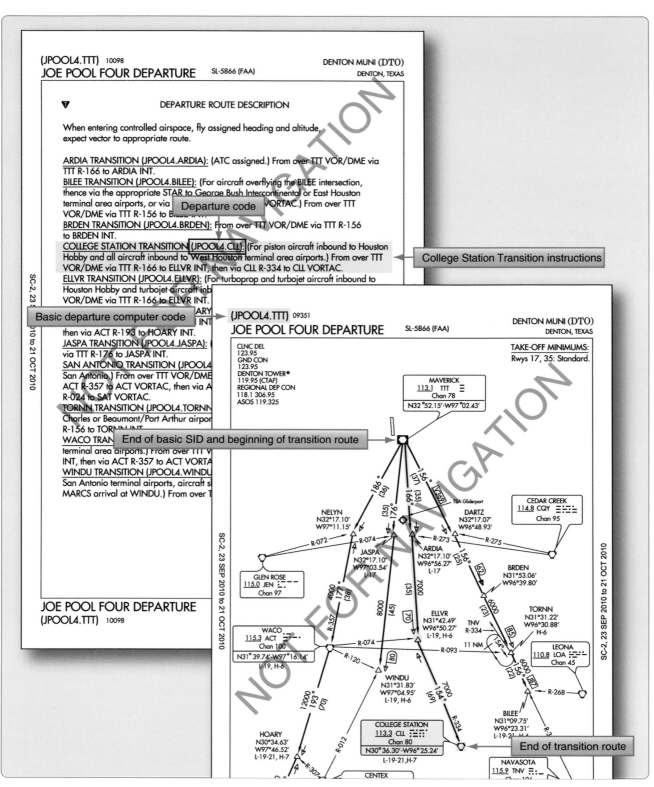

(JPOOL4.TTT) 10098
JOE POOL FOUR DEPARTURE   SL-5866 (FAA)

DENTON MUNI (DTO)
DENTON, TEXAS

▼   DEPARTURE ROUTE DESCRIPTION

When entering controlled airspace, fly assigned heading and altitude, expect vector to appropriate route.

ARDIA TRANSITION (JPOOL4.ARDIA): (ATC assigned.) From over TTT VOR/DME via TTT R-166 to ARDIA INT.

BILEE TRANSITION (JPOOL4.BILEE): (For aircraft overflying the BILEE intersection, thence via the appropriate STAR to George Bush Intercontinental or East Houston terminal area airports, or via `Departure code` VORTAC.) From over TTT VOR/DME via TTT R-156 to BILEE INT.

BRDEN TRANSITION (JPOOL4.BRDEN): From over TTT VOR/DME via TTT R-156 to BRDEN INT.

COLLEGE STATION TRANSITION (JPOOL4.CLL): (For piston aircraft inbound to Houston Hobby and all aircraft inbound to West Houston terminal area airports.) From over TTT VOR/DME via TTT R-166 to ELLVR INT, then via CLL R-334 to CLL VORTAC.   `College Station Transition instructions`

ELLVR TRANSITION (JPOOL4.ELLVR): (For turboprop and turbojet aircraft inbound to Houston Hobby and turbojet aircraft inb... VOR/DME via TTT R-166 to ELLVR INT.

`Basic departure computer code` ...ARY... INT. then via ACT R-193 to HOARY INT.

JASPA TRANSITION (JPOOL4.JASPA): (via TTT R-176 to JASPA INT.

SAN ANTONIO TRANSITION (JPOOL4... San Antonio.) From over TTT VOR/DME... ACT R-357 to ACT VORTAC, then via A... R-024 to SAT VORTAC.

TORNN TRANSITION (JPOOL4.TORN... Charles or Beaumont/Port Arthur airpor... R-156 to TORNN INT.

WACO TRAN... `End of basic SID and beginning of transition route` terminal area airports.) From over TTT V... INT, then via ACT R-357 to ACT VORTA...

WINDU TRANSITION (JPOOL4.WINDU... San Antonio terminal airports, aircraft s... MARCS arrival at WINDU.) From over T...

JOE POOL FOUR DEPARTURE
(JPOOL4.TTT) 10098

SC-2, 23 SEP 2010 to 21 OCT 2010

(JPOOL4.TTT) 09351
JOE POOL FOUR DEPARTURE   SL-5866 (FAA)

DENTON MUNI (DTO)
DENTON, TEXAS

CLNC DEL
123.95
GND CON
123.95
DENTON TOWER*
119.95 (CTAF)
REGIONAL DEP CON
118.1 306.95
ASOS 119.325

TAKE-OFF MINIMUMS:
Rwys 17, 35: Standard.

MAVERICK
113.1 TTT
Chan 78
N32°52.15'-W97°02.43'

CEDAR CREEK
114.8 CQY
Chan 95

`End of transition route`

Figure 1-23. Transition routes as depicted on SID.

1-25

while providing seamless transitions to the en route structure. ATC clearance must be received prior to flying a SID. SIDs also provide additional benefits to both the airspace capacity and the airspace users by reducing radio congestion, allowing more efficient airspace use, and simplifying departure clearances. All of the benefits combine to provide effective, efficient terminal operations, thereby increasing the overall capacity of the NAS.

If you cannot comply with a SID, if you do not possess the charted SID procedure, or if you simply do not wish to use SIDs, include the statement "NO SIDs" in the remarks section of your flight plan. Doing so notifies ATC that they cannot issue you a clearance containing a SID, but instead will clear you via your filed route to the extent possible, or via a Preferential Departure Route (PDR). It should be noted that SID usage not only decreases clearance delivery time, but also greatly simplifies your departure, easing you into the IFR structure at a desirable location and decreases your flight management load. While you are not required to depart using a SID, it may be more difficult to receive an "as filed" clearance when departing busy airports that frequently use SID routing.

SIDs are always charted graphically and are located in the TPP after the last approach chart for an airport. The SID may be one or two pages in length, depending on the size of the graphic and the amount of space required for the departure description. Each chart depicts the departure route, navigational fixes, transition routes, and required altitudes. The departure description outlines the particular procedure for each runway. [Figure 1-22]

Transition Routes

Charted transition routes allow pilots to transition from the end of the basic SID to a location in the en route structure. Typically, transition routes fan out in various directions from the end of the basic SID to allow pilots to choose the transition route that takes them in the direction of intended departure. A transition route includes a course, a minimum altitude, and distances between fixes on the route. When filing a SID for a specific transition route, include the transition in the flight plan, using the correct departure and transition code. ATC also assigns transition routes as a means of putting the flight on course to the destination. In any case, the pilot must receive an ATC clearance for the departure and the associated transition, and the clearance from ATC will include both the departure name and transition (e.g., Joe Pool Nine Departure, College Station Transition). [Figure 1-23]

The SID is designed to allow the pilot to provide his or her own navigation with minimal radio communication.

This type of procedure usually contains an initial set of departure instructions followed by one or more transition routes. A SID may include an initial segment requiring radar vectors to help the flight join the procedure, but the majority of the navigation remains the pilot's responsibility. [Figure 1-24]

A radar SID usually requires ATC to provide radar vectors from just after takeoff (ROC is based on a climb to 400 feet above the DER elevation before making the initial turn) until reaching the assigned route or a fix depicted on the SID chart. Radar SIDs do not include departure routes or transition routes because independent pilot navigation is not involved. The procedure sets forth an initial set of departure instructions that typically include an initial heading and altitude. ATC must have radar contact with the aircraft to be able to provide vectors. ATC expects you to immediately comply with radar vectors, and they expect you to notify them if you are unable to fulfill their request. ATC also expects you to make contact immediately if an instruction causes you to compromise safety due to obstructions or traffic.

It is prudent to review radar SID charts prior to use because this type of procedure often includes nonstandard lost communication procedures. If you were to lose radio contact while being vectored by ATC, you would be expected to comply with the lost communication procedure as outlined on the chart, not necessarily those procedures outlined in the AIM. [Figure 1-25]

SID Flight Planning Considerations

Take into consideration the departure paths included in the SIDs, and determine if you can use a standardized departure procedure. You have the opportunity to choose the SID that best suits your flight plan. During the flight planning phase, you can investigate each departure, and determine which procedure allows you to depart the airport in the direction of your intended flight. Also consider how a climb gradient to a specific altitude affects the climb time and fuel burn portions of the flight plan. Notes giving procedural requirements are listed on the graphic portion of a departure procedure, and they are mandatory in nature. [Figure 1-26] Mandatory procedural notes may include:

- Aircraft equipment requirements (DME, ADF, etc.)

- ATC equipment in operation (radar)

- Minimum climb requirements

- Restrictions for specific types of aircraft (turbojet only)

- Limited use to certain destinations

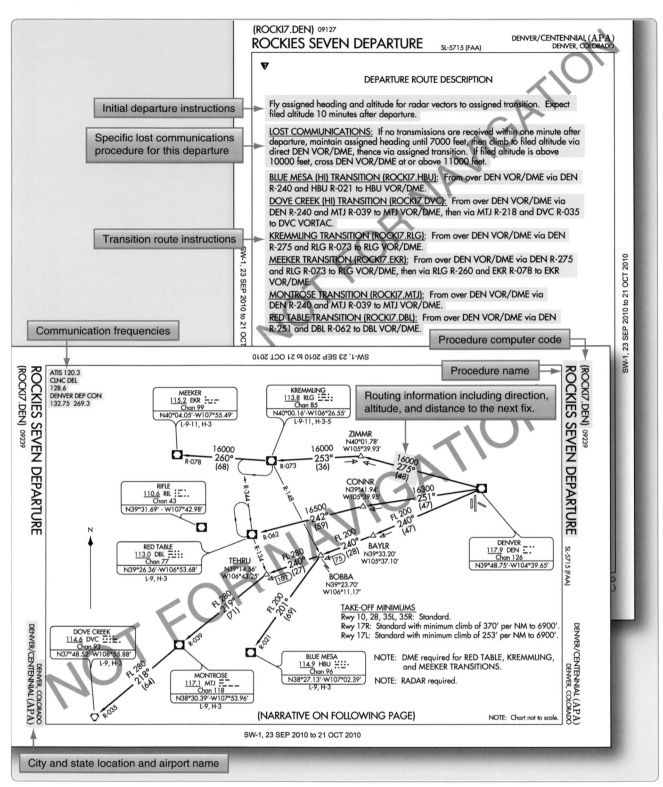

Figure 1-24. Example of a common SID at Denver, Colorado.

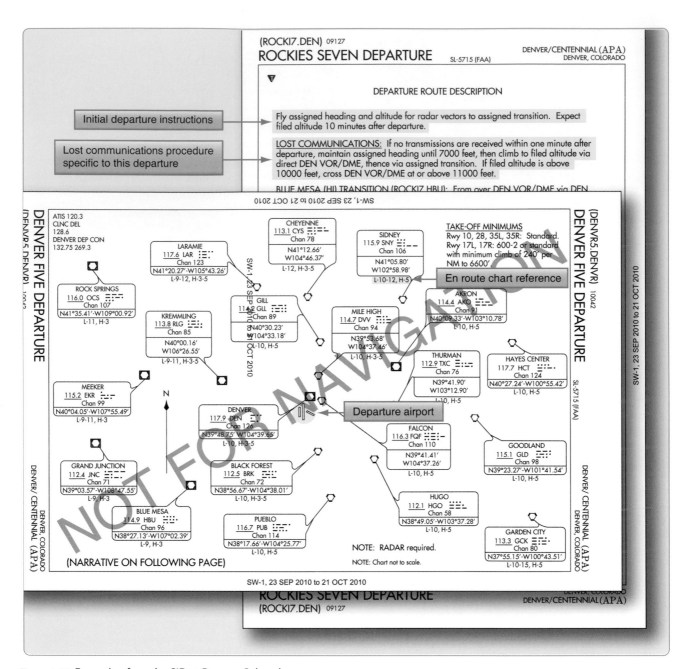

Figure 1-25. Example of a radar SID at Denver, Colorado.

Cautionary statements may also be included on the procedure to notify you of specific activity, but these are strictly advisory. [Figure 1-26] If you are unable to comply with a specific requirement, you must not file the procedure as part of your flight plan. If ATC assigns you a SID, you may need to quickly recalculate your all-engines-operating performance numbers. If you cannot comply with the climb gradient in the SID, you should not accept a clearance for that SID and furthermore, you must not accept the procedure if ATC assigns it.

A clearance for a SID which contains published altitude restrictions may be issued using the phraseology "climb via." Climb via is an abbreviated clearance that requires compliance with the procedure lateral path, associated speed and altitude restrictions along the cleared route or procedure. Expanded procedures for "Climb via" can be found in the Aeronautical Information Manual (AIM).

ATC can assign SIDs or radar vectors as necessary for traffic management and convenience. To fly a SID, you must receive approval to do so in a clearance. In order to accept a clearance that includes a SID, you must have the charted SID procedure in your possession at the time of departure. It is your responsibility as pilot in command to accept or reject the issuance of a SID by ATC. You must accept or reject the clearance based on:

• The ability to comply with the required performance.

1-28

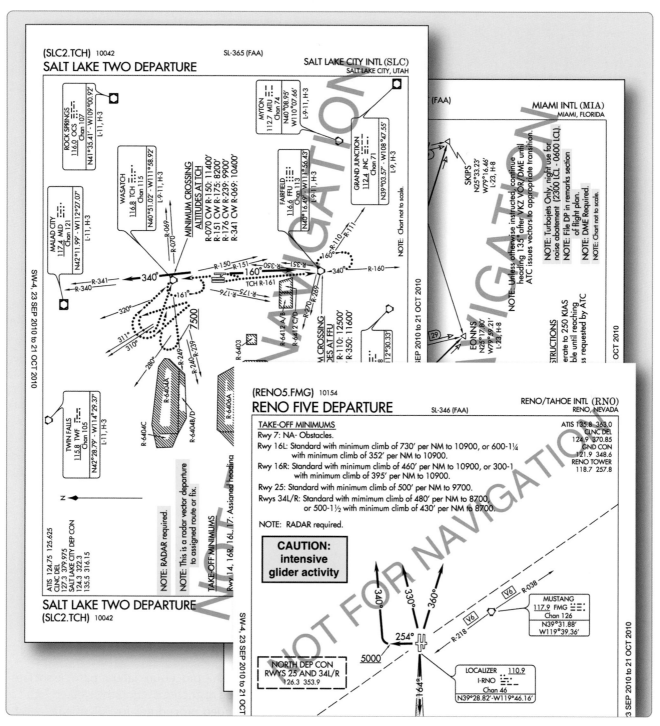

Figure 1-26. Departure procedure notes and cautionary statements.

- The ability to navigate to the degree of accuracy required for the procedure.

- Possession of the charted SID procedure.

- Personal understanding of the SID in its entirety.

When you accept a clearance to depart using a SID or radar vectors, ATC is responsible for traffic separation. When departing with a SID, ATC expects you to fly the procedure as charted because the procedure design considers obstacle clearance. It is also expected that you will remain vigilant in scanning for traffic when departing in visual conditions. Furthermore, it is your responsibility to notify ATC if your clearance would endanger your safety or the safety of others. DPs are also categorized by equipment requirements as follows:

- Non-RNAV DP—established for aircraft equipped with conventional avionics using ground-based

NAVAIDs. These DPs may also be designed using dead reckoning navigation. Some flight management systems (FMS) are certified to fly a non-RNAV DP if the FMS unit accepts inputs from conventional avionics sources, such as DME, VOR, and localizer (LOC). These inputs include radio tuning and may be applied to a navigation solution one at a time or in combination. Some FMS provide for the detection and isolation of faulty navigation information.

- RNAV DP—established for aircraft equipped with RNAV avionics (e.g., GPS, VOR/DME, DME/DME). Automated vertical navigation is not required, and all RNAV procedures not requiring GPS must be annotated with the note: "RADAR REQUIRED." Prior to using TSO-C129 GPS equipment for RNAV departures, approach receiver autonomous integrity monitoring (RAIM) availability should be checked for that location.

- Radar DP—radar may be used for navigation guidance for SID design. Radar SIDs are established when ATC has a need to vector aircraft on departure to a particular ATS Route, NAVAID, or fix. A fix may be a ground-based NAVAID, a waypoint, or defined by reference to one or more radio NAVAIDs. Not all fixes are waypoints since a fix could be a VOR or VOR/DME, but all waypoints are fixes. Radar vectors may also be used to join conventional or RNAV navigation SIDs. SIDs requiring radar vectors must be annotated "RADAR REQUIRED."

Area Navigation (RNAV) Departures

Historically, departure procedures were built around existing ground-based technology and were typically designed to accommodate lower traffic volumes. Often, departure and arrival routes use the same NAVAIDs creating interdependent, capacity diminishing routes. RNAV is a method of navigation that permits aircraft operation on any desired flight path within the coverage of ground- or spaced- based NAVAIDs or within the limits of the capability of self-contained aids or a combination of these. In the future, there will be an increased dependence on the use of RNAV in lieu of routes defined by ground-based NAVAIDs. As a part of the evolving RNAV structure, the FAA has developed departure procedures for pilots flying aircraft equipped with some type of RNAV technology. RNAV allows for the creation of new departure routes that are independent of present fixes and NAVAIDs. RNAV routing is part of the National Airspace Redesign (NAR) and is expected to reduce complexity and increase efficiency of terminal airspace.

When new RNAV departure procedures are designed, they will require minimal vectoring and communications between pilots and ATC. Usually, each departure procedure includes position, time, and altitude, which increase the ability to predict what the pilot will actually do. RNAV departure procedures have the ability to increase the capacity of terminal airspace by increasing on-time departures, airspace utilization, and improved predictability.

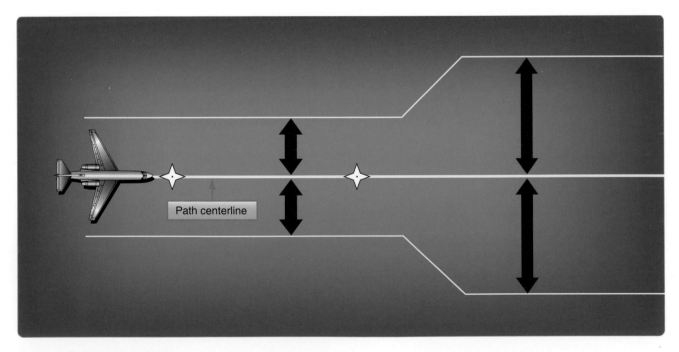

Path centerline

Figure 1-27. RNP departure levels.

FAA Flight Plan Aircraft Suffixes	
Suffix	Equipment Capability
	NO DME
/X	No transponder
/T	Transponder with no Mode C
/U	Transponder with Mode C
	DME
/D	No transponder
/B	Transponder with no Mode C
/A	Transponder with Mode C
	TACAN ONLY
/M	No transponder
/N	Transponder with no Mode C
/P	Transponder with Mode C
	AREA NAVIGATION (RNAV)
/Y	LORAN, VOR/DME, or INS with no transponder
/C	LORAN, VOR/DME, or INS, transponder with no Mode C
/I	LORAN, VOR/DME, or INS, transponder with Mode C
	ADVANCED RNAV WITH TRANSPONDER AND MODE C (If an aircraft is unable to operate with a transponder and/or Mode C, it will revert to the appropriate code listed above under Area Navigation.)
/E	Flight Management System (FMS) with DME/DME and IRU position updating
/F	Flight Management System (FMS) with DME/DME position updating
/G	Global Navigation Satellite System (GNSS), including GPS or WAAS, with en route and terminal capability.
/R	Required Navigational Performance (RNP). The aircraft meets the RNP type prescribed for the route segment(s), route(s), and/or area concerned.
	Reduced Vertical Separation Minimum (RVSM). Prior to conducting RVSM operations within the U.S., the operator must obtain authorization from the FAA or from the responsible authority, as appropriate.
/J	/E with RVSM
/K	/F with RVSM
/L	/G with RVSM
/Q	/R with RVSM
/W	RVSM

ICAO Flight Plan Equipment Codes	
Radio communication, navigation and approach aid equipment and capabilities	
"N" if no COM/NAV/approach aid equipment for the route to be flown is carried, or the equipment is unserviceable, OR "S" if standard COM/NAV/ approach aid equipment for the route to be flown is carried and serviceable (see Note 1), AND/OR INSERT one or more of the following letters to indicate the serviceable COM/NAV/approach aid equipment and capabilities available:	
A	GBAS landing system
B	LPV (APV with SBAS)
C	LORAN C
D	DME
E1	FMC WPR ACARS
E2	D-FIS ACARS
E3	PDC ACARS
F	ADF
G	GNSS (See Note 2)
H	HF RTF
I	Intertial Navigation
J1	CPDLC ATN VDL Mode 2 (See Note 3)
J2	CPDLC FANS 1/A HFDL
J3	CPDLC FANS 1/A VDL Mode 4
J4	CPDLC FANS 1/A VDL Mode 2
J5	CPDLC FANS 1/A SATCOM (INMARSAT)
J6	CPDLC FANS 1/A SATCOM (MTSAT)
J7	CPDLC FANS 1/A SATCOM (Iridium)
K	MLS
L	ILS
M1	OmegaATC RTF SATCOM
M2	ATC RTF (MTSAT)
M3	ATC RTF (Iridium)
O	VOR
P1 - P9	Reserved for RCP
R	PBN approved (see Note 4)
T	TACAN
U	UHF RTF
V	VHF RTF
W	RVSM approved
X	MNPS approved
Y	VHF with 8.33 kHz channel spacing capability
Z	Other equipment carried or other capabilities (see Note 5)

Any alphanumeric characters not indicated above are reserved. Effective September 1, 2005

Figure 1-28. Flight plan equipment codes (continued on next page).

ICAO Flight Plan Equipment Codes

Surveillance equipment and capabilities

INSERT N if no surveillance equipment for the route to be flown is carried, or the equipment is unserviceable, OR INSERT one or more of the following descriptors, to a maximum of 20 characters, to describe the serviceable surveillance equipment carriedand/or capabilities on board:

	SSR Modes A and C
A	Transponder - Mode A (4 digits - 4,096 codes)
C	Transponder - Mode A (4 digits - 4,096 codes) and Mode C

	SSR Modes S
E	Transponder—Mode S, including aircraft identification, pressure-altitude and extended squitter (ADS-B) capability
H	Transponder—Mode S, including aircraft identification, pressure-altitude and enhanced surveillance capability
I	Transponder—Mode S, including aircraft identification, but no pressure-altitude capability
L	Transponder—Mode S, including aircraft identification, pressure-altitude, extended squitter (ADS-B) and enhanced surveillance capability
P	Transponder—Mode S, including pressure-altitude, but no aircraft identification capability
S	Transponder—Mode S, including both pressure altitude and aircraft identification capability
X	Transponder — Mode S with neither aircraft identification nor pressure-altitude capability

Note: Enhanced surveillance capability is the ability of the aircraft to down-link aircraft derived data via a Mode S transponder.

	ADS-B
B1	ADS-B with dedicated 1090 MHz ADS-B "out" capability
B2	ADS-B with dedicated 1090 MHz ADS-B "out" and "in" capability
U1	ADS-B "out" capability using UAT
U2	ADS-B "out" and "in" capability using UAT
V1	ADS-B "out" capability using VDL Mode 4
V2	ADS-B "out" and "in" capability using VDL Mode 4
D1	D1 ADS-C with FANS 1/A capabilities
G1	G1 ADS-C with ATN capabilities

Alphanumeric characters not indicated above are reserved.

Example: ADE3RV/HB2U2V2G1

Note: Additional surveillance application should be listed in Item 18 following the indicator SUR/ .

NOTE	
Note 1	If the letter S is used, standard equipment is considered to be VHF RTF, VOR and ILS, unless another combination is prescribed by the appropriate ATS authority.
Note 2	If the letter G is used, the types of external GNSS augmentation, if any, are specified in Item 18 following the indicator NAV/ and separated by a space.
Note 3	See RTCA/EUROCAE Interoperability Requirements Standard For ATN Baseline 1 (ATN B1 INTEROP Standard DO-280B/ED-110B) for data link services air traffic control clearance and information/air traffic control communications management/air traffic control microphone check.
Note 4	Information on navigation capability is provided to ATC for clearance and routing purposes.
Note 5	If the letter Z is used, specify in Item 18 the other equipment carried or other capabilities, preceded by COM/, NAV/ and/or DAT, as appropriate.
Note 6	If the letter R is used, the performance based navigation levels that can be met are specified in Item 18 following the indicator PBN/. Guidance material on the application of performance based navigation to a specific route segment, route or area is contained in the Performance-Based Navigation Manual (Doc 9613).
Note 7	RNAV equipped aircraft capable of flying RNAV SIDs, putting "NO SID" in the remarks section will not always result in a clearance via a Preferential Departure Route (PDR). The Pilot/Dispatcher must amend Field 18 NAV from D1 to D0 and remove PBRN RNAV1 Code (D1-D4).
Note 8	If a RNAV DP is filed, an ICAO flight plan must be used. In Field 18, Pilots/Dispatchers must file a D1 or D2 depending on the RNAV DP. Additionally, Field 18 should include PBN/D1-D4 depending on the navigation update source. See AIM/PANS ATM 4444 for ICAO filing procedures.
Note 9	RNAV Q-routes require en route RNAV 2, corresponding NAV/E2 code and PBN/C1-C4 based on navigation system update source.
Note 10	If an aircraft does not meet the requirements for RVSM, then the W filed in ICAO flight plan Field 10A must be removed and STS/NONRVSM must be annotated in Field 18.
Note 11	Filing requirements for RNAV STARS. Field 18 of the ICAO flight plan must have a NAV/A1 or A2 assigned to the RNAV STAR. Additionally, PBN/D1-D4 for RNAV1 or C1-C4 for RNAV2 should be filed. If unable to accept the RNAV STAR, the flight plan must be amended to change the NAV/A1 or A2 to A0.

Figure 1-28. Flight plan equipment codes (continued).

All public RNAV SIDs and graphic ODPs are RNAV 1. These procedures generally start with an initial RNAV or heading leg near the departure end runway. In addition, these procedures require system performance currently met by GPS, DME/DME/Inertial Reference Unit (IRU) RNAV systems that satisfy the criteria discussed in AC 90-100, U.S. Terminal and En Route Area Navigation (RNAV) Operations. RNAV departures are identifiable by the inclusion of the term RNAV in the title of the departure. From a required navigation performance (RNP) standpoint, RNAV departure routes are designed with 1 or 2 NM performance standards as listed below. This means you as the pilot and your aircraft equipment must be able to maintain the aircraft within 1 or 2 NM either side of route centerline. [Figure 1-27]

- RNAV 1 procedures require that the aircraft's total system error remain bounded by ±1 NM for 95 percent of the total flight time.

- RNAV 2 requires a total system error of not more than 2 NM for 95 percent of the total flight time.

RNP is RNAV with on-board monitoring and alerting; RNP is also a statement of navigation performance necessary for operation within defined airspace. RNP 1 (in-lieu-of RNAV 1) is used when a DP that contains a constant radius to a fix (RF) leg or when surveillance (radar) monitoring is not desired for when DME/DME/IRU is used. These procedures are annotated with a standard note: "RNP 1."

If unable to comply with the requirements of an RNAV or RNP procedure, pilots need to advise ATC as soon as possible. For example, "N1234, failure of GPS system, unable RNAV, request amended clearance." Pilots are not authorized to fly a published RNAV or RNP procedure unless it is retrievable by the procedure name from the navigation database and conforms to the charted procedure. No other modification of database waypoints or creation of user-defined waypoints on published RNAV or RNP procedures is permitted, except to change altitude and/or airspeed waypoint constraints to comply with an ATC clearance/instruction, or to insert a waypoint along the published route to assist in complying with an ATC instruction. For example, "Climb via the WILIT departure except cross 30 north of CHUCK at/ or above FL 210." This is limited only to systems that allow along track waypoint construction.

Pilots of aircraft utilizing DME/DME for primary navigation updating should ensure any required DME stations are in service as determined by NOTAM, ATIS, or ATC advisory. DME/DME navigation system updating may require specific DME facilities to meet performance standards. Based on DME availability evaluations at the time of publication, current DME coverage is not sufficient to support DME/DME RNAV operations everywhere without IRU augmentation or

use of GPS. [Figure 1-28] DP chart notes may also include operational information for certain types of equipment, systems, and performance requirements, in addition to the type of RNAV departure procedure.

While operating on RNAV segments, pilots are encouraged to use the flight director in lateral navigation mode. RNAV terminal procedures may be amended by ATC issuing radar vectors and/or clearances direct to a waypoint. Pilots should avoid premature manual deletion of waypoints from their active "legs" page to allow for rejoining procedures. While operating on RNAV segments, pilots operating /R aircraft should adhere to any flight manual limitation or operating procedure required to maintain the RNP value specified for the procedure. In 2008, the FAA implemented the use of en route host automation ICAO flight plan (FP) processing for requesting assignment of RNAV SID, Standard Terminal Arrivals (STARs), or RNAV routes U.S. domestic airspace. This is part of a risk reduction strategy for introduction of the En Route Automation Modernization (ERAM) system in October 2008. ERAM also will use ICAO FP processing and as a result aircrews should be aware that as the FAA updates to ERAM the standard FAA flight plan equipment suffix codes will change to the ICAO flight plan equipment suffix codes.

For procedures requiring GPS and/or aircraft approvals requiring GPS, if the navigation system does not automatically alert the flight crew of a loss of GPS, aircraft operators must develop procedures to verify correct GPS operation. If not equipped with GPS, or for multi-sensor systems with GPS that do not alert upon loss of GPS, aircraft must be capable of navigation system updating using DME/DME or DME/DME/IRU for type 1 and 2 procedures. AC 90-100 may be used as operational guidance for RNAV ODPs. Pilots of FMS-equipped aircraft who are assigned an RNAV DP procedure and subsequently receive a change of runway, transition, or procedure, must verify that the appropriate changes are loaded and available for navigation.[Figures 1-29 and 1-30]

Additionally, new waypoint symbols are used in conjunction with RNAV charts. There are two types of waypoints currently in use: fly-by (FB) and fly-over (FO). A FB waypoint typically is used in a position at which a change in the course of procedure occurs. Charts represent them with four-pointed stars. This type of waypoint is designed to allow you to anticipate and begin your turn prior to reaching the waypoint, thus providing smoother transitions. Conversely, RNAV charts show a FO waypoint as a four-pointed star enclosed in a circle. This type of waypoint is used to denote a missed approach point, a missed approach holding point, or other specific points in space that must be flown over. [Figure 1-31] Pilots should not change any database waypoint type from a FB to FO, or vice versa.

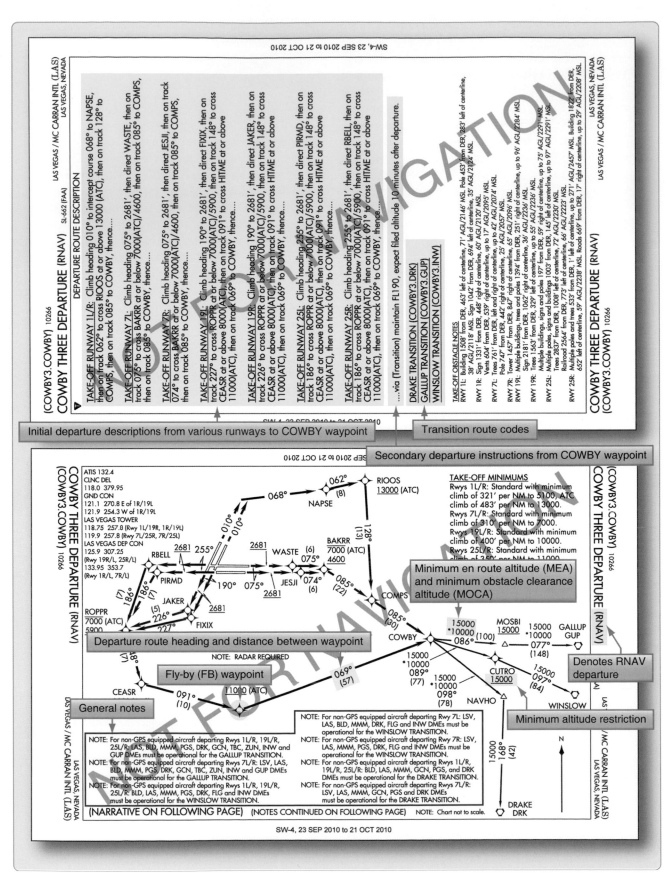

Figure 1-29. Examples of RNAV SID.

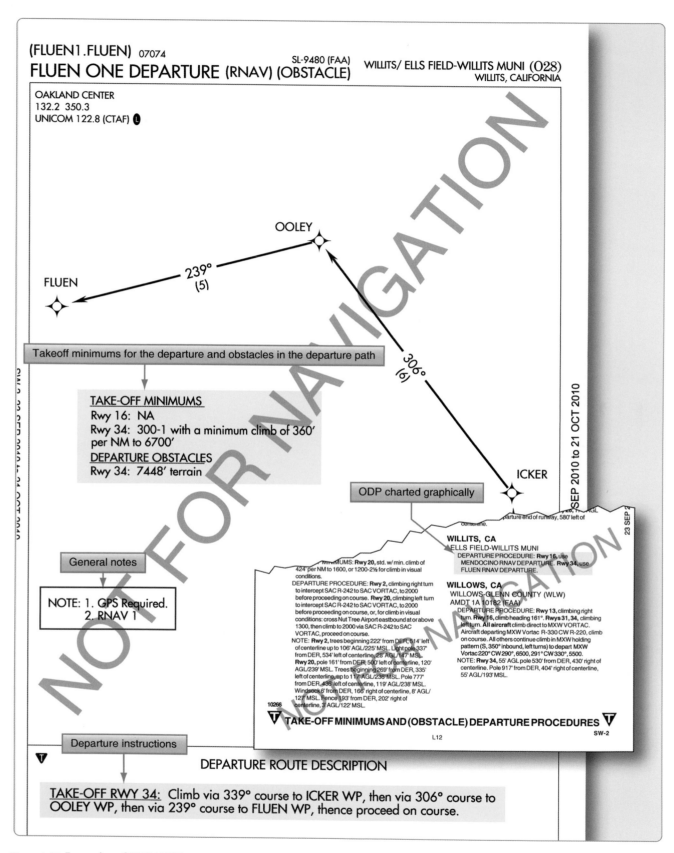

Figure 1-30. Examples of RNAV ODP.

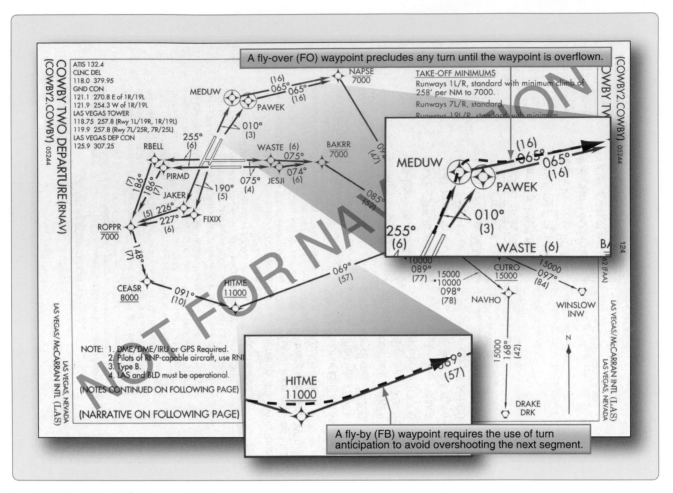

Figure 1-31. Fly-over and fly-by waypoints.

There are specific requirements, however, that must be met before using RNAV procedures. Every RNAV departure chart lists general notes and may include specific equipment and performance requirements, as well as the type of RNAV departure procedure in the chart plan view. New aircraft equipment suffix codes are used to denote capabilities for advanced RNAV navigation for flight plan filing purposes. [Figure 1-32]

SID Altitudes

SID altitudes can be charted in four different ways. The first are mandatory altitudes, the second, minimum altitudes, the third, maximum altitudes and the fourth is a combination of minimum and maximum altitudes or also referred to as block altitudes. Below are examples of how each will be shown on a SID approach plate.

- Mandatory altitudes – $\overline{\underline{5500}}$

- Minimum altitudes – $\underline{2300}$

- Maximum altitudes – $\overline{3300}$

- Combination of minimum and maximum – $\overline{7000}$
  $\underline{4600}$

Some SIDs may still have "(ATC)" adjacent to a crossing

altitude as shown in Figure 1-33 which implies that the crossing altitude is there to support an ATC requirement. A new charting standard has begun a process to remove, over a period of time, the ATC annotation. The Cowboy Four Departure (RNAV) shown in Figure 1-34 depicts the new charting standard without ATC annotations. When necessary, ATC may amend or delete SID crossing altitude restrictions; when doing so, ATC assumes responsibility for obstacle clearance until the aircraft is re-established laterally and vertically on the published SID route.

Pilot Responsibility for Use of RNAV Departures

RNAV usage brings with it multitudes of complications as it is being implemented. It takes time to transition, to disseminate information, and to educate current and potential users. As a current pilot using the NAS, you need to have a clear understanding of the aircraft equipment requirements for operating in a given RNP environment. You must understand the type of navigation system installed in your aircraft, and furthermore, you must know how your system operates to ensure that you can comply with all RNAV requirements. Operational information should be included in your AFM or its supplements. Additional information concerning how

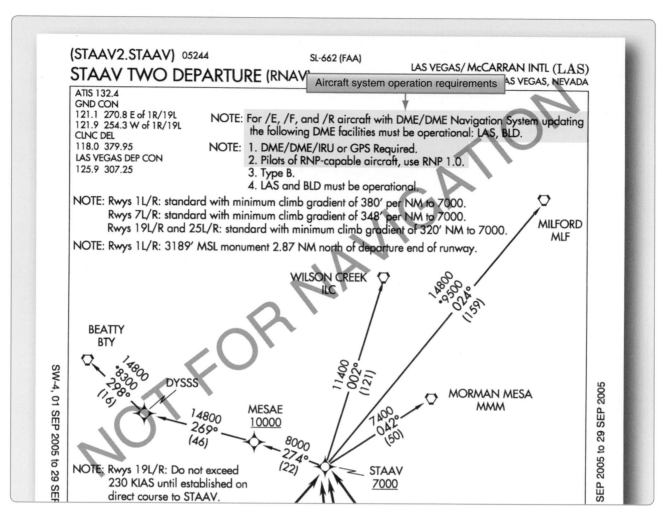

Figure 1-32. Operation requirements for RNAV.

to use your equipment to its fullest capacity, including "how to" training, may be gathered from your avionics manufacturer. If you are in doubt about the operation of your avionics system and its ability to comply with RNAV requirements, contact the FAA directly through your local Flight Standards District Office (FSDO). RNAV departure procedures are being developed at a rapid pace to provide RNAV capabilities at all airports. With every chart revision cycle, new RNAV departures are being added for small and large airports. These departures are flown in the same manner as traditional navigation-based departures; pilots are provided headings, altitudes, navigation waypoint, and departure descriptions. RNAV SIDs are found in the TPP with traditional departure procedures.

Radar Departures

A radar departure is another option for departing an airport on an IFR flight. You might receive a radar departure if the airport does not have an established departure procedure, if you are unable to comply with a departure procedure, or if you request "No SIDs" as a part of your flight plan. Expect ATC to issue an initial departure heading if you are being radar vectored after takeoff, however, do not expect to be given a purpose for the specific vector heading. Rest assured that the controller knows your flight route and will vector you into position. By nature of the departure type, once you are issued your clearance, the responsibility for coordination of your flight rests with ATC, including the tower controller and, after handoff, the departure controller who will remain with you until you are released on course and allowed to "resume own navigation."

For all practical purposes, a radar departure is the easiest type of departure to use. It is also a good alternative to a published departure procedure, particularly when none of the available departure procedures are conducive to your flight route. However, it is advisable to always maintain a detailed awareness of your location while you are being radar vectored by ATC. If for some reason radar contact is lost, you will be asked to provide position reports in order for ATC to monitor your flight progress. Also, ATC may release you to "resume own navigation" after vectoring you off course momentarily for a variety of reasons, including weather or traffic.

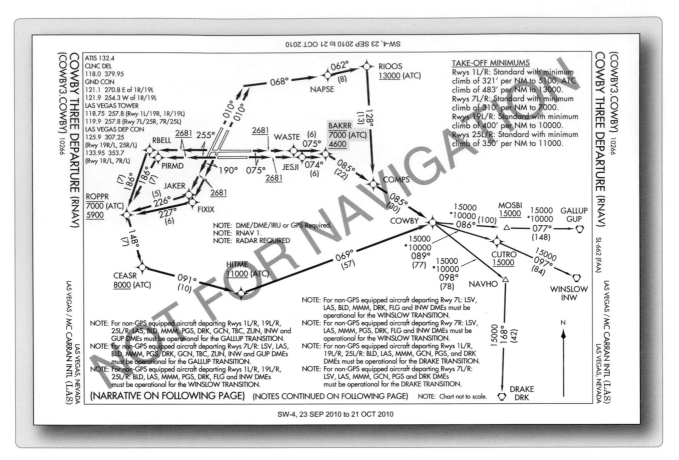

Figure 1-33. Crossing altitude is there to support an ATC requirement.

Upon initial contact, state your aircraft or flight number, the altitude you are climbing through, and the altitude to which you are climbing. The controller will verify that your reported altitude matches that emitted by your transponder. If your altitude does not match, or if you do not have Mode C capabilities, you will be continually required to report your position and altitude for ATC.

The controller is not required to provide terrain and obstacle clearance just because ATC has radar contact with your aircraft. It remains your responsibility until the controller begins to provide navigational guidance in the form of radar vectors. Once radar vectors are given, you are expected to promptly comply with headings and altitudes as assigned. Question any assigned heading if you believe it to be incorrect or if it would cause a violation of a regulation, then advise ATC immediately and obtain a revised clearance.

Diverse Vector Area

ATC may establish a minimum vectoring altitude (MVA) around certain airports. This altitude, based on terrain and obstruction clearance, provides controllers with minimum altitudes to vector aircraft in and around a particular location. However, at times, it may be necessary to vector aircraft below this altitude to assist in the efficient flow of departing traffic. For this reason, an airport may have an established Diverse Vector Area (DVA). This DVA may be established below the MVA or Minimum IFR Altitude (MIA) in a radar environment at the request of Air Traffic. This type of DP meets the TERPs criteria for diverse departures, obstacles and terrain avoidance in which random radar vectors below the MVA/MIA may be issued to departing traffic.

The existence of a DVA will be noted in the Takeoff Minimums and Obstacle Departure Procedures section of the U.S. Terminal Procedure Publication (TPP). The Takeoff Departure procedure will be listed first, followed by any applicable DVA. Pilots should be aware that Air Traffic facilities may utilize a climb gradient greater than the standard 200 FPNM within a DVA. This information will be identified in the DVA text for pilot evaluation against the aircraft's performance. Pilots should note that the DVA has been assessed for departures which do not follow a specified ground track, but will remain within the specified area. ATC may also vector an aircraft off a previously assigned DP. In all cases, the minimum 200 FPNM climb gradient is assumed unless a higher is specified on the departure, and obstacle clearance is not provided by ATC until the controller begins to provide navigational guidance

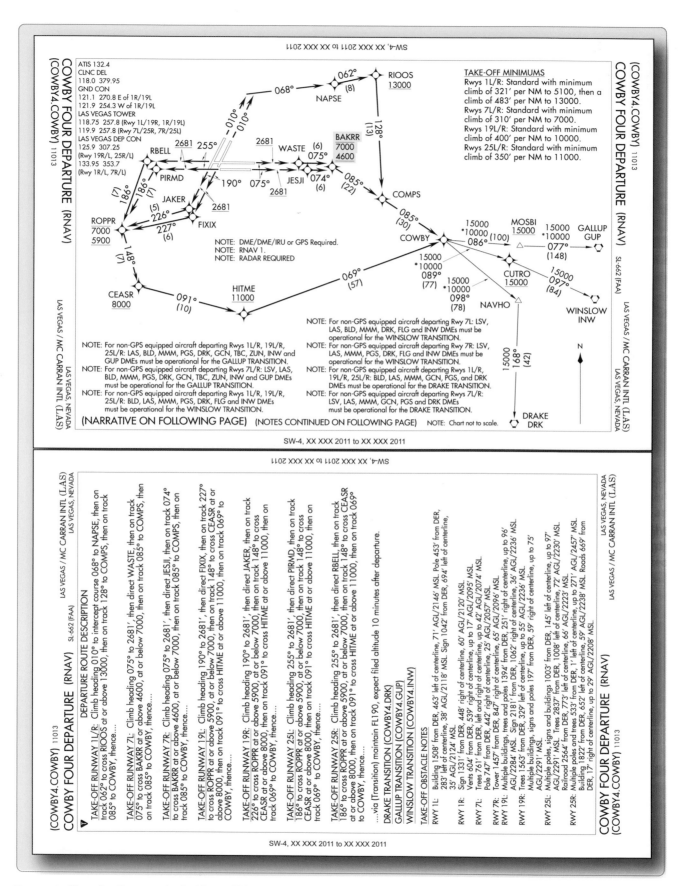

Figure 1-34. New charting standard without ATC annotations.

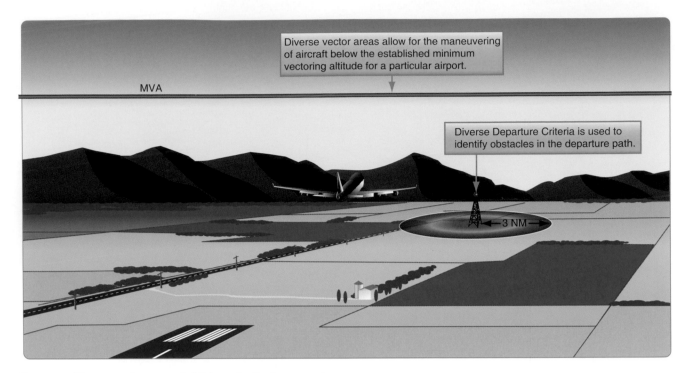

Diverse vector areas allow for the maneuvering of aircraft below the established minimum vectoring altitude for a particular airport.

MVA

Diverse Departure Criteria is used to identify obstacles in the departure path.

←— 3 NM —→

Figure 1-35. Diverse vector area establishment criteria.

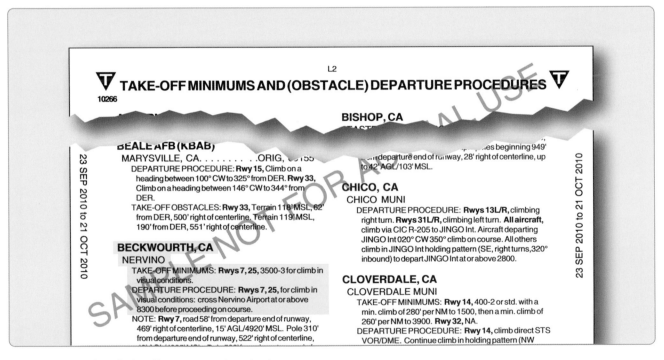

L2

TAKE-OFF MINIMUMS AND (OBSTACLE) DEPARTURE PROCEDURES

10266

BISHOP, CA

BEALE AFB (KBAB)

MARYSVILLE, CA . . . . .ORIG, 09155
DEPARTURE PROCEDURE: **Rwy 15,** Climb on a heading between 100° CW to 325° from DER. **Rwy 33,** Climb on a heading between 146° CW to 344° from DER.
TAKE-OFF OBSTACLES: **Rwy 33,** Terrain 118' MSL, 62' from DER, 500' right of centerline. Terrain 119' MSL, 190' from DER, 551' right of centerline.

...departure end of runway, 28' right of centerline, up to 42' AGL/103' MSL.

CHICO, CA

CHICO MUNI
DEPARTURE PROCEDURE: **Rwys 13L/R,** climbing right turn. **Rwys 31L/R,** climbing left turn. **All aircraft,** climb via CIC R-205 to JINGO Int. Aircraft departing JINGO Int 020° CW 350° climb on course. All others climb in JINGO Int holding pattern (SE, right turns, 320° inbound) to depart JINGO Int at or above 2800.

BECKWOURTH, CA

NERVINO
TAKE-OFF MINIMUMS: **Rwys 7, 25,** 3500-3 for climb in visual conditions.
DEPARTURE PROCEDURE: **Rwys 7, 25,** for climb in visual conditions: cross Nervino Airport at or above 8300 before proceeding on course.
NOTE: **Rwy 7,** road 58' from departure end of runway, 469' right of centerline, 15' AGL/4920' MSL. Pole 310' from departure end of runway, 522' right of centerline,

CLOVERDALE, CA

CLOVERDALE MUNI
TAKE-OFF MINIMUMS: **Rwy 14,** 400-2 or std. with a min. climb of 280' per NM to 1500, then a min. climb of 260' per NM to 3900. **Rwy 32,** NA.
DEPARTURE PROCEDURE: **Rwy 14,** climb direct STS VOR/DME. Continue climb in holding pattern (NW

23 SEP 2010 to 21 OCT 2010

23 SEP 2010 to 21 OCT 2010

Figure 1-36. Examples of takeoff minimums obstacle clearance.

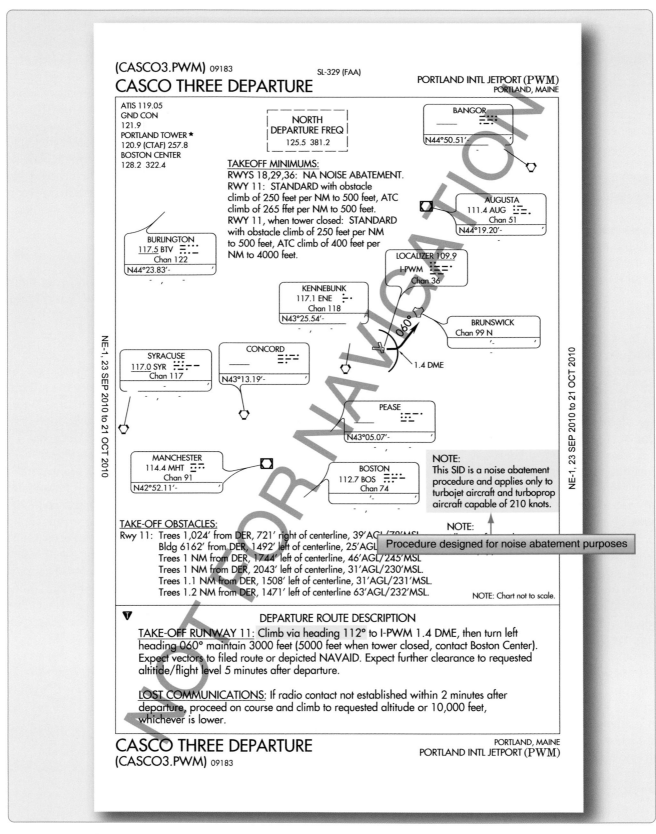

(CASCO3.PWM) 09183   SL-329 (FAA)   PORTLAND INTL JETPORT (PWM)
CASCO THREE DEPARTURE   PORTLAND, MAINE

ATIS 119.05
GND CON
121.9
PORTLAND TOWER ★
120.9 (CTAF) 257.8
BOSTON CENTER
128.2 322.4

NORTH
DEPARTURE FREQ
125.5 381.2

BANGOR
N44°50.51'-

TAKEOFF MINIMUMS:
RWYS 18,29,36: NA NOISE ABATEMENT.
RWY 11: STANDARD with obstacle
climb of 250 feet per NM to 500 feet, ATC
climb of 265 ffet per NM to 500 feet.
RWY 11, when tower closed: STANDARD
with obstacle climb of 250 feet per NM
to 500 feet, ATC climb of 400 feet per
NM to 4000 feet.

AUGUSTA
111.4 AUG
Chan 51
N44°19.20'-

BURLINGTON
117.5 BTV
Chan 122
N44°23.83'-

LOCALIZER 109.9
I-PWM
Chan 36

KENNEBUNK
117.1 ENE
Chan 118
N43°25.54'-

060°

BRUNSWICK
Chan 99 N
'-   '

1.4 DME

SYRACUSE
117.0 SYR
Chan 117

CONCORD
N43°13.19'-

PEASE
N43°05.07'-

MANCHESTER
114.4 MHT
Chan 91
N42°52.11'-

BOSTON
112.7 BOS
Chan 74
'-   '

NOTE:
This SID is a noise abatement
procedure and applies only to
turbojet aircraft and turboprop
aircraft capable of 210 knots.

NOTE:

Procedure designed for noise abatement purposes

TAKE-OFF OBSTACLES:
Rwy 11: Trees 1,024' from DER, 721' right of centerline, 39'AGL/78'MSL.
        Bldg 6162' from DER, 1492' left of centerline, 25'AGL.
        Trees 1 NM from DER, 1744' left of centerline, 46'AGL/245'MSL.
        Trees 1 NM from DER, 2043' left of centerline, 31'AGL/230'MSL.
        Trees 1.1 NM from DER, 1508' left of centerline, 31'AGL/231'MSL.
        Trees 1.2 NM from DER, 1471' left of centerline 63'AGL/232'MSL.

NOTE: Chart not to scale.

▼   DEPARTURE ROUTE DESCRIPTION

TAKE-OFF RUNWAY 11: Climb via heading 112° to I-PWM 1.4 DME, then turn left
heading 060° maintain 3000 feet (5000 feet when tower closed, contact Boston Center).
Expect vectors to filed route or depicted NAVAID. Expect further clearance to requested
altitide/flight level 5 minutes after departure.

LOST COMMUNICATIONS: If radio contact not established within 2 minutes after
departure, proceed on course and climb to requested altitude or 10,000 feet,
whichever is lower.

CASCO THREE DEPARTURE   PORTLAND, MAINE
(CASCO3.PWM) 09183   PORTLAND INTL JETPORT (PWM)

NE-1, 23 SEP 2010 to 21 OCT 2010
NE-1, 23 SEP 2010 to 21 OCT 2010

Figure 1-37. Noise abatement SID.

(vectors). Lastly, pilots should understand ATC instructions take precedence over an ODP [Figure 1-35].

Visual Climb Over Airport (VCOA)

A visual climb over airport (VCOA) is a departure option for an IFR aircraft, operating in VMC equal to or greater than the specified visibility and ceiling, to visually conduct climbing turns over the airport to the published "climb-to" altitude from which to proceed with the instrument portion of the departure. A VCOA is a departure option developed when obstacles farther than 3 SM from the airport require a CG of more than 200 FPNM.

These procedures are published in the Take-Off Minimums and (Obstacle) Departure Procedures section of the TPP. [Figure 1-36] Prior to departure, pilots are required to notify ATC when executing the VCOA.

Noise Abatement Procedures

As the aviation industry continues to grow and air traffic increases, so does the population of people and businesses around airports. As a result, noise abatement procedures have become commonplace at most of the nation's airports. 14 CFR Part 150 specifies the responsibilities of the FAA to investigate the recommendations of the airport operator in a noise compatibility program and approve or disapprove the noise abatement suggestions. This is a crucial step in ensuring that the airport is not unduly inhibited by noise requirements and that air traffic workload and efficiency are not significantly impacted, all while considering the noise problems addressed by the surrounding community.

While most DPs are designed for obstacle clearance and workload reduction, there are some SIDs that are developed solely to comply with noise abatement requirements. Portland International Jetport is an example of an airport where a SID was created strictly for noise abatement purposes as noted in the DP. [Figure 1-36] Typically, noise restrictions are incorporated into the main body of the SID. These types of restrictions require higher departure altitudes, larger climb gradients, reduced airspeeds, and turns to avoid specific areas. Noise restrictions may also be evident during a radar departure. ATC may require you to turn away from your intended course or vector you around a particular area. While these restrictions may seem burdensome, it is important to remember that it is your duty to comply with written and spoken requests from ATC.

Additionally, when required, departure instructions specify the actual heading to be flown after takeoff, as is the case in Figure 1-37 under the departure route description, "Climb via heading 112 degrees..." Some existing procedures specify, "Climb runway heading." Over time, both of these departure instructions will be updated to read, "Climb heading 112 degrees...." Runway heading is the magnetic direction that corresponds with the runway centerline extended (charted on the airport diagram), not the numbers painted on the runway. Pilots cleared to "fly or maintain runway heading" are expected to fly or maintain the published heading that corresponds with the extended centerline of the departure runway (until otherwise instructed by ATC), and are not to apply drift correction (e.g., RWY 11, actual magnetic heading of the runway centerline 112.2 degrees, "fly heading 112 degrees"). In the event of parallel departures, this prevents a loss of separation caused by only one aircraft applying a wind drift.

Procedural Notes

An important consideration to make during your flight planning is whether or not you are able to fly your chosen departure procedure as charted.

DP Responsibilities

Responsibility for the safe execution of DPs rests on the shoulders of both ATC and the pilot. Without the interest and attention of both parties, the IFR system cannot work in harmony, and achievement of safety is impossible.

ATC, in all forms, is responsible for issuing clearances appropriate to the operations being conducted, assigning altitudes for IFR flight above the minimum IFR altitudes for a specific area of controlled airspace, ensuring the pilot has acknowledged the clearance or instructions, and ensuring the correct read back of instructions. Specifically related to departures, ATC is responsible for specifying the direction of takeoff or initial heading when necessary, obtaining pilot concurrence that the procedure complies with local traffic patterns, terrain, and obstruction clearance, and including DP as part of the ATC clearance when pilot compliance for separation is necessary.

The pilot has a number of responsibilities when simply operating in conjunction with ATC or when using DPs under an IFR clearance:

- Acknowledge receipt and understanding of an ATC clearance.

- Read back any part of a clearance that contains "hold short" instructions.

- Request clarification of clearances.

- Request an amendment to a clearance if it is unacceptable from a safety perspective.

- Promptly comply with ATC requests. Advise ATC immediately if unable to comply with a clearance.

- You are required to contact ATC if you are unable to comply with all-engines-operating climb gradients and climb rates. It is also expected that you are capable of maintaining the climb gradient outlined in either a standard or non-standard DP. If you cannot maintain a standard climb gradient or the climb gradient specified in an ODP, you must wait until you can depart under VMC.

When planning for a departure, pilots should:

- Consider the type of terrain and other obstructions in the vicinity of the airport.

- Determine if obstacle clearance can be maintained visually, or if they need to make use of a DP.

- Determine if an ODP or SID is available for the departure airport.

- Determine what actions allow for a safe departure out of an airport that does not have any type of affiliated DPs.

By simply complying with DPs in their entirety as published, obstacle clearance is guaranteed. Depending on the type of departure used, responsibility for terrain clearance and traffic separation may be shared between pilots and controllers.

Departures From Tower-Controlled Airports

Departing from a tower-controlled airport is relatively simple in comparison to departing from non-towered airport. Normally you request your IFR clearance through ground control or clearance delivery. Communication frequencies for the various controllers are listed on departure, approach, and airport charts, as well as the A/FD. At some airports, you may have the option of receiving a pre-taxi clearance. This program allows you to call ground control or clearance delivery no more than 10 minutes prior to beginning taxi operations and receive your IFR clearance. A pre-departure clearance (PDC) program that allows pilots to receive a clearance via data link from a dispatcher or a data link communications service provider, e.g. ARINC, is available for Part 121 and 135 operators. A clearance is given to the dispatcher, who in turn, relays it to the crew, enabling the crew to bypass communication with clearance delivery, thus reducing frequency congestion. Once you have received your clearance, it is your responsibility to comply with the instructions as given, and notify ATC if you are unable to comply with the clearance. If you do not understand the clearance, or if you think that you have missed a portion of the clearance, contact ATC immediately for clarification.

Departures From Airports Without an Operating Control Tower

There are hundreds of airports across the United States that operate successfully every day without the benefit of a control tower. While a tower is certainly beneficial when departing IFR, most other departures can be made with few challenges. As usual, you must file your flight plan at least 30 minutes in advance. During your planning phase, investigate the departure airport's method for receiving an instrument clearance. You can contact the Automated Flight Service Station (AFSS) on the ground by telephone, and they will request your clearance from ATC. Typically, when a clearance is given in this manner, the clearance includes a void time. You must depart the airport before the clearance void time; if you fail to depart, you must contact ATC by a specified notification time, which is within 30 minutes of the original void time. After the clearance void time, your reserved space within the IFR system is released for other traffic.

There are several other ways to receive a clearance at a non-towered airport. If you can contact the AFSS or ATC on the radio, you can request your departure clearance. However, these frequencies are typically congested, and they may not be able to provide you with a clearance via the radio. You also can use a Remote Communications Outlet (RCO) to contact an AFSS if one is located nearby. Some airports have licensed UNICOM operators that can also contact ATC on your behalf and, in turn, relay your clearance from ATC. You are also allowed to depart the airport VFR, if conditions permit, and contact the controlling authority to request your clearance in the air. As technology improves, new methods for delivery of clearances at non-towered airports are being created.

Ground Communication Outlet

Ground Communication Outlets (GCO), have been developed in conjunction with the FAA to provide pilots flying in and out of non-towered airports with the capability to contact ATC and AFSS via very high frequency (VHF) radio to a telephone connection. This lets pilots obtain an instrument clearance or close a VFR/IFR flight plan. You can use four key clicks on your VHF radio to contact the nearest ATC facility and six key clicks to contact the local AFSS, but it is intended to be used only as a ground operational tool. A GCO is an unstaffed, remote controlled ground-to-ground communication facility that is relatively inexpensive to install and operate. Installations of these types of outlets are scheduled at instrument airports around the country.

GCOs are manufactured by different companies including ARINC and AVTECH, each with different operating characteristics but with the ability to accomplish the same goal. This latest technology has proven to be an incredibly useful tool for communicating with the appropriate authorities when departing IFR from a non-towered airport. The GCO should help relieve the need to use the telephone to call ATC and the need to depart into marginal conditions just to achieve radio contact. GCO information is listed on airport charts and instrument approach charts with other communications frequencies. Signs may also be located on an airport to notify you of the frequency and proper usage.

See and Avoid Techniques

Meteorological conditions permitting, you are required to use "see and avoid" techniques to avoid traffic, terrain, and other obstacles. To avoid obstacles during a departure, the takeoff minimums may include a non-standard ceiling and visibility minimum. These are given to pilots so they can depart an airport without being able to meet the established climb gradient. Instead, they must see and avoid obstacles in the departure path. In these situations, ATC provides radar traffic information for radar-identified aircraft outside controlled airspace, workload permitting, and safety alerts to pilots believed to be within an unsafe proximity to obstacles or aircraft.

VFR Departures

There may be times when you need to fly an IFR flight plan due to the weather you will encounter at a later time (or if you simply wish to fly IFR to remain proficient), but the weather outside is clearly VFR. It may be that you can depart VFR, but you need to get an IFR clearance shortly after departing the airport. A VFR departure can be used as a tool that allows you to get off the ground without having to wait for a time slot in the IFR system, however, departing VFR with the intent of receiving an IFR clearance in the air can also present serious hazards worth considering.

A VFR departure dramatically changes the takeoff responsibilities for you and for ATC. Upon receiving clearance for a VFR departure, you are cleared to depart; however, you must maintain separation between yourself and other traffic. You are also responsible for maintaining terrain and obstruction clearance, as well as remaining in VFR weather conditions. You cannot fly in IMC without first receiving your IFR clearance. Likewise, a VFR departure relieves ATC of these duties and basically requires them only to provide you with safety alerts as workload permits.

Maintain VFR until you have obtained your IFR clearance and have ATC approval to proceed on course in accordance with your clearance. If you accept this clearance and are below the minimum IFR altitude for operations in the area, you accept responsibility for terrain/obstruction clearance until you reach that altitude.

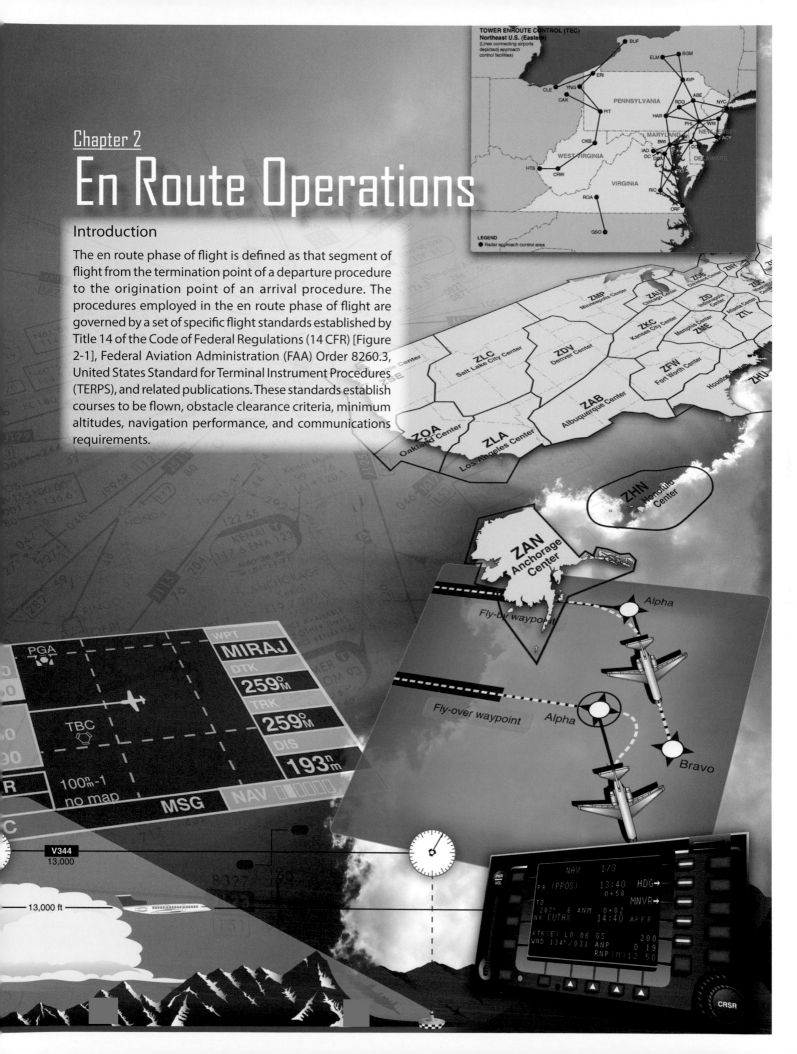

Chapter 2
En Route Operations

Introduction

The en route phase of flight is defined as that segment of flight from the termination point of a departure procedure to the origination point of an arrival procedure. The procedures employed in the en route phase of flight are governed by a set of specific flight standards established by Title 14 of the Code of Federal Regulations (14 CFR) [Figure 2-1], Federal Aviation Administration (FAA) Order 8260.3, United States Standard for Terminal Instrument Procedures (TERPS), and related publications. These standards establish courses to be flown, obstacle clearance criteria, minimum altitudes, navigation performance, and communications requirements.

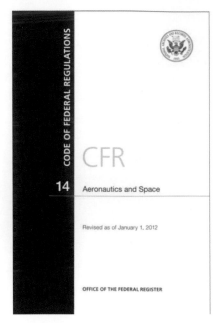

Figure 2-1. Code of Federal Regulations, Title 14 Aeronautics and Space.

En Route Navigation

En route instrument flight rules (IFR) navigation is evolving from the ground-based navigational aid (NAVAID) airway system to a sophisticated satellite and computer-based system that can generate courses to suit the operational requirements of almost any flight. The FAA Global Navigation Satellite System (GNSS) provides satellite-based positioning, navigation, and timing services in the United States to enable performance-based operations for all phases of flight, to include en route navigation.

14 CFR Part 91, section 91.181, is the basis for the course to be flown. Unless authorized by ATC, to operate an aircraft within controlled airspace under IFR, pilots must either fly along the centerline when on a Federal airway or, on routes other than Federal airways, along the direct course between NAVAIDs or fixes defining the route. The regulation allows maneuvering to pass well clear of other air traffic or, if in visual meteorogical conditions (VMC), to clear the flightpath both before and during climb or descent.

Airways

Airway routing occurs along pre-defined pathways called airways. [Figure 2-2] Airways can be thought of as three-dimensional highways for aircraft. In most land areas of the world, aircraft are required to fly airways between the departure and destination airports. The rules governing airway routing, Standard Instrument Departures (SID) and Standard Terminal Arrival (STAR), are published flight procedures that cover altitude, airspeed, and requirements for entering and leaving the airway. Most airways are eight nautical miles (14 kilometers) wide, and the airway flight levels keep aircraft separated by at least 500 vertical feet from aircraft on the flight level above and below when operating under VFR. When operating under IFR, between the surface and an altitude of Flight Level (FL) 290, no aircraft should come closer vertically than 1,000 feet. Above FL290 , no aircraft should come closer than 2,000 feet except in airspace where Reduced Vertical Separation Minima (RVSM) can be applied in which case the vertical separation is reduced to 1,000 feet. Airways usually intersect at NAVAIDs that designate the allowed points for changing from one airway to another. Airways have names consisting of one or more letters followed by one or more digits (e.g., V484 or UA419).

The en route airspace structure of the National Airspace System (NAS) consists of three strata. The first stratum low

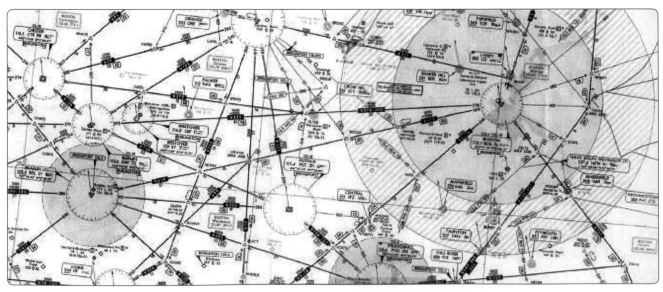

Figure 2-2. Airways depicted on an aeronautical chart.

altitude airways in the United States can be navigated using NAVAIDs, have names that start with the letter V, and are called Victor Airways. [Figure 2-3] They cover altitudes from approximately 1,200 feet above ground level (AGL) up to, but not including 18,000 feet above mean sea level (MSL). The second stratum high altitude airways in the United States all have names that start with the letter J, and are called Jet Routes. [Figure 2-4] These routes run from 18,000 feet to 45,000 feet. The third stratum allows random operations above flight level (FL) 450. The altitude separating the low and high airway structure varies from county to country. For example, in Switzerland it is 19,500 feet and 25,000 feet in Egypt.

Air Route Traffic Control Centers

The FAA defines an Air Route Traffic Control Center (ARTCC) as a facility established to provide air traffic control (ATC) service to aircraft operating on IFR flight plans within controlled airspace, principally during the en route phase of flight. When equipment capabilities and controller workload permit, certain advisory/assistance services may be provided to VFR aircraft.

ARTCCs, usually referred to as Centers, are established primarily to provide air traffic service to aircraft operating on IFR flight plans within the controlled airspace, and principally during the en route phase of flight. There are 21 ARTCC's in the United States. [Figure 2-5] Any aircraft

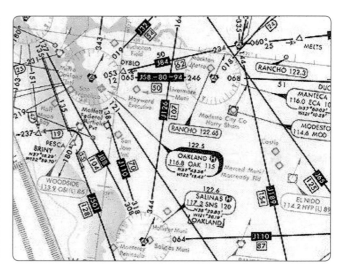

Figure 2-3. Victor airways.

Figure 2-4. Jet routes.

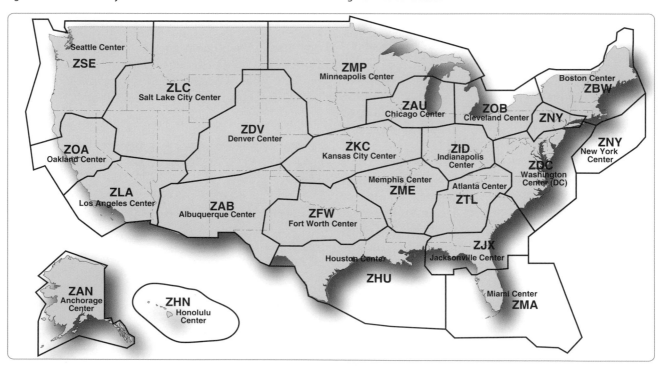

Figure 2-5. Air Route Traffic Control Centers.

operating under IFR within the confines of an ARTCC's airspace is controlled by air traffic controllers at the Center. This includes all sorts of different types of aircraft: privately owned single engine aircraft, commuter airlines, military jets, and commercial airlines.

The largest component of the NAS is the ARTCC. Each ARTCC covers thousands of square miles encompassing all or part of several states. ARTCCs are built to ensure safe and expeditious air travel. All Centers operate 7-days a week, 24-hours a day, and employ a combination of several hundred ATC specialists, electronic technicians, computer system specialists, environmental support specialists, and administrative staff. Figure 2-6 is an example of the Boston ARTCC. The green lines mark the boundaries of the Boston Center area, and the red lines mark the boundaries of Military Operations Areas (MOAs), Prohibited, Restricted, Alert, and Warning Areas.

Safe Separation Standards

The primary means of controlling aircraft is accomplished by using highly sophisticated computerized radar systems. In addition, the controller maintains two-way radio communication with aircraft in his or her sector. In this way, the specialist ensures that the aircraft are separated by the following criteria:

- Laterally—5 miles
- Vertically—
-   1,000 feet (if the aircraft is below FL290, or between FL290 and FL410 for RVSM compliant aircraft)
-   2,000 feet (if the aircraft is at FL290 or above)

The controllers can accomplish this separation by issuing instructions to the pilots of the aircraft involved. Altitude assignments, speed adjustments, and radar vectors are examples of instructions that might be issued to aircraft.

En route control is handled by pinpointing aircraft positions through the use of flight progress strips. These strips are pieces of printed paper containing pertinent information extracted from the pilot's flight plan. These strips are printed 20 minutes prior to an aircraft reaching each Center's sector. A flight progress strip tells the controller everything needed to direct that aircraft. If the flight progress strips of each aircraft approaching a sector are arranged properly, it is possible to determine potential conflicts long before the aircraft are even visible on the Center controller's display. In areas where radar coverage is not available, this is the sole means of separating aircraft.

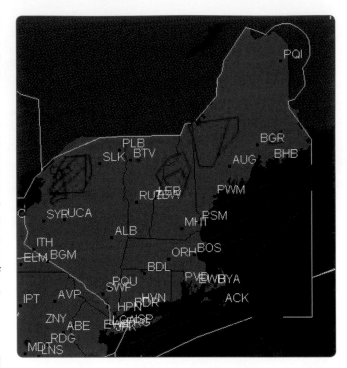

Figure 2-6. Boston Air Route Traffic Control Center.

The strips, one for each en route point from which the pilot reports his or her position, are posted on a slotted board in front of the air traffic controller. [Figure 2-7] At a glance, he or she is able to see certain vital data: the type of airplane and who is flying it (airline, business, private, or military pilot), aircraft registration number or flight number, route, speed, altitude, airway designation, and the estimated time of arrival (ETA) at destination. As the pilot calls in the aircraft's position and time at a predetermined location, the strips are removed from their slots and filed. Any change from the original flight plan is noted on the strips as the flight continues. Thus, from a quick study of the flight progress board, a controller can assess the overall traffic situation and can avoid possible conflicts.

Figure 2-7. Flight progress strips.

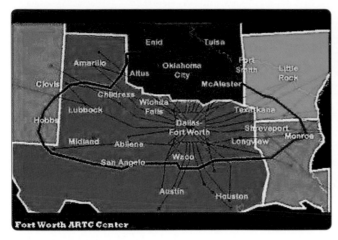

Figure 2-8. Fort Worth Air Route Traffic Control Center.

Figure 2-8 shows the Fort Worth, Texas Air Route Traffic Control Center (ZFW) and the geographical area that it covers. The Center has approximately 350 controllers. Most are certified and some are in on-the-job training.

Sectors

The airspace controlled by a Center may be further administratively subdivided into smaller, manageable pieces of airspace called sectors. A few sectors extend from the ground up, but most areas are stratified into various levels to accommodate a wide variety of traffic. Each sector is staffed by a set of controllers and has a unique radio frequency that the controller uses to communicate with the pilots. As aircraft transition from one sector to another, they are instructed to change to the radio frequency used by the next sector. Each sector also has secure landline communications with adjacent sectors, approach controls, areas, ARTCCs, flight service centers, and military aviation control facilities.

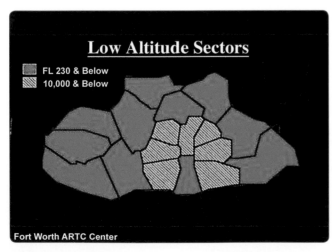

Figure 2-9. Low altitude sectors.

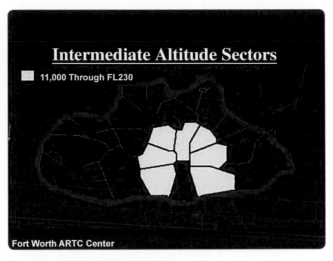

Figure 2-10. Intermediate altitude sectors.

Figure 2-11. High altitude sectors.

Figure 2-12. Ultra high altitude sectors.

The ARTCC at Fort Worth, Texas is subdivided into sectors that are categorized as follows:

- Eighteen low altitude sectors. [Figure 2-9]

2-5

- Seven intermediate altitude sectors. [Figure 2-10]
- Sixteen high altitude sectors. [Figure 2-11]
- One ultra high altitude sector. [Figure 2-12]

From one to three controllers may work a sector, depending upon the amount of air traffic. Each controller is assigned to work the positions within an area of specialization. Controllers have direct communication with pilots, with surrounding sectors and Centers, plus the towers and Flight Service Stations (FSS) under their jurisdiction. Each control position is equipped with computer input and readout devices for aircraft flight plan data.

The Center controllers have many decision support tools (computer software programs) that provide vital information to assist the controllers in maintaining safe separation distances for all aircraft flying through their sector. For example, one tool available allows the controller to display the extended route of any aircraft on the radar screen called a vector line. This line projects where the aircraft will be within a specified number of minutes, assuming the aircraft does not change its course. This is a helpful tool to determine if aircraft flying intersecting routes pass safely within the separation standard, or if they conflict with each other. In addition to vector lines, the controller can also display a route line for any given aircraft on his or her radar screen. This tells the controller where

a particular aircraft is in specified number of minutes, as well as the path the aircraft will fly to get there. Decision support tools such as these help each controller look ahead and avoid conflicts.

In-flight Requirements and Instructions

The CFRs require the pilot in command under IFR in controlled airspace to continuously monitor an appropriate Center or control frequency. When climbing after takeoff, an IFR flight is either in contact with a radar-equipped local departure control or, in some areas, an ARTCC facility. As a flight transitions to the en route phase, pilots typically expect a handoff from departure control to a Center frequency if not already in contact with the Center.

The FAA National Aeronautical Information Systems (AIS) publishes en route charts depicting Centers and sector frequencies. [Figure 2-13] During handoff from one Center to another, the previous controller assigns a new frequency. In cases where flights may be out of range, the Center frequencies on the face of the chart are very helpful. In Figure 2-13, notice the boundary between Memphis, Tennessee and Atlanta, Georgia Centers, and the remote sites with discrete very high frequency (VHF) and ultra high frequency (UHF) for communicating with the appropriate ARTCC. These Center frequency boxes can be used for finding the nearest frequency within the aircraft range.

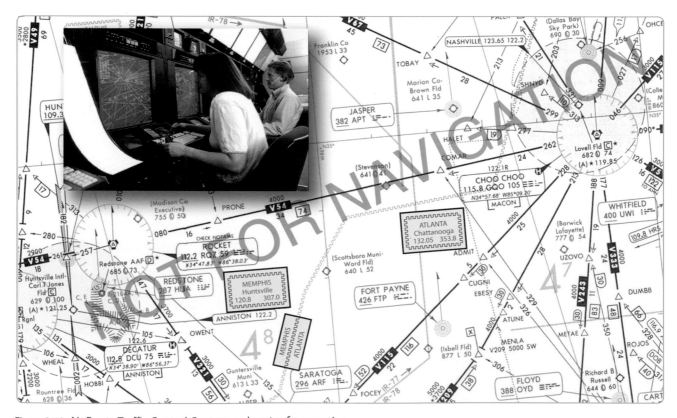

Figure 2-13. Air Route Traffic Control Centers and sector frequencies.

They also can be used for making initial contact with the Center for clearances. The exact location for the Center transmitter is not shown, although the frequency box is placed as close as possible to the known location.

During the en route phase, as a flight transitions from one Center facility to the next, a handoff or transfer of control is required as previously described. The handoff procedure is similar to the handoff between other radar facilities, such as departure or approach control. During the handoff, the controller whose airspace is being vacated issues instructions that include the name of the facility to contact, appropriate frequency, and other pertinent remarks.

Accepting radar vectors from controllers does not relieve pilots of their responsibility for safety of flight. Pilots must maintain a safe altitude and keep track of their position, and it is their obligation to question controllers, request an amended clearance, or, in an emergency, deviate from their instructions if they believe that the safety of flight is in doubt. Keeping track of altitude and position when climbing, and during all other phases of flight, are basic elements of situational awareness (SA). Aircraft equipped with an enhanced ground proximity warning system (EGPWS), terrain awareness and warning system (TAWS), or traffic alert and collision avoidance system (TCAS) help pilots detect and/or correct for potential unsafe proximities to other aircraft and increases pilot(s) situational awareness. Regardless of equipment, pilots must always maintain SA regarding their location and the location of traffic in their vicinity.

High Altitude Area Navigation Routing

Special high altitude routes allow pilots routing options for flight within the initial high altitude routing (HAR) Phase I expansion airspace. Pilots are able to fly user-preferred routes, referred to as non-restrictive routing (NRR), between specific fixes described by pitch (entry into) and catch (exit out of) fixes in the HAR airspace. Pitch points indicate an end of departure procedures, preferred IFR routings, or other established routing programs where a flight can begin a segment of NRR. The catch point indicates where a flight ends a segment of NRR and joins published arrival procedures, preferred IFR routing, or other established routing programs.

The HAR Phase I expansion airspace is defined as that airspace at and above FL 350 in fourteen of the western and southern ARTCCs. The airspace includes Minneapolis (ZMP), Chicago (ZAU), Kansas City (ZKC), Denver (ZDV), Salt Lake City (ZLC), Oakland (ZOA), Seattle Centers (ZSE), Los Angeles (ZLA), Albuquerque (ZAB), Fort Worth (ZFW), Memphis (ZME), and Houston (ZHU). Jacksonville (ZJX)

and Miami (ZMA) are included for east-west routes only. To develop a flight plan, select pitch and catch points which can be found in the Airport/Facility Directory (A/FD) based upon your desired route across the Phase I airspace. Filing requirements to pitch points, and from catch points, remain unchanged from current procedures. For the portion of the route between the pitch and catch points, NRR is permitted. Where pitch points for a specific airport are not identified, aircraft should file an appropriate departure procedure (DP), or any other user preferred routing prior to the NRR portion of their routing. Where catch points for a specific airport are not identified aircraft should file, after the NRR portion of their routing, an appropriate arrival procedure or other user preferred routing to their destination.

Additionally, information concerning the location and schedule of special use airspace (SUA) and Air Traffic Control Assigned Airspace (ATCAA) can be found at http://sua.faa.gov. ATCAA refers to airspace in the high altitude structure supporting military and other special operations. Pilots are encouraged to file around these areas when they are scheduled to be active, thereby avoiding unplanned reroutes around them.

In conjunction with the HAR program, area navigation (RNAV) routes have been established to provide for a systematic flow of air traffic in specific portions of the en route flight environment. The designator for these RNAV routes begin with the letter Q, for example, Q-501. Where those routes aid in the efficient orderly management of air traffic, they are published as preferred IFR routes.

Preferred IFR Routes

Preferred IFR routes are established between busier airports to increase system efficiency and capacity. They normally extend through one or more ARTCC areas and are designed to achieve balanced traffic flows among high density terminals. IFR clearances are issued on the basis of these routes except when severe weather avoidance procedures or other factors dictate otherwise. Preferred IFR routes are listed in the A/FD and can also be found on www.fly.faa.gov, which requires entering the following data: departure airport designator, destination, route type, area, aircraft types, altitude, route string, direction, departure ARTCC, and arrival ARTCC. [Figure 2-14] If a flight is planned to or from an area having such routes but the departure or arrival point is not listed in the A/FD, pilots may use that part of a preferred IFR route that is appropriate for the departure or arrival point listed. Preferred IFR routes are correlated with departure procedures (DPs) and STARs and may be defined by airways, jet routes, direct routes between NAVAIDs, waypoints, NAVAID radials/ distance measuring equipment (DME), or any combinations thereof.

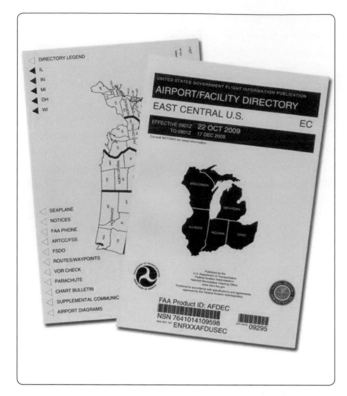

Figure 2-14. Airport/Facility Directory.

Preferred IFR routes are published in the A/FD for the low and high altitude stratum. If they begin or end with an airway number, it indicates that the airway essentially overlies the airport and flights normally are cleared directly on the airway. Preferred IFR routes beginning or ending with a fix indicate that pilots may be routed to or from these fixes via a SID route, radar vectors, or a STAR. Routes for major terminals are listed alphabetically under the name of the departure airport. Where several airports are in proximity, they are listed under the principal airport and categorized as a metropolitan area (e.g., New York Metro Area). One way preferred IFR routes are listed is numerically, showing the segment fixes and the direction and times effective. Where more than one route is listed, the routes have equal priority for use. Official location identifiers are used in the route description for very high frequency omnidirectional ranges (VORs) and very high frequency omnidirectional ranges/ tactical air navigation (VORTACs), and intersection names are spelled out. The route is direct where two NAVAIDs, an intersection and a NAVAID, a NAVAID and a NAVAID radial and distance point, or any navigable combination of these route descriptions follow in succession.

A system of preferred IFR routes helps pilots, flight crews, and dispatchers plan a route of flight to minimize route changes, and to aid in the efficient, orderly management of air traffic using Federal airways. Preferred IFR routes are designed to serve the needs of airspace users and to provide for a systematic flow of air traffic in the major terminal and en route flight environments. Cooperation by all pilots in filing preferred routes results in fewer air traffic

PREFERRED IFR ROUTES

Terminals	Route	Effective Times (UTC)
	(60–170 incl 210 kts plus, non–turbojet) V14 CEDOR DNY051 DNY V449 LHY V93 LVZ V613 FJC PTW..	1100–0300X
	or	
	(70–170 turbojets only) V14 CEDOR DNY051 DNY SLATT–STAR..	
Trenton (TTN)	(90–170, non–turbojet) V14 CEDOR DNY051 DNY LHY LVZ V613 FJC V149 MAZIE ARD	1100–0300X
	or	
	(90–170, turbojet) V14 CEDOR DNY051 DNY LHY LVZ V29 ETX V30 V149 MAZIE ARD	1100–0300X
BALTIMORE (BWI)—See Washington/Baltimore Metro		
BOSTON METRO AREA (BOS)		
Cleveland (CLE)	(60–170) MHT V490 UCA V2 SYR V84 GEE V464 V115 TDT V72 V232 CXR	1000–0300X
Kennedy (JFK)	(110–170, jets) LUCOS SEY067 SEY PARCH CCC ROBER	1100–0300X
	or	
	(110–170, Props) LUCOS SEY067 SEY HTO V46 DPK	
	or	
	(AOB 100) BOSOX V419 V14 ORW V16 DPK	

Figure 2-15 Preferred IFR routes.

delays and better efficiency for departure, en route, and arrival air traffic service. [Figure 2-15]

Substitute Airway or Route Structures

ARTCCs are responsible for specifying essential substitute airway or route segments (sub-routes) and fixes for use during scheduled or unscheduled VOR/VORTAC shutdowns. Scheduled shutdowns of navigational facilities require planning and coordination to ensure an uninterrupted flow of air traffic. AIS (or AJV-5), in coordination with the ARTCCs, determine when the length of outages or other factors require publication of sub- routes and Flight Inspection Services (AJW-3) provides flight inspection services, obstacle clearance verification, certification, and final approval of substitute routes.

Substitute Airway En Route Flight Procedures

A schedule of proposed facility shutdowns within the region is maintained and forwarded as far in advance as possible to enable the substitute routes to be published. Substitute routes are normally based on VOR/VORTAC facilities established and published for use in the appropriate altitude strata. In the case of substitute routes in the upper airspace stratum, it may be necessary to establish routes by reference to VOR/VORTAC facilities used in the low altitude system. Non-directional (radio) beacon (NDB) facilities may only be used where VOR/VORTAC coverage is inadequate and ATC requirements necessitate use of such NAVAIDs. Where operational necessity dictates, NAVAIDs may be used beyond their standard service volume (SSV) limits that define the reception limits of unrestricted NAVAIDs, which are usable for random/unpublished route navigation, provided that the routes can be given adequate frequency protection.

The centerline of substitute routes must be contained within controlled airspace [Figure 2-16], although substitute

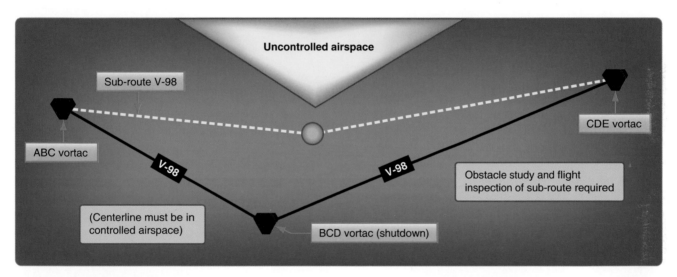

Figure 2-16 14 CFR Part 95 sub-routes.

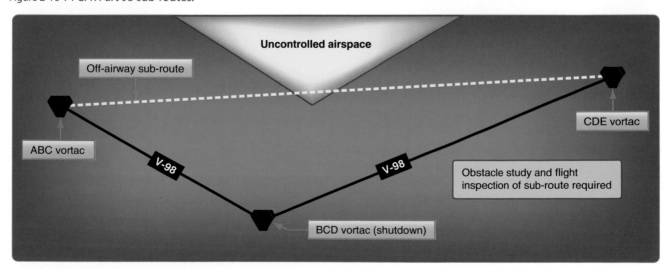

Figure 2-17 Non-Part 95 sub-routes.

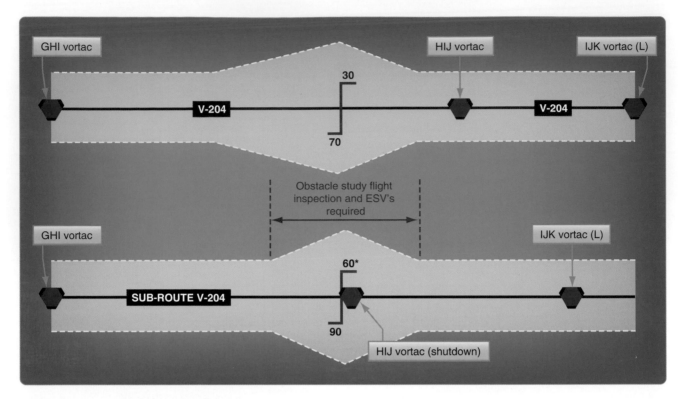

Figure 2-18 Sub-route wider than existing route.

routes for off-airway routes may not be in controlled airspace. [Figure 2-17] Substitute routes are flight inspected to verify clearance of controlling obstacles and to check for satisfactory facility performance. If substitute routes do not overlie existing routes, or are wider than existing routes, map studies are required to identify controlling obstacles. [Figure 2-18] The format for describing substitute routes is from navigational fix to navigational fix. A minimum en route altitude (MEA) and a maximum authorized altitude (MAA) are provided for each route segment. Temporary reporting points may be substituted for the out-of-service facility and only those other reporting points that are essential for ATC. Normally, temporary reporting points over intersections are not necessary where Center radar coverage exists. A minimum reception altitude (MRA) is established for each temporary reporting point.

Tower En Route Control

Tower en route control (TEC) is an ATC program available to pilots that provides a service to aircraft proceeding to and from metropolitan areas. It links designated approach control areas by a network of identified routes made up of the existing airway structure of the NAS, which makes it possible to fly an IFR flight without leaving approach control airspace. [Figure 2-19] This service is designed to help expedite air traffic and reduces ATC and pilot communication requirements. The program is generally used by non-turbojet aircraft operating at and below 10,000 feet but a few facilities, such as Milwaukee and

Chicago, have allowed turbojets to proceed between city pairs. Participating flights are relatively short with a duration of 2 hours or less.

TEC is referred to as tower en route, or tower-to-tower, and allows flight beneath the en route structure. TEC reallocates airspace both vertically and geographically to allow flight planning between city pairs while remaining with approach control airspace. All users are encouraged to use the TEC route descriptions located in the A/FD when filing flight plans. [Figure 2-20] All published TEC routes are designed to avoid en route airspace, and the majority is within radar coverage.

Tower En Route Control Route Descriptions

The graphic depiction of TEC routes located in the A/FD is not to be used for navigation or for detailed flight planning because not all city pairs are depicted. The information is intended to show geographic areas connected by TEC. [Figure 2-19] Pilots should refer to the route descriptions for specific flight planning.

As shown in Figure 2-20, the route description contains four columns of information. The first column is the approach control area within which the departure airport is located, which are listed alphabetically. The second column shows the specific route, airway, or radial that is to be used. The third column shows the highest altitude allowed for the route, and the fourth shows the destination airport,

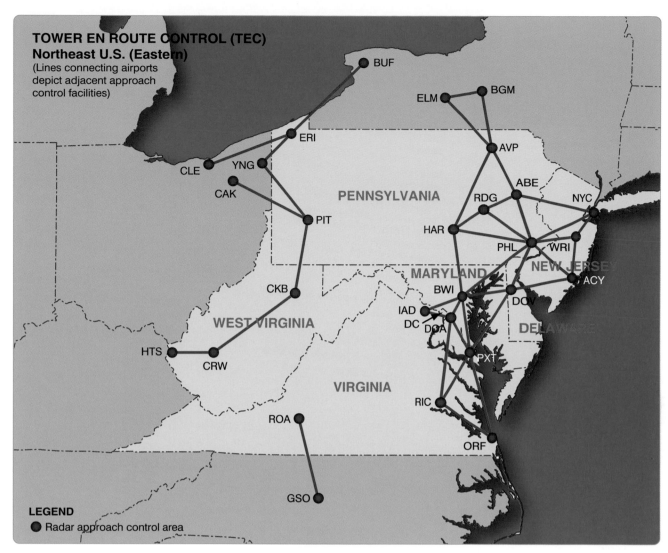

Figure 2-19. Tower En Route Control (TEC) Northeast U.S. (Eastern).

which are also listed alphabetically. When flight planning, it is important to always check current publications for information about the departure and destination airport. Routes are effective only during each respective terminal facilities normal operating hours. Always check NOTAMs to ensure that appropriate terminal facilities are operating for the planned flight time. Altitudes are always listed in thousands of feet. ATC may request that the pilot changes altitude while in flight in order to maintain the flight within approach control airspace. ATC provides radar monitoring and, if necessary, course guidance if the highest altitude assigned is below the MEA.

Shown in Figure 2-21, under the second column, the word "Direct" appears as the route when radar vectors are used or no airway exists. This also indicates that a SID or STAR may be assigned by ATC. When a NAVAID or intersection identifier appears with no airway immediately preceding or following the identifier, the routing is understood to be direct to or from that point unless otherwise cleared by ATC. Routes beginning and ending with an airway indicate that the airway essentially overflies the airport, or radar vectors are issued. [Figure 2-21] Where more than one route is listed to the same destination, ensure that the correct route for the type of aircraft classification has been filed. These are denoted after the route in the altitude column using J (jet powered), M (turbo props/special, cruise speed 190 knots or greater), P (non-jet, cruise speed 190 knots or greater), or Q (non-jet, cruise speed 189 knots or less). [Figure 2-22] Although all airports are not listed under the destination column, IFR flights may be planned to satellite airports in the proximity of major airports via the same routing. When filing flight plans, the coded route identifier (i.e., BURL 1, VTUL4, or POML3) may be used in lieu of the route of flight.

Airway and Route System

There are three fixed route systems established for air navigation purposes. They are the Federal airway

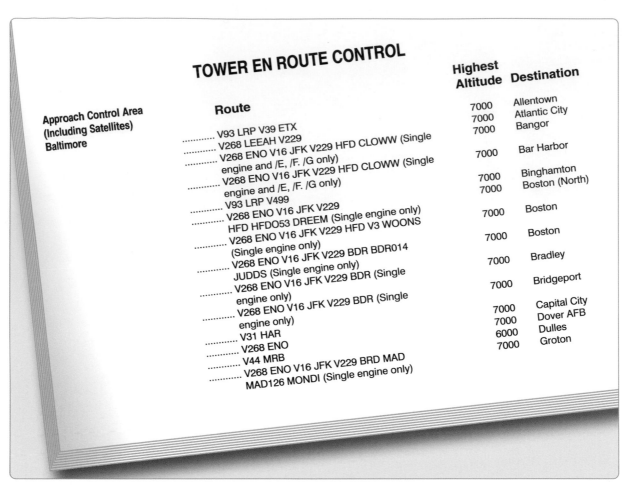

TOWER EN ROUTE CONTROL

Approach Control Area (Including Satellites)	Route	Highest Altitude	Destination
Baltimore			
	V93 LRP V39 ETX	7000	Allentown
	V268 LEEAH V229	7000	Atlantic City
	V268 ENO V16 JFK V229 HFD CLOWW (Single engine and /E, /F. /G only)	7000	Bangor
	V268 ENO V16 JFK V229 HFD CLOWW (Single engine and /E, /F. /G only)	7000	Bar Harbor
	V93 LRP V499	7000	Binghamton
	V268 ENO V16 JFK V229 HFD HFDO53 DREEM (Single engine only)	7000	Boston (North)
	V268 ENO V16 JFK V229 HFD V3 WOONS (Single engine only)	7000	Boston
	V268 ENO V16 JFK V229 BDR BDR014 JUDDS (Single engine only)	7000	Boston
	V268 ENO V16 JFK V229 BDR (Single engine only)	7000	Bradley
	V268 ENO V16 JFK V229 BDR (Single engine only)	7000	Bridgeport
	V31 HAR	7000	Capital City
	V268 ENO	7000	Dover AFB
	V44 MRB	6000	Dulles
	V268 ENO V16 JFK V229 BRD MAD MAD126 MONDI (Single engine only)	7000	Groton

Figure 2-20. Baltimore Airport/Facility Directory, Tower En Route Control route descriptions.

TOWER EN ROUTE CONTROL

Approach Control Area (Including Satellites)	Route	Highest Altitude	Destination
Allentown			
	EJC V149 LHY	8000	Albany
	ETX LHY	8000	Albany
	V149 MAZIE ARD CYN	5000	Atlantic City
	V93 LRP	8000	Baltimore
	EXT V162 DUMMR V93 LRP	6000	Baltimore
	V39 LRP	8000	Baltimore
	V130	10000	Bradley
	Direct	10000	Bradley
	FJC STW	5000	Caldwell
	(2) EXT V30 SBJ	5000	Farmingdale
	ETX V162 HAR	8000	Harrisburg
	Direct	10000	Hartford
	EXT ETX004 WEISS	4000	Hazleton
	EXT V39	4000	Lancaster

Figure 2-21. Allentown Airport/Facility Directory, Tower En Route Control route descriptions.

2-12

Approach Control Area (Including Satellites)	TOWER EN ROUTE CONTROL Route	Highest Altitude	Destination
Atlantic City	V229 DIXIE V276 ARD	6000	Allentown
	V1 DIXIE V276 ARD (Single engine only)	6000	Allentown
	V1 ATR V308 OTT	4000	Andrews, AFB
	LEEAH V268 BAL	4000	Baltimore
	V1 JFK V229 HFD CLOWM (Single engine and /E, /F, /G only)	6000	Bangor
	V1 JFK V229 HFD CLOWM (Single engine and /E, /F, /G only)	6000	Bar Harbor
	V1 JFK V229 HFD HFD053 DREEM (Single (Single engine only)	6000	Boston (North)
	V1 JFK V229 HFD V3 WOONS (Single engine only	6000	Boston
	V1 JFK V229 HFD FOSTY WOONS (Single engine only)	6000	Boston
	V1 JFK V229 BDR BDR14 JUDDS (Single engine only)	6000	Bradley
	V184 ZIGGI JFK 210 JFK V229 BDR (Twins only, n/a between 1400-2100)	6000	Bridgeport
	HOWIE V1 JFK V229 BDR (Single engine only)	6000	Bridgeport
	V184 00D DQO V469 HAR	4000	Capital City

Figure 2-22. Atlantic City Airport/Facility Directory, Tower En Route Control route descriptions.

consisting of VOR (low victor airways, high jet routes), NDB (low or medium frequency) and the RNAV route system. To the extent possible, these route systems are aligned in an overlying manner to facilitate transition between each. The majority of the airways are made up of victor airways, jet routes, and RNAV, but some low/medium frequency (L/MF) airways and routes are still being used in Alaska and one other that is located off the coast of North Carolina and is called Green 13 (G13). [Figure 2-23]

Airway/Route Depiction

IFR en route charts show all IFR radio NAVAIDs that have been flight-checked by the FAA and are operational. The FAA, AIS publishes and distributes U.S. Government Civil Aeronautical Charts and flight information publications. IFR en route navigation information is provided on three charts: IFR en route low altitude chart, IFR en route high altitude chart, and Terminal Area Chart (TAC). [Figure 2-24A and B]

IFR En Route Low Altitude Chart

En route low altitude charts provide aeronautical information for navigation under IFR conditions below 18,000 feet MSL. Low altitude charts [Figure 2-25] include the following information:

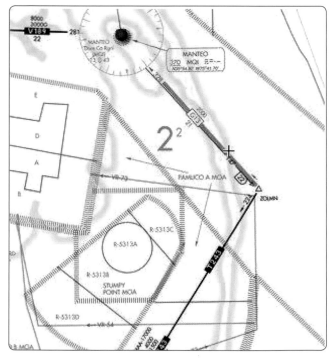

Figure 2-23. Low frequency airway G13.

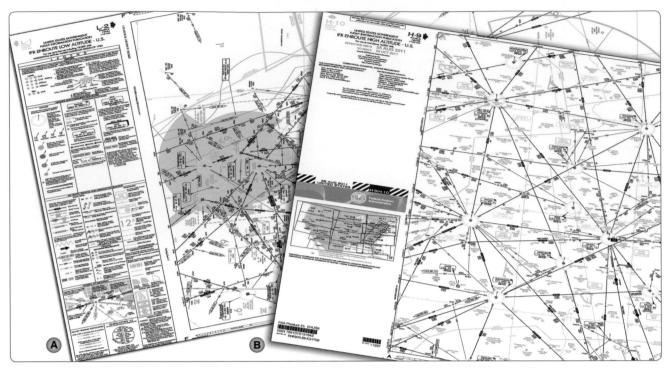

Figure 2-24. IFR en route low altitude (left) and high altitude (right) charts.

- Airways[Figure 2-25A]
- RNAV routes [Figure 2-25B]
- Limits of controlled airspace [Figure 2-25C]
- VHF radio aids to navigation (frequency, identification, channel, geographic coordinates) [Figure 2-25D]
- Airports that have an instrument approach procedure or a minimum 3,000 foot hard surface runway [Figure 2-25E]
- Off-route obstruction clearance altitudes (OROCA) [Figure 2-25F]
- Reporting points [Figure 2-25G]
- Special use airspace areas [Figure 2-25H]
- Military training routes [Figure 2-25I]

IFR aeronautical charts depict VOR airways (airways based on VOR or VORTAC NAVAIDs) in black, identified by a "V" (Victor) followed by the route number (e.g., V12). [Figure 2-26] LF/MF airways (airways based on LF/MF NAVAIDs) are sometimes referred to as colored airways because they are identified by color name and number (e.g., Amber One, charted as A1). Green and red airways are plotted east and west, and amber and blue airways are plotted north and south. Regardless of their color identifier, LF/MF airways are depicted in brown. [Figure 2-27]

Airway/route data, such as the airway identifications, bearings or radials, mileages, and altitude (e.g., MEA),

minimum obstacle clearance altitude (MOCA), and MAA, are shown aligned with the airway and in the same color as the airway. [Figure 2-26]

All airways/routes that are predicated on VOR or VORTAC NAVAIDs are defined by the outbound radial from the NAVAID. Airways/routes that are predicated on LF/MF NAVAIDs are defined by the inbound bearing.

New low altitude RNAV routes have been created by the FAA. RNAV routes provide more direct routing for IFR aircraft and enhance the safety and efficiency of the NAS. In order to utilize these routes, aircraft must be equipped with IFR approved GNSS. In Alaska, when using RNAV routes, the aircraft must be equipped with Technical Standing Order (TSO)-145a and 146a equipment.

Low altitude RNAV only routes are identified by the letter "T" prefix, followed by a three digit number (T-200 to T-500). RNAV routes are depicted in aeronautical blue, as well as the RNAV route data, which includes the following [Figure 2-28]:

- Route line
- Identification boxes
- Mileages
- Waypoints
- Waypoint names
- Magnetic reference bearings
- MEAs

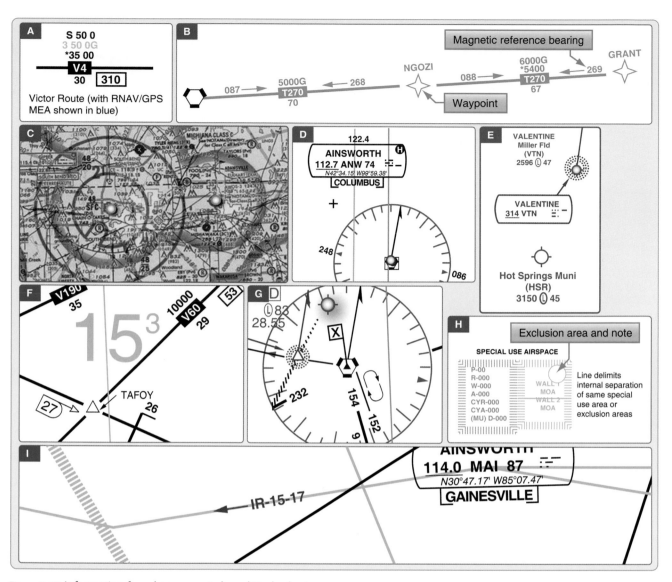

Figure 2-25. Information found on en route low altitude charts.

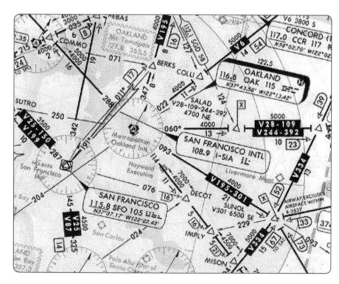

Figure 2-26. Victor airways.

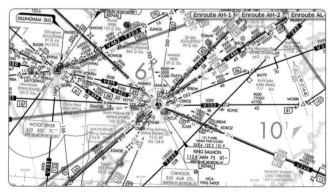

Figure 2-27. LF/MF airways.

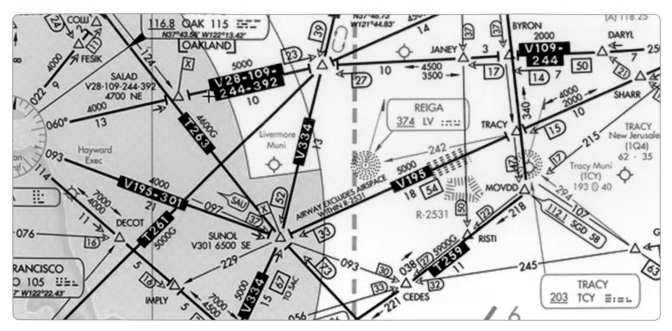

Figure 2-28. Low altitude RNAV routes.

Magnetic reference bearings are shown originating from a waypoint, fix/reporting point, or NAVAID. A GNSS MEA for each segment is established to ensure obstacle clearance and communications reception. All MEAs are identified with a "G" suffix. [Figure 2-29]

Joint Victor/RNAV routes are depicted using black for the victor airways and blue for the RNAV routes, and the identification boxes for each are shown adjacent to one another. Magnetic reference bearings are not shown. MEAs are stacked in pairs or in two separate columns, GNSS and

Victor. On joint routes, RNAV specific information is printed in blue. [Figure 2-30]

IFR En Route High Altitude Chart

En route high altitude charts provide aeronautical information for navigation under IFR conditions at and above FL180. [Figure 2-31] High altitude charts include the following information:

*   Jet route structure
*   RNAV Q-routes

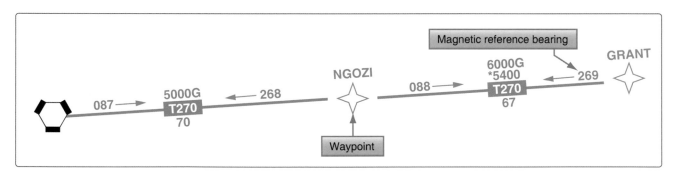

Figure 2-29. Low altitude RNAV route data.

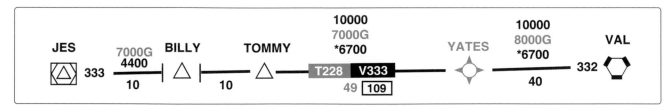

Figure 2-30. Joint Victor/RNAV airway.

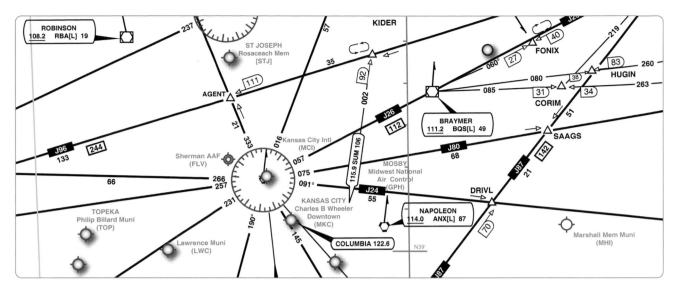

Figure 2-31. IFR en route high altitude chart.

- VHF radio aids to navigation (frequency, ID, channel, geographic coordinates)

- Selected airports

- Reporting points

- Navigation reference system (NRS) waypoints [Figure 2-32]

Jet routes are depicted in black with a "J" identifier followed by the route number (e.g., "J12") and are based on VOR or VORTAC NAVAIDs. [Figure 2-33] RNAV "Q" Route MEAs are shown when other than 18,000 feet. [Figure 2-34] MEAs for GNSS RNAV aircraft are identified with a "G" suffix. MEAs for DME/DME/IRU RNAV aircraft do not have a "G" suffix. All RNAV routes and associated data is charted in aeronautical blue and magnetic reference bearings are shown originating from a waypoint, fix/reporting point, or NAVAID. When joint Jet/RNAV routes are depicted, the route identification boxes are located adjacent to each

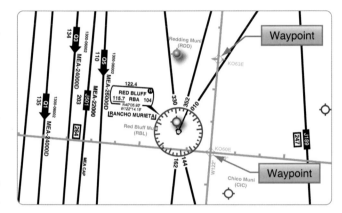

Figure 2-32. Navigation reference system (NRS) waypoints.

other with the route charted in black. [Figure 2-35] With the exception of "Q" routes in the Gulf of Mexico, GNSS or DME/DME/IRU RNAV equipment is required along with radar monitoring capabilities. For aircraft that have DME/DME/IRU RNAV equipment, refer to the A/FD for specific DME information.

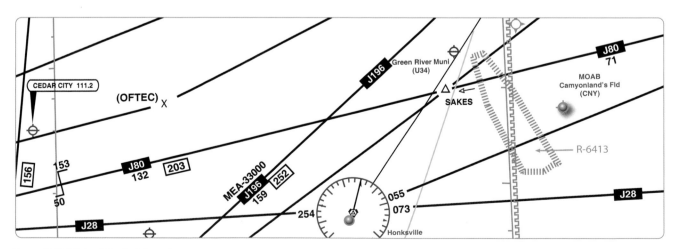

Figure 2-33. High altitude jet routes.

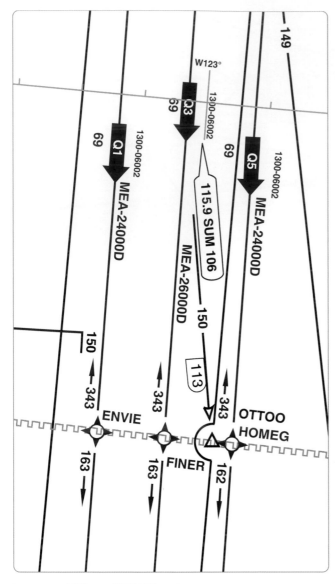

Figure 2-34. MEAs on RNAV (Q) routes.

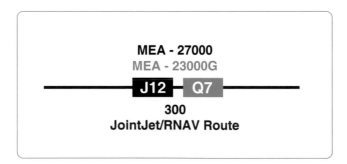

Figure 2-35. Joint jet/RNAV routes.

VHF Airways

Victor airways are a system of established routes that run along specified VOR radials, from one VOR station to another. The purpose is to make flight planning easier and they help ATC to organize and regulate the air traffic

flow. Almost all commercial flights are routed along these airways but they are available for use by any pilot provided that the proper altitudes are employed.

Victor Airway Navigation Procedures

The procedure for getting established on a victor airway is to either fly directly to a nearby VOR or to intercept an airway radial along the route of flight. Once the pilot is established on an airway, it is important to follow the procedures and guidelines put in place to ensure air traffic separation and optimal safety on the airway. When using victor airways for navigation, procedures do not allow the pilot to jump from one VOR to another, but must navigate from one to the next by using the alternating outbound/inbound procedure of linking VORs. For example, when departing from Zanesville VOR on V-214, the pilot selects the 090° radial with a FROM indication on the course deviation indicator (CDI) and should correct as necessary to continuously maintain track on the centerline of the airway. [Figure 2-36] The pilot should continue on this course until it is time to change over to the inbound course to the Bellaire VOR.

LF/MF Airways

The basic LF/MF airway width is 4.34 nautical miles (NM) on each side of the centerline; the width expands by five degrees when the distance from the facility providing course guidance is greater than 49.66 NM. [Figure 2-37]

En Route Obstacle Clearance Areas

All published routes in the NAS are based on specific obstacle clearance criteria. An understanding of en route obstacle clearance areas helps with SA and may help avoid controlled flight into terrain (CFIT). Obstacle clearance areas for the en route phase of flight are identified as primary, secondary, and turning areas.

The primary and secondary area obstacle clearance criteria, airway and route widths, and the ATC separation procedures for en route segments are a function of safety and practicality in flight procedures. These flight procedures are dependent upon the pilot, the aircraft, and the navigation system being used, resulting in a total VOR system accuracy factor along with an associated probability factor. The pilot/aircraft information component of these criteria includes pilot ability to track the radial and the flight track resulting from turns at various speeds and altitudes under different wind conditions. The navigation system information includes navigation facility radial alignment displacement, transmitter monitor tolerance, and receiver accuracy. All of these factors were considered during development of en route criteria. From this analysis, the

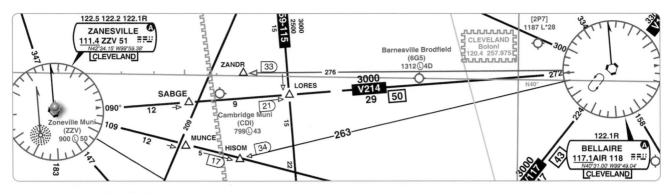

Figure 2-36. Zanesville VOR/Victor Airway 214.

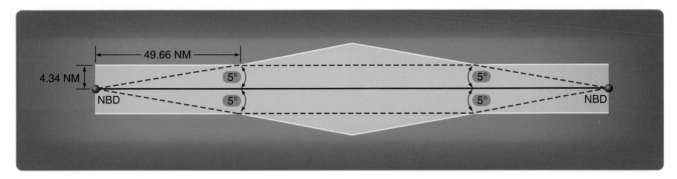

Figure 2-37. LF/MR airway width.

computations resulted in a total system accuracy of ±4.5° 95 percent of the time and ±6.7° 99 percent of the time. The 4.5° value became the basis for primary area obstacle clearance criteria, airway and route widths, and the ATC separation procedures. The 6.7° value provides secondary obstacle clearance area dimensions.

Primary and Secondary En Route Obstacle Clearance Areas

The primary obstacle clearance area has a protected width of 8 NM with 4 NM on each side of the centerline. The primary area has widths of route protection based upon system accuracy of a ±4.5° angle from the NAVAID. These 4.5° lines extend out from the NAVAID and intersect the boundaries of the primary area at a point approximately

51 NM from the NAVAID. Ideally, the 51 NM point is where pilots would change over from navigating away from the facility, to navigating toward the next facility, although this ideal is rarely achieved. [Figure 2-38]

If the distance from the NAVAID to the change-over point (COP) is more than 51 NM, the outer boundary of the primary area extends beyond the 4 NM width along the 4.5° line when the COP is at midpoint. This means the primary area, along with its obstacle clearance criteria, is extended out into what would have been the secondary area. Additional differences in the obstacle clearance area result in the case of the effect of an offset COP or dogleg segment. For protected en route areas, the minimum obstacle clearance in the primary area, not designated as

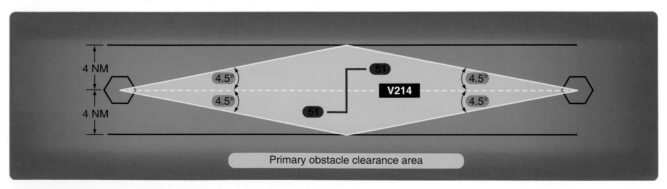

Figure 2-38. Primary obstacle clearance area.

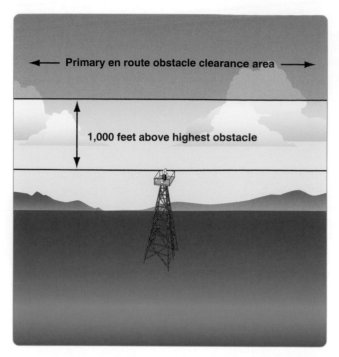

Figure 2-39. Non-mountainous obstacle clearance in the primary area.

mountainous under 14 CFR Part 95—IFR altitude, is 1,000 feet over the highest obstacle. [Figure 2-39] The secondary obstacle clearance area extends along a line 2 NM on each side of the primary area. Navigation system accuracy in the secondary area has widths of route protection of a ±6.7° angle from the NAVAID. These 6.7° lines intersect the outer boundaries of the secondary areas at the same point as primary lines, 51 NM from the NAVAID. If the distance from the NAVAID to the COP is more than 51 NM, the secondary area extends along the 6.7° line when the COP is at mid-point. [Figure 2-40] In all areas, mountainous and non-mountainous, obstacles that are located in secondary areas are considered as obstacles to air navigation if they extend above the secondary obstacle clearance plane. This plane begins at a point 500 feet above the obstacles

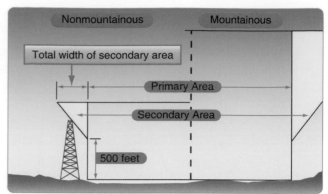

Figure 2-41. Primary and secondary obstacle clearance area.

(natural or man-made) upon which the primary obstacle clearance area is based, and slants upward at an angle that causes it to intersect the outer edge of the secondary area at a point 500 feet higher. [Figure 2-41]

Changeover Points

When flying airways, pilots normally change frequencies midway between NAVAIDs, although there are times when this is not practical. If the navigation signals cannot be received from the second VOR at the midpoint of the route, a COP is depicted and shows the distance in NM to each NAVAID. [Figure 2-42] COPs indicate the point where a frequency change is necessary to receive course guidance from the facility ahead of the aircraft instead of the one behind. These COPs divide an airway or route segment and ensure continuous reception of navigation signals at the prescribed minimum en route IFR altitude. They also ensure that other aircraft operating within the same portion of an airway or route segment receive consistent azimuth signals from the same navigation facilities regardless of the direction of flight.

Where signal coverage from two VORs overlaps at the MEA, the COP normally is designated at the midpoint. Where

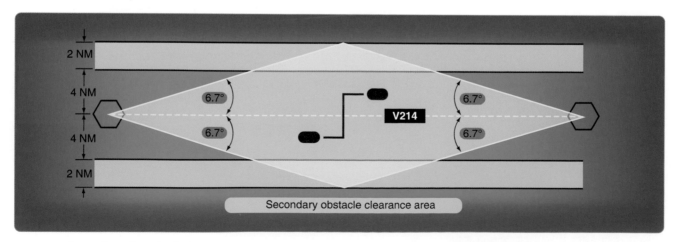

Figure 2-40. Secondary obstacle clearance area.

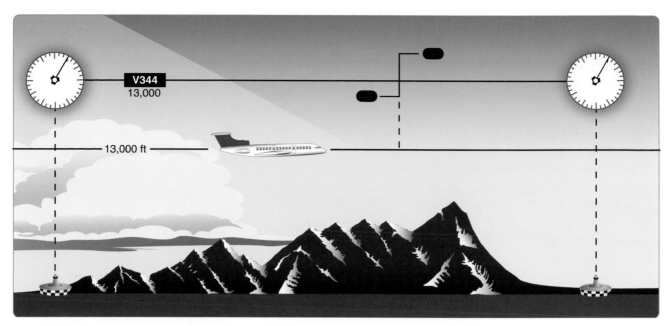

Figure 2-42. Changeover points.

radio frequency interference or other navigation signal problems exist, the COP is placed at the optimum location, taking into consideration the signal strength, alignment error, or any other known condition that affects reception. The COP has an effect on the primary and secondary obstacle clearance areas. On long airway or route segments, if the distance between two facilities is over 102 NM and the COP is placed at the midpoint, the system accuracy lines extend beyond the minimum widths of 8 and 12 NM, and a flare or spreading outward results at the COP. [Figure 2-43] Offset COP and dogleg segments on airways or routes can also result in a flare at the COP.

Direct Route Flights

Direct route flights are flights that are not flown on the radials or courses of established airways or routes. Direct route flights must be defined by indicating the radio fixes over which the flight passes. Fixes selected to define the route should be those over which the position of the aircraft can be accurately determined. Such fixes automatically become compulsory reporting points for the flight, unless advised otherwise by ATC. Only those NAVAIDs established for use in a particular structure (i.e., in the low or high structures) may be used to define the en route phase of a direct flight within that altitude structure.

Figure 2-44 shows a straight line on a magnetic course from SCRAN intersection of 270° direct to the Fort Smith Regional Airport in Arkansas that passes just north of restricted areas R-2401A and B and R-2402. Since the airport and the restricted areas are precisely plotted, there is an assurance that you will stay north of the restricted areas. From a practical standpoint, it might be better to fly direct

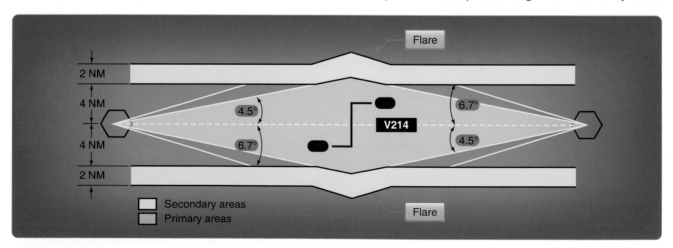

Figure 2-43. Changeover point effect on long airway or route segment.

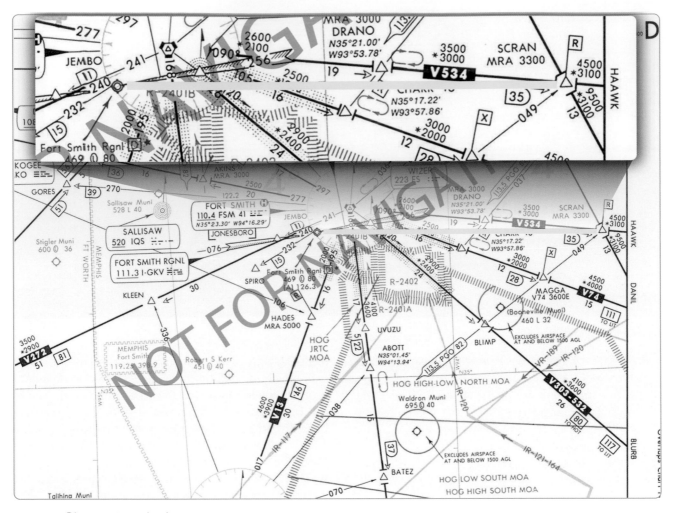

Figure 2-44. Direct route navigation.

to the Wizer NDB. This route goes even further north of the restricted areas and places you over the final approach fix to Runway 25 at Fort Smith.

The azimuth feature of VOR aids and the azimuth and distance (DME) features of VORTAC and TACAN aids are assigned certain frequency protected areas of airspace that are intended for application to established airway and route use and to provide guidance for planning flights outside of established airways or routes. These areas of airspace are expressed in terms of cylindrical service volumes of specified dimensions called class limits or categories.

An operational service volume has been established for each class in which adequate signal coverage and frequency protection can be assured. To facilitate use of VOR, VORTAC, or TACAN aids, consistent with their operational service volume limits, pilot use of such aids for defining a direct route of flight in controlled airspace should not exceed the following:

1. Operations above FL 450—use NAVAIDs not more than 200 NM apart. These aids are depicted on en route high altitude charts.

2. Operation off established routes from 18,000 feet MSL to FL 450—use NAVAIDs not more than 260 NM apart. These aids are depicted on en route high altitude charts.

3. Operation off established airways below 18,000 feet MSL—use NAVAIDs not more than 80 NM apart. These aids are depicted on en route low altitude charts.

4. Operation off established airways between 14,500 feet MSL and 17,999 feet MSL in the conterminous United States—(H) facilities not more than 200 NM apart may be used.

Increasing use of self-contained airborne navigational systems that do not rely on the VOR/VORTAC/TACAN system has resulted in pilot requests for direct routes that exceed NAVAID service volume limits. These direct route requests are approved only in a radar environment with approval based on pilot responsibility for navigation on the authorized direct route. Radar flight following is provided by ATC for ATC purposes. At times, ATC initiates a direct route in a radar environment that exceeds NAVAID service volume limits. In such cases, ATC provides radar monitoring and navigational assistance as necessary.

When filing for a direct route flight, airway or jet route numbers, appropriate to the stratum in which operation is conducted, may also be included to describe portions of the route to be flown. The following is an example of how a direct route flight would be written.

MDW V262 BDF V10 BRL STJ SLN GCK

Spelled out: from Chicago Midway Airport via Victor 262 to Bradford, Victor 10 to Burlington, Iowa, direct St. Joseph, Missouri, direct Salina, Kansas, direct Garden City, Kansas.

NOTE: When route of flight is described by radio fixes, the pilot is expected to fly a direct course between the points named.

Pilots should keep in mind that they are responsible for adhering to obstruction clearance requirements on those segments of direct routes that are outside of controlled airspace. The MEAs and other altitudes shown on low altitude IFR en route charts pertain to those route segments within controlled airspace, and those altitudes may not meet obstruction clearance criteria when operating off those routes.

Published RNAV Routes

Published RNAV routes are fixed, permanent routes that can be flight planned and flown by aircraft with RNAV capability. These are being expanded worldwide as new RNAV routes are developed, and existing charted, conventional routes are being designated for RNAV use. It is important to be alert to the rapidly changing application of RNAV techniques being applied to conventional en route airways. Published RNAV routes may potentially be found on any en route chart. The published RNAV route designation may be obvious, or, on the other hand, RNAV route designations may be less obvious, as in the case where a published route shares a common flight track with a conventional airway.

NOTE: The use of RNAV is dynamic and rapidly changing, therefore, en route charts are continuously being updated for information changes, and you may find some differences between charts.

Basic designators for air traffic service (ATS) routes and their use in voice communications have been established. One of the main purposes of a system of route designators is to allow both pilots and ATC to make unambiguous reference to RNAV airways and routes. Basic designators for ATS routes consist of a maximum of five, and in no case to exceed six, alpha/numeric characters in order to be usable by both ground and airborne automation systems. The designator indicates the type of the route, such as high/low altitude, specific airborne navigation equipment requirements, such as RNAV, and the aircraft type using the route primarily and exclusively. The basic route designator consists of one or two letter(s) followed by a number from 1 to 999.

Composition of Designators

The prefix letters that pertain specifically to RNAV designations are included in the following list:

1. The basic designator consists of one letter of the alphabet followed by a number from 1 to 999. The letters may be:

    a. A, B, G, R—for routes that form part of the regional networks of ATS route and are not RNAV routes;

    b. L, M, N, P—for RNAV routes that form part of the regional networks of ATS routes;

    c. H, J, V, W—for routes that do not form part of the regional networks of ATS routes and are not RNAV routes;

    d. Q, T, Y, Z—for RNAV routes that do not form part of the regional networks of ATS routes.

2. Where applicable, one supplementary letter must be added as a prefix to the basic designator as follows:

    a. K—to indicate a low level route established for use primarily by helicopters;

    b. U—to indicate that the route or portion thereof is established in the upper airspace;

    c. S—to indicate a route established exclusively for use by supersonic airplanes during acceleration/deceleration and while in supersonic flight.

3. Where applicable, a supplementary letter may be added after the basic designator of the ATS route as a suffix as follows:

    a. F—to indicate that on the route or portion thereof advisory service only is provided;

    b. G—to indicate that on the route or portion thereof flight information services only is provided;

    c. Y—for RNP 1 routes at and above FL 200 to indicate that all turns on the route between 30° and 90° must be made within the tolerance of a tangential arc between the straight leg segments defined with a radius of 22.5 NM;

d. Z—for RNP 1 routes at and below FL 190 to indicate that all turns on the route between 30° and 90° should be made within the tolerance of a tangential arc between the straight leg segments defined with a radius of 15 NM.

NOTE: RNAV Q-routes require en route RNAV 2, corresponding NAV/E2 code and PBN/C1-C4 based on navigation system update source.

Use of Designators in Communications

In voice communications, the basic letter of a designator should be spoken in accordance with the International Civil Aviation Organization (ICAO) spelling alphabet. Where the prefixes K, U, or S, previously mentioned, are used in voice communications, they should be pronounced as:

K—Kopter

U—Upper, as in the English language

S—Supersonic

Where suffixes F, G, Y or Z specified in above, are used, the flight crew should not be required to use them in voice communications. Below is an example of how the letters and numbers are spoken.

A11—Alpha Eleven

UR5—Upper Romeo Five

KB34—Kopter Bravo Thirty Four

UW456—Upper Whiskey Four Fifty Six

The en route chart excerpt depicts three published RNAV jet routes: J804R, J888R, and J996R. [Figure 2-45] The R suffix is a supplementary route designator denoting an RNAV route. The overlapping symbols for the AMOTT intersection and waypoint indicate that AMOTT can be identified by conventional navigation or by latitude and longitude coordinates. Although coordinates were originally included for aircraft equipped with an inertial navigation system (INS), they are now a good way to cross check between the coordinates on the chart and in the flight management system (FMS) or global positioning system (GPS) databases

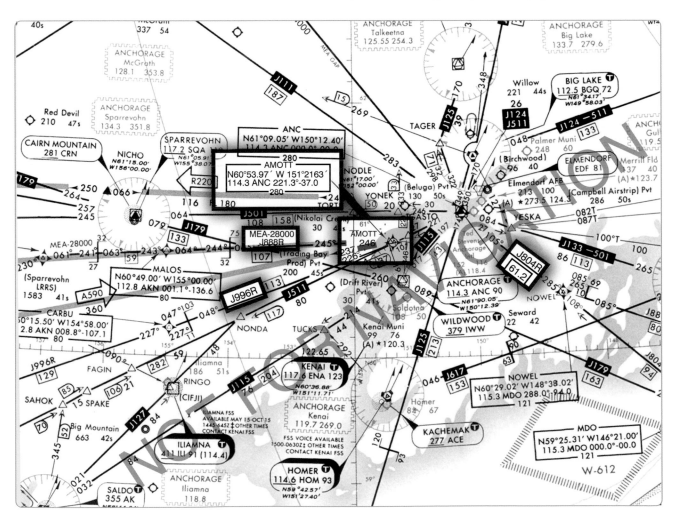

Figure 2-45. Published RNAV jet routes.

to ensure you are tracking on your intended en route course. The AMOTT RNAV waypoint includes bearing and distance from the Anchorage VORTAC.

Random RNAV Routes

Random RNAV routes are direct routes that are based on RNAV capability between waypoints defined in terms of latitude or longitude coordinates, degree-distance fixes, or offsets from established routes or airways at a specified distance and direction. Radar monitoring by ATC is required on all random RNAV routes. Random RNAV routes can only be approved in a radar environment. Factors that are considered by ATC when approving random RNAV routes include the capability to provide radar monitoring and compatibility with traffic volume and flow. ATC radar monitor each flight; however, navigation on the random RNAV route is the responsibility of the pilot.

Pilots flying aircraft that are equipped with approved area navigation equipment may file for RNAV routes throughout the NAS and may be filed for in accordance with the following procedures:

1. File airport-to-airport flight plans.

2. File the appropriate RNAV capability certification suffix in the flight plan.

3. Plan the random route portion of the flight plan to begin and end over appropriate arrival and departure transition fixes or appropriate NAVAIDs for the altitude stratum within which the flight is conducted. The use of normal preferred DPs and STAR, where established, is recommended.

4. File route structure transitions to and from the random route portion of the flight.

5. Define the random route by waypoints. File route description waypoints by using degree distance fixes based on navigational aids that are appropriate for the altitude stratum.

6. File a minimum of one route description waypoint for each ARTCC through whose area the random route is flown. These waypoints must be located within 200 NM of the preceding center's boundary.

7. File an additional route description waypoint for each turnpoint in the route.

8. Plan additional route description waypoints as required to ensure accurate navigation via the filed route of flight. Navigation is the pilot's responsibility unless ATC assistance is requested.

9. Plan the route of flight so as to avoid prohibited and restricted airspace by 3 NM unless permission has been obtained to operate in that airspace and the appropriate ATC facilities are advised.

NOTE: To be approved for use in the NAS, RNAV equipment must meet the appropriate system availability, accuracy, and airworthiness standards. For additional guidance on equipment requirements, see Advisory Circular (AC) 20-138C, Airworthiness Approval of Positioning and Navigation Systems. For airborne navigation database, see AC 90-105, Approval Guidance for RNP Operations and Barometric Vertical Navigation in the U.S. National Airspace System.

Pilots flying aircraft that are equipped with latitude/longitude coordinate navigation capability, independent of VOR/TACAN references, may file for random RNAV routes at and above FL 390 within the conterminous United States using the following procedures:

1. File airport-to-airport flight plans prior to departure.

2. File the appropriate RNAV capability certification suffix in the flight plan.

3. Plan the random route portion of the flight to begin and end over published departure/arrival transition fixes or appropriate NAVAIDs for airports without published transition procedures. The use of preferred departure and arrival routes, such as DP and STAR where established, is recommended.

4. Plan the route of flight so as to avoid prohibited and restricted airspace by 3 NM unless permission has been obtained to operate in that airspace and the appropriate ATC facility is advised.

5. Define the route of flight after the departure fix, including each intermediate fix (turnpoint) and the arrival fix for the destination airport in terms of latitude/longitude coordinates plotted to the nearest minute or in terms of Navigation Reference System (NRS) waypoints. For latitude/longitude filing, the arrival fix must be identified by both the latitude/ longitude coordinates and a fix identifier as shown in the example below.
MIA[1] SRQ[2] 3407/10615[3] 3407/11546 TNP[4] LAX[5]

> [1]Departure airport
>
> [2]Departure fix
>
> [3]Intermediate fix (turning point)
>
> [4]Arrival fix
>
> [5]Destination airport
>
> Or:

ORD[1] IOW[2] KP49G[3] KD34U[4] KL160[5] OAL[6] MOD2[7] SFO[8]

> [1]Departure airport
>
> [2]Transition fix (pitch point)

[3]Minneapolis ARTCC waypoint

[4]Denver ARTCC waypoint

[5]Los Angeles ARTCC waypoint (catch point)

[6]Transition fix

[7]Arrival

[8]Destination airport

6. Record latitude/longitude coordinates by four figures describing latitude in degrees and minutes followed by a solidus and five figures describing longitude in degrees and minutes.

7. File at FL 390 or above for the random RNAV portion of the flight.

8. Fly all routes/route segments on Great Circle tracks.

9. Make any in-flight requests for random RNAV clearances or route amendments to an en route ATC facility.

Off-Airway Routes

14 CFR Part 95 prescribes altitudes governing the operation of aircraft under IFR on Federal airways, jet routes, RNAV low or high altitude routes, and other direct routes for which a MEA is designated. In addition, it designates mountainous areas and COPs. Off-airway routes are established in the same manner and in accordance with the same criteria as airways and jet routes. If a pilot flies for a scheduled air carrier or operator for compensation or hire, any requests for the establishment of off-airway routes are initiated by the company through the principal operations inspector (POI) who works directly with the company and coordinates FAA

Authorized areas of en route operation	Limitations, provisions, and reference paragraphs
The 48 contiguous United States and the District of Columbia	Note 1
Canada, excluding Canadian MNPS airspace and the areas of magnetic unreliability as established in the Canadian AIP	Note 3

SPECIAL REQUIREMENTS:
Note 1 - B-737 Class II navigation operations with a single long-range system is authorized only within this area of en route operation.

Note 3 - Only B-747 and DC-10 operations authorized in these areas.

Figure 2-46. Excerpt of authorized areas of en route operation.

approval. Air carrier authorized routes should be contained in the company's Operations Specifications (OpSpecs) under the auspices of the air carrier operating certificate. [Figure 2-46]

Off-airway routes predicated on public navigation facilities and wholly contained within controlled airspace are published as direct Part 95 routes. Off-airway routes predicated on privately owned navigation facilities or not contained wholly within controlled airspace are published as off-airway non- Part 95 routes. In evaluating the adequacy of off-airway routes, the following items are considered: the type of aircraft and navigation systems used; proximity to military bases, training areas, low level military routes; and the adequacy of communications along the route.

Commercial operators planning to fly off-airway routes should have specific instructions in the company's OpSpecs that address en route limitations and provisions regarding en route authorizations to use the GPS or other RNAV systems in the NAS. The company's manuals and checklists should include practices and procedures for long-range navigation and training on the use of long range navigation equipment. Minimum equipment lists (MELs) and maintenance programs must address the long range navigation equipment. Examples of other selected areas requiring specialized en route authorization include the following:

• Class I navigation in the United States Class A airspace using area of long range navigation system.

• Class II navigation using multiple long range navigation systems.

• Operations in central East Pacific airspace.

• North Pacific operations.

• Operations within North Atlantic (NAT) minimum navigation performance specifications (MNPS) airspace.

• Operations in areas of magnetic unreliability.

• North Atlantic operation (NAT/OPS) with two engine airplanes under 14 CFR Part 121.

• Extended range operations (ER-OPS) with two engine airplanes under 14 CFR Part 121.

• Special fuel reserves in international operations.

• Planned in-flight re-dispatch or re-release en route.

• Extended over water operations using a single long-range communication system.

- Operations in reduced vertical separation minimum (RVSM) airspace.

Off-Route Obstruction Clearance Altitude

An off-route obstruction clearance altitude (OROCA) is an off-route altitude that provides obstruction clearance with a 1,000-foot buffer in non-mountainous terrain areas and a 2,000-foot buffer in designated mountainous areas within the United States. This altitude may not provide signal coverage from ground-based NAVAIDs, ATC radar, or communications coverage. OROCAs are intended primarily as a pilot tool for emergencies and SA. OROCAs depicted on en route charts do not provide the pilot with an acceptable altitude for terrain and obstruction clearance for the purposes of off-route, random RNAV direct flights in either controlled or uncontrolled airspace. OROCAs are not subject to the same scrutiny as MEAs, minimum vectoring altitude (MVAs), MOCAs, and other minimum IFR altitudes. Since they do not undergo the same obstruction evaluation, airport airspace analysis procedures, or flight inspection, they cannot provide the same level of confidence as the other minimum IFR altitudes.

When departing an airport VFR intending to or needing to obtain an IFR clearance en route, you must be aware of the position of your aircraft relative to terrain and obstructions. When accepting a clearance below the MEA, MIA, MVA, or the OROCA, you are responsible for your own terrain/obstruction clearance until reaching the MEA, MIA, or MVA. If unable to visually maintain terrain/obstruction clearance, pilots should advise ATC and state intentions of the flight. [Figure 2-47]

For all random RNAV flights, there needs to be at least one waypoint in each ARTCC area through which you intend to fly. One of the biggest problems in creating an RNAV direct route is determining if the route goes through special use airspace. For most direct routes, the chances of going through prohibited, restricted, or special use airspace are good. In the United States, all direct routes should be planned to avoid prohibited or restricted airspace by at least 3 NM. If a bend in a direct route is required to avoid special use airspace, the turning point needs to be part

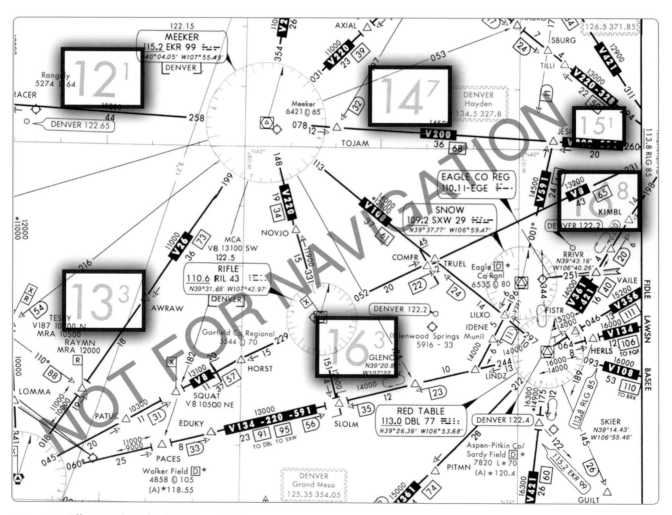

Figure 2-47. Off-route obstacle clearance altitude.

Figure 2-48. Random RNAV route.

of the flight plan. Two of the most prominent long range navigation systems today include FMS with integrated GPS and stand-alone GPS. The following example is a simplified overview showing how the RNAV systems might be used to fly a random RNAV route.

Shown in Figure 2-48, the aircraft is northeast of Tuba City VORTAC at FL 200 using RNAV (showing both GPS and FMS), RNAV direct on a southwesterly heading to Lindbergh Regional Airport in Winslow. As the pilot is monitoring his or her position and cross-checking the avionics against the high altitude en route chart, he or she receives a company message instructing to divert to Las Vegas, requiring a change in the flight plan as highlighted on the depicted chart excerpt.

During the flight deck review of the high and low altitude en route charts, the pilot determines that the best course of action is to fly direct to the MIRAJ waypoint, 28 DME northeast of the Las Vegas VORTAC on the 045° radial. This places the aircraft 193 NM out on a 259° magnetic course inbound, and may help to avoid diverting north, allowing to bypass the more distant originating and intermediate fixes feeding into Las Vegas. The pilot requests an RNAV random route clearance direct MIRAJ to expedite the flight. Denver Center comes back with the following amended flight plan and initial clearance into Las Vegas:

"Marathon five sixty four, turn right heading two six zero, descend and maintain one six thousand, cleared present position direct MIRAJ."

The latitude and longitude coordinates of the aircraft's present position on the high altitude chart is N36 19.10 and W110 40.24 as the course is changed. Notice the GPS moving map (upper left), the FMS control display unit (below the GPS), and FMS map mode navigation displays (to the right of the GPS) as the flight is rerouted to Las Vegas. For SA, the pilot makes note that the altitude is well above any of the OROCAs on the direct route as the flight arrives in the Las Vegas area using the low altitude chart.

Monitoring of Navigation Facilities

VOR, VORTAC, and instrument landing system (ILS) facilities, as well as most NDBs and marker beacons installed by the FAA, are provided with an internal monitoring feature. Internal monitoring is provided at the facility through the use of equipment that causes a facility shutdown if performance deteriorates below established tolerances. A remote status indicator also may be provided through the use of a signal- sampling receiver, microwave link, or telephone circuit. Older FAA NDBs and some non-Federal NDBs do not have the internal feature, and monitoring is accomplished by manually checking the operation at least once each hour. FAA facilities, such as automated flight service stations (AFSSs) and ARTCCs/sectors, are usually the control point for NAVAID facility status. Pilots can query the appropriate FAA facility if they have questions in flight regarding NAVAID status, in addition to checking NOTAMs prior to flight, since NAVAIDs and associated monitoring equipment are continuously changing.

Navigational Gaps

A navigational course guidance gap, referred to as an MEA gap, describes a distance along an airway or route segment where a gap in navigational signal coverage exists. The navigational gap may not exceed a specific distance that varies directly with altitude, from 0 NM at sea level to 65 NM at 45,000 feet MSL and not more than one gap may exist in the airspace structure for the airway or route segment. Additionally, a gap usually does not occur at any airway or route turning point. To help ensure the maximum amount of continuous positive course guidance available when flying, there are established en route criteria for both straight and turning segments. Where large gaps exist that require altitude changes, MEA "steps" may be

established at increments of not less than 2,000 feet below 18,000 feet MSL, or not less than 4,000 feet at 18,000 MSL and above, provided that a total gap does not exist for the entire segment within the airspace structure. MEA steps are limited to one step between any two facilities to eliminate continuous or repeated changes of altitude in problem areas. The allowable navigational gaps pilots can expect to see are determined, in part, by reference to the graph depicted in Figure 2-49. Notice the en route chart excerpt depicting that the MEA is established with a gap in navigation signal coverage northwest of the Carbon VOR/DME on V134 . At the MEA of 13,000, the allowable navigation course guidance gap is approximately 18.5 NM, as depicted in Figure 2-49. The navigation gap area is not identified on the chart by distances from the navigation facilities. Proper flight planning will help pilots prepare for MEA gaps by insuring that appropriate maps are available as they may need to dead reckon through the gap. Calculating the ground track (with adjustments for winds) before and after the gap will also help to stay on course when navigational course guidance is not available.

NAVAID Accuracy Check

The CFRs and good judgment dictate that the equipment of aircraft flying under IFR be within a specified tolerance before taking off. When approved procedures are available, they should be used for all equipment inspections.

VOR Accuracy

VOR accuracy can be checked by using any of the following methods: VOR test facility signal (VOT), VOR checkpoint signs, dual VOR check, or airborne VOR check.

VOT

The VOT is an approved test signal and is located on an airport. This enables the pilot to check the VOR accuracy from the flight deck before takeoff. Listed below are the steps used for a VOT:

1. Tune the VOR receiver to the VOT frequency. VOT frequencies can be found in the A/FD. [Figure 2-50] These frequencies are coded with a series of Morse code dots or a continuous 1020-cycle tone.

2. On the VOR, set the course selector to 0° and the track bar (TB) indicator should read center. The TO-FROM indicator should read FROM.

3. Set the course selector to 180° and the TO-FROM indicator should read TO and the TB should then be centered.

NOTE: Determining the exact error in the receiver is done by turning the track selector until the TB is centered and noting the degrees difference between 180° or 0°. The

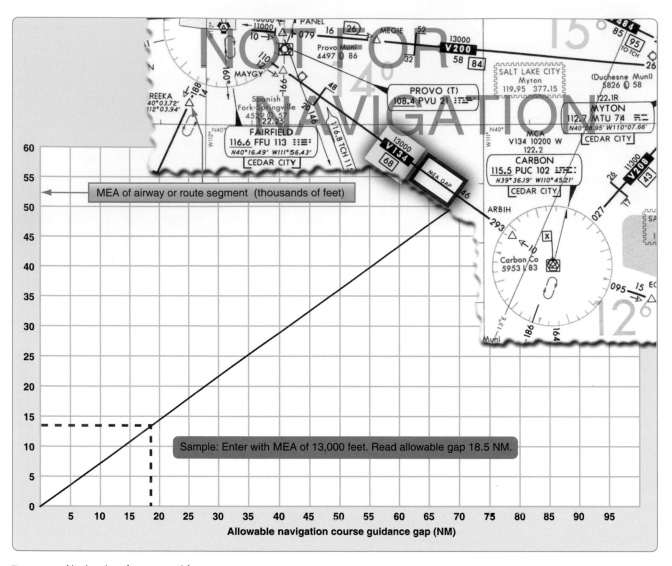

Figure 2-49. Navigational course guidance gaps.

Sample: Enter with MEA of 13,000 feet. Read allowable gap 18.5 NM.

MEA of airway or route segment (thousands of feet)

Allowable navigation course guidance gap (NM)

VOR test facilities (VOT)		
Facility Name .... (Airport Name) ....Frequency	Type VOT Facility	Remarks
Bradlay Intl 111.40	G	
Bridgeport 109.25	G	
Groton 110.25	G	
Hartford 108.20	G	

Figure 2-50. VOR test facilities (VOT) frequencies.

VOR 116.4
147° 4.1 NM

DME and VOR check radial

Figure 2-51. VOR checkpoint signs.

maximum bearing error with the VOT system check is plus or minus 4° and apparent errors greater than 4° indicate that the VOR receiver is beyond acceptable tolerance.

VOR Checkpoint Signs

Many aerodromes have VOR checkpoint signs that are located beside the taxiways. [Figure 2-51] These signs indicate the exact point on the aerodrome that there is sufficient signal strength from a VOR to check the aircraft's VOR receiver against the radial designated on the sign. Listed below are the steps to use at a VOR checkpoint:

1. Tune the proper VOR frequency.

2. Identify the VOR frequency.

3. Set the published radial on the course deviation indicator (CDI).

4. Confirm that the TB is centered.

5. Check the needle sensitivity by changing the omnibearing select (OBS) 10° each way.

6. Set the reciprocal of the radial and check the TO-FROM flag change.

7. The maximum permissible difference between aircraft equipment and the designated radial is 4° and 0.5 NM of the posted distance.

Dual VOR Check

If a VOT or VOR checkpoint is not available and the aircraft is equipped with dual VORs, the equipment may be checked against one another by tuning both sets to the VOR facility at the same time and noting the indicated bearings to that station. [Figure 2-52] A difference greater than 4° between the two VORs indicates that one of the receivers may be out of tolerance.

Airborne VOR Check

VOR equipment can also be checked for accuracy while in flight by flying over a fix or landmark located on a published radial and noting the indicated radial. Variances of more than 6° from the published radial should be considered out of tolerance and not be used for IFR navigation.

NDB Accuracy Check

The pilot must identify an NDB before using it for navigation, and continuously monitor it while using it for an instrument approach. The lack of an IDENT may indicate that the NDB is out of service, even though it may still be transmitting (for instance for maintenance or test purposes). If an incorrect IDENT is heard, then the NDB should not be used.

RNAV Accuracy Check

RNAV accuracy checks may differ depending on the different type of equipment and manufacturer. When available, all written procedures should be followed. Below is a list of generic checks that should be used when checking the accuracy of the system prior to flight.

1. System initialization—pilots should confirm that the navigation database is current and verify that the aircrafts present position has been entered correctly.

2. Active flight plan check—the active flight plan should be checked by comparing the aeronautical charts, departure and arrival procedures, and other applicable documents with the map display.

Figure 2-52. Instrument panel with dual VORs.

3. Prior to takeoff—ensure that the RNAV system is available. If possible, check to see that the system is updating when aircraft position is changing.

NOTE: While in flight, continue to verify system accuracy by displaying bearing/range to a VOR/DME on the RNAV system and compare it to the actual RMI reading of that particular NAVAID.

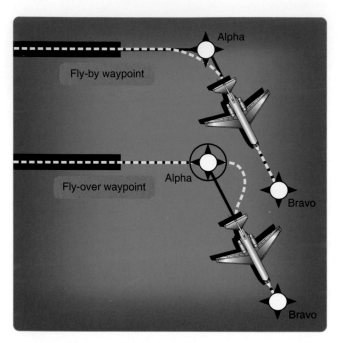

Figure 2-53. Fly-by and fly-over waypoints.

Waypoints

Waypoints are predetermined geographical locations that are defined in terms of latitude/longitude coordinates or fixes, used to define an RNAV route or the flight path of an aircraft employing RNAV. Waypoints may be a simple named point in space or may be associated with existing NAVAIDs, intersections, or fixes. A waypoint is most often used to indicate a change in direction, speed, or altitude along the desired path. Aviation RNAV procedures make use of both fly-over and fly-by waypoints. A fly-over waypoint is a waypoint that must be crossed vertically by an aircraft. A fly-by waypoint is a waypoint that marks the intersection of two straight paths, with the transition from one path to another being made by the aircraft using a precisely calculated turn that flies by but does not vertically cross the waypoint. [Figure 2-53]

User-Defined Waypoints

Pilots typically create user-defined waypoints for use in their own random RNAV direct navigation. They are newly established, unpublished airspace fixes that are designated geographic locations/positions that help provide positive course guidance for navigation and a means of checking progress on a flight. They may or may not be actually plotted by the pilot on en route charts, but would normally be communicated to ATC in terms of bearing and distance or latitude/longitude. An example of user-defined waypoints typically includes those generated by various means including keyboard input, and even electronic map mode functions used to establish waypoints with a cursor on the display.

Another example is an offset phantom waypoint, which is a point-in-space formed by a bearing and distance from NAVAIDs, such as VORTACs and tactical air navigation (TACAN) stations, using a variety of navigation systems. When specifying unpublished waypoints in a flight plan, they can be communicated using the frequency/bearing/distance format or latitude and longitude, and they automatically become compulsory reporting points unless otherwise advised by ATC. All airplanes with latitude and longitude navigation systems flying above FL 390 must use latitude and longitude to define turning points.

Floating Waypoints

Floating waypoints, or reporting points, represent airspace fixes at a point in space not directly associated with a conventional airway. In many cases, they may be established for such purposes as ATC metering fixes, holding points, RNAV-direct routing, gateway waypoints, STAR origination points leaving the en route structure, and SID terminating points joining the en route structure. In the top example of Figure 2-54, a low altitude en route chart depicts three floating waypoints that have been highlighted: SCORR, FILUP, and CHOOT. Notice that waypoints are named with five-letter identifiers that are unique and pronounceable. Pilots must be careful of similar waypoint names. Notice on the high altitude en route chart excerpt in the bottom example, the similar sounding and spelled floating waypoint named SCOOR, rather than SCORR. This emphasizes the importance of correctly entering waypoints into database-driven navigation systems. One waypoint character incorrectly entered into your navigation system could adversely affect your flight. The SCOOR floating reporting point also is depicted on a Severe Weather Avoidance Plan (SWAP) en route chart. These waypoints and SWAP routes assist pilots and controllers when severe weather affects the East Coast.

Computer Navigation Performance

An integral part of RNAV using en route charts typically involves the use of airborne navigation databases.

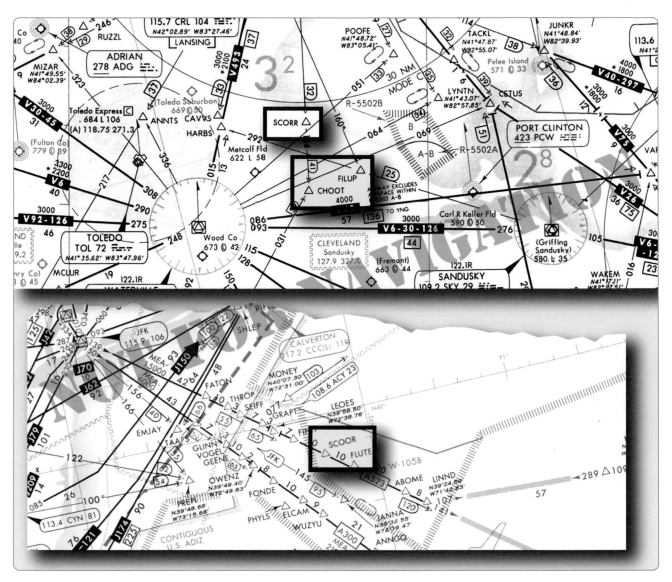

Figure 2-54. Floating waypoints.

Because GPS receivers are basically "to-to" navigators, they must always be navigating to a defined point. On overlay approaches, if no pronounceable five-character name is published for an approach waypoint or fix, it has been given a database identifier consisting of letters and numbers. These points appear in the list of waypoints in the approach procedure database, but may not appear on the approach chart. A point used for the purpose of defining the navigation track for an airborne computer system (i.e., GPS or FMS) is called a Computer Navigation Fix (CNF). CNFs include unnamed DME fixes, beginning and ending points of DME arcs, and sensor final approach fixes (FAFs) on some GPS overlay approaches.

To aid in the approach chart/database correlation process, the FAA has begun a program to assign five-letter names to CNFs and to chart CNFs on various National Oceanic

Service aeronautical products. [Figure 2-55] These CNFs are not to be used for any ATC application, such as holding for which the fix has not already been assessed. CNFs are charted to distinguish them from conventional reporting points, fixes, intersections, and waypoints. A CNF name is enclosed in parenthesis, e.g., (MABEE) and is placed next to the CNF it defines. If the CNF is not at an existing point defined by means such as crossing radials or radial/DME, the point is indicated by an X. The CNF name is not used in filing a flight plan or in aircraft/ATC communications. Use current phraseology (e.g., facility name, radial, distance) to describe these fixes.

Many of the RNAV systems available today make it all too easy to forget that en route charts are still required and necessary for flight. As important as databases are, they really are onboard the airplane to provide navigation guidance and situational awareness (SA); they are not

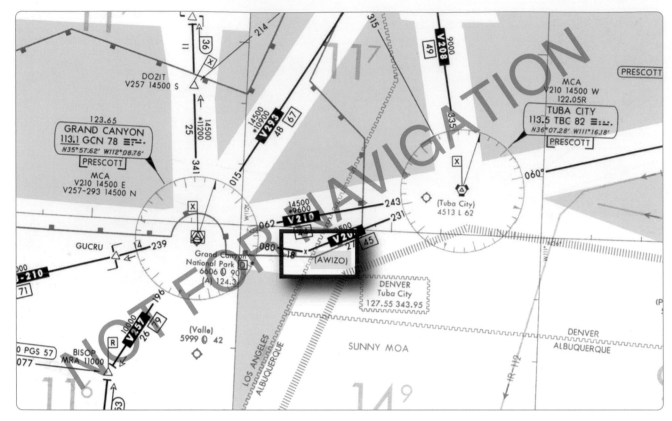

Figure 2-55. Computer navigation fix.

intended as a substitute for paper charts. When flying with GPS, FMS, or planning a flight with a computer, it is critical to understand the limitations of the system you are using, for example, incomplete information, unloadable procedures, complex procedures, and database storage limitations.

Required Navigation Performance

Required navigation performance (RNP) is RNAV with onboard navigation monitoring and alerting. RNP is also a statement of navigation performance necessary for operation within a defined airspace. A critical component of RNP is the ability of the aircraft navigation system to monitor its achieved navigation performance, and to identify for the pilot whether the operational requirement is, or is not being met during an operation. This onboard performance monitoring and alerting capability, therefore, allows a lessened reliance on ATC intervention

(via radar monitoring, automatic dependent surveillance-broadcast (ADS-B), multilateration, communications), and/or route separation to achieve the overall safety of the operation. RNP capability of the aircraft is a major component in determining the separation criteria to ensure that the overall containment of the operation is met.

The RNP capability of an aircraft varies depending upon the aircraft equipment and the navigation infrastructure. For example, an aircraft may be equipped and certified for RNP 1.0, but may not be capable of RNP 1.0 operations due to limited NAVAID coverage.

RNP Levels

An RNP level or type is applicable to a selected airspace, route, or procedure. As defined in the Pilot/Controller Glossary, the RNP level or type is a value typically expressed

RNP Level	Typical Application	Primary Route Width (NM) - Centerline to Boundary
0.1 to 1.0	RNP AR Approach Segments	0.1 to 1.0
0.3 to 1.0	RNP Approach Segments	0.3 to 1.0
1	Terminal and En Route	1.0
2	En Route	2.0

Figure 2-56. U.S. standard RNP levels.

as a distance in nautical miles from the intended centerline of a procedure, route, or path. RNP applications also account for potential errors at some multiple of RNP level (e.g., twice the RNP level).

Standard RNP Levels

United States standard values supporting typical RNP airspace are shown in Figure 2-56. Other RNP levels as identified by ICAO, other states, and the FAA may also be used.

Application of Standard RNP Levels

United States standard levels of RNP typically used for various routes and procedures supporting RNAV operations may be based on use of a specific navigational system or sensor, such as GPS, or on multi-sensor RNAV systems having suitable performance.

NOTE: The performance of navigation in RNP refers not only to the level of accuracy of a particular sensor or aircraft navigation system, but also to the degree of precision with which the aircraft is flown. Specific required flight procedures may vary for different RNP levels.

IFR En Route Altitudes

Minimum En Route Altitudes (MEAs), Minimum Reception Altitudes (MRAs), Maximum Authorized Altitudes (MAAs), Minimum Obstacle Clearance Altitudes (MOCAs), Minimum Turning Altitudes (MTAs) and Minimum Crossing Altitudes (MCAs) are established by the FAA for instrument flight along Federal airways, as well as some off-airway routes. The altitudes are established after it has been determined that the NAVAIDs to be used are adequate and so oriented on the airways or routes that signal coverage is acceptable, and that flight can be maintained within prescribed route widths.

For IFR operations, regulations require that pilots operate their aircraft at or above minimum altitudes. Except when necessary for takeoff or landing, pilots may not operate an aircraft under IFR below applicable minimum altitudes, or if no applicable minimum altitude is prescribed, in the case of operations over an area designated as mountainous, an altitude of 2,000 feet above the highest obstacle within a horizontal distance of 4 NM from the course to be flown. In any other case, an altitude of 1,000 feet above the highest obstacle within a horizontal distance of 4 NM from the course to be flown must be maintained as a minimum altitude. If both a MEA and a MOCA are prescribed for a particular route or route segment, pilots may operate an aircraft below the MEA down to, but not below, the MOCA, only when within 22 NM of the VOR. When climbing to

a higher minimum IFR altitude (MIA), pilots must begin climbing immediately after passing the point beyond which that minimum altitude applies, except when ground obstructions intervene, the point beyond which that higher minimum altitude applies must be crossed at or above the applicable MCA for the VOR.

If on an IFR flight plan, but cleared by ATC to maintain VFR conditions on top, pilots may not fly below minimum en route IFR altitudes. Minimum altitude rules are designed to ensure safe vertical separation between the aircraft and the terrain. These minimum altitude rules apply to all IFR flights, whether in IFR or VFR weather conditions, and whether assigned a specific altitude or VFR conditions on top.

Minimum En Route Altitude (MEA)

The MEA is the lowest published altitude between radio fixes that assures acceptable navigational signal coverage and meets obstacle clearance requirements between those fixes. The MEA prescribed for a Federal airway or segment, RNAV low or high route, or other direct route applies to the entire width of the airway, segment, or route between the radio fixes defining the airway, segment, or route. MEAs for routes wholly contained within controlled airspace normally provide a buffer above the floor of controlled airspace consisting of at least 300 feet within transition areas and 500 feet within control areas. MEAs are established based upon obstacle clearance over terrain and manmade objects, adequacy of navigation facility performance, and communications requirements.

RNAV Minimum En Route Altitude

RNAV MEAs are depicted on some IFR en route low altitude charts, allowing both RNAV and non-RNAV pilots to use the same chart for instrument navigation.

Minimum Reception Altitude (MRA)

MRAs are determined by FAA flight inspection traversing an entire route of flight to establish the minimum altitude the navigation signal can be received for the route and for off-course NAVAID facilities that determine a fix. When the MRA at the fix is higher than the MEA, an MRA is established for the fix and is the lowest altitude at which an intersection can be determined.

Maximum Authorized Altitude (MAA)

An MAA is a published altitude representing the maximum usable altitude or flight level for an airspace structure or route segment. [Figure 2-57] It is the highest altitude on a Federal airway, jet route, RNAV low or high route, or other direct route for which an MEA is designated at which adequate reception of navigation signals is assured.

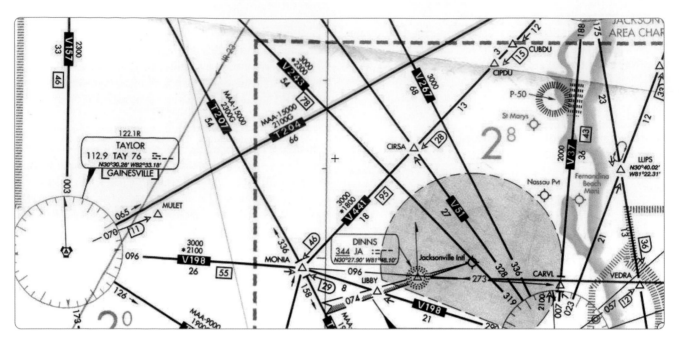

Figure 2-57. Maximum authorized altitude (MAA).

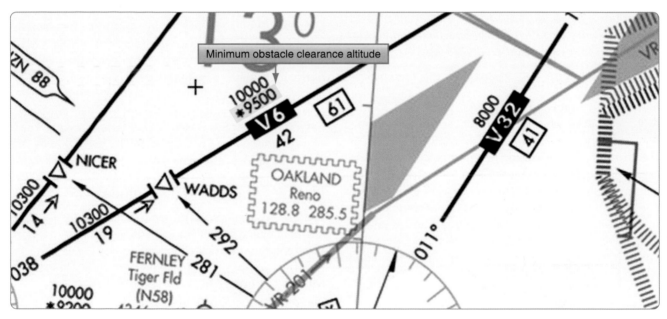

Figure 2-58. Minimum obstacle clearance altitude (MOCA).

MAAs represent procedural limits determined by technical limitations or other factors, such as limited airspace or frequency interference of ground-based facilities.

Minimum Obstruction Clearance Altitude (MOCA)

The MOCA is the lowest published altitude in effect between fixes on VOR airways, off-airway routes, or route segments that meets obstacle clearance requirements for the entire route segment. [Figure 2-58] This altitude also assures acceptable navigational signal coverage only within 22 NM of a VOR. The MOCA seen on the en route chart may have been computed by adding the required obstacle clearance (ROC) to the controlling obstacle in the primary area or computed by using a TERPS chart if the controlling obstacle is located in the secondary area. This figure is then rounded to the nearest 100 foot increment (i.e., 2,049 feet becomes 2,000, and 2,050 feet becomes 2,100 feet). An extra 1,000 feet is added in mountainous areas, in most cases.

ATC controllers have an important role in helping pilots remain clear of obstructions. Controllers are instructed to

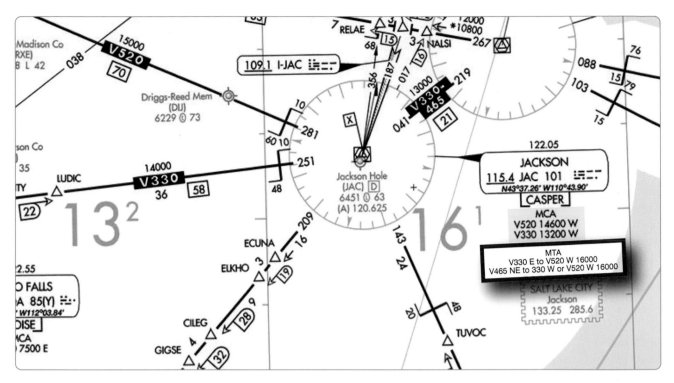

Figure 2-59. Minimum turning altitude (MTA).

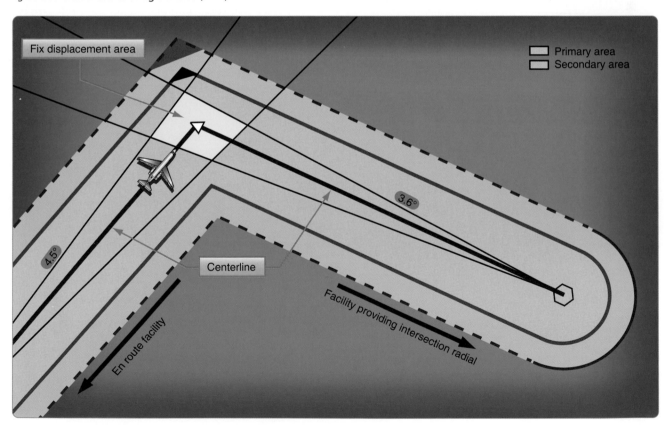

Figure 2-60. Turning area at the intersection fix with NAVAID distance less than 51 NM.

issue a safety alert if the aircraft is in a position that, in their judgment, places the pilot in unsafe proximity to terrain, obstructions, or other aircraft. Once pilots inform ATC of action being taken to resolve the situation, the controller may discontinue the issuance of further alerts. A typical terrain/obstruction alert may sound like this: "(Aircraft call sign), Low altitude alert. Check your altitude immediately. The MOCA in your area is 12,000."

Minimum Turning Altitude (MTA)

Minimum turning altitude (MTA) is a charted altitude providing vertical and lateral obstruction clearance based on turn criteria over certain fixes, NAVAIDs, waypoints, and on charted route segments. [Figure 2-59] When a VHF airway or route terminates at a NAVAID or fix, the primary area extends beyond that termination point. When a change of course on VHF airways and routes is necessary, the en route obstacle clearance turning area extends the primary and secondary obstacle clearance areas to accommodate the turn radius of the aircraft. Since turns at or after fix passage may exceed airway and route boundaries, pilots are expected to adhere to airway and route protected airspace by leading turns early before a fix. The turn area provides obstacle clearance for both turn anticipation (turning prior to the fix) and flyover protection (turning after crossing the fix). This does not violate the requirement to fly the centerline of the airway. Many factors

enter into the construction and application of the turning area to provide pilots with adequate obstacle clearance protection. These may include aircraft speed, the amount of turn versus NAVAID distance, flight track, curve radii, MEAs, and MTA. [Figure 2-60]

Due to increased airspeeds at 10,000 feet MSL or above, an expanded area in the vicinity of the turning fix is examined to ensure the published MEA is sufficient for obstacle clearance. In some locations (normally mountainous), terrain/obstacles in the expanded search area may obviate the published MEA and necessitate a higher minimum altitude while conducting the turning maneuver. Turning fixes requiring a higher MTA are charted with a flag along with accompanying text describing the MTA restriction. [Figure 2-59]

An MTA restriction normally consists of the ATS route leading to the turning fix, the ATS route leading from the turning fix, and an altitude (e.g., MTA V330 E TO V520 W 16000). When an MTA is applicable for the intended route of flight, pilots must ensure they are at or above the charted MTA prior to beginning the turn and maintain at or above the MTA until joining the centerline of the ATS route following the turn. Once established on the centerline following the turning fix, the MEA/MOCA determines the minimum altitude available for assignment.

Transmittal of Airways/Route Data										
Airway number or route	From / To	Routine or docket number	Controlling @ terrain/Obstruction and coordinates	MRA / MOCA	MAA / MEA	GNSS MEA	Change over point	Fix MRA/MCA	Remarks	Flight inspection dates
V330	Idaho Falls, ID VOR/DME		Tree 6177 @ 432912.00N/1114118.00W	8000	17500			*9500E	300 MTN ROC RED DEL MCA ATIDA COME ADD MCA AT OSITY DEC MOCA	
	*Osity, ID		Terrain 6077 432912.00N/1114118.00W	--7900--	8000					
									DEL directional MEA MEA CARONIAL ALT	
V330	Osity, ID		AAO 12138 (SEC) @ 434118.30N/1104858.30W	14000	17500		JAC 10	# MTA * 13400W	INC MCA PRECIP TER DEC MOCA MEA CARDINAL ALT	
	# Jackson, WY VOR/DME		Terrain 11132 433900.00N/1105057.00W	--13600--	14000					
									JAC R-251 UNUSABLE BYD 10 # CHART: MTA V330 E TO VS20W 16000	
Date	Office AJW-3773	Title Manager				Signature Ray Nussear				

Figure 2-61. Minimum turning altitude information located in the remarks section of FAA Form 8260-16 Transmittal of Airways/Route Data.

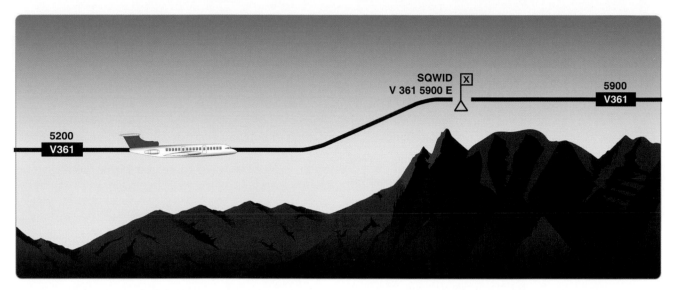

Figure 2-62. Minimum crossing altitude (MCA).

An MTA may also preclude the use of a specific altitude or a range of altitudes during a turn. For example, the MTA may restrict the use of 10,000 through 11,000 feet MSL. In this case, any altitude greater than 11,000 feet MSL is unrestricted, as are altitudes less than 10,000 feet MSL provided MEA/MOCA requirements are satisfied.

All MTA information associated with the airway/route inbound to the turn fix/facility is put in the remarks section of FAA Form 8260-16, Transmittal of Airways/Route Data, using the following format [Figure 2-61]:

#CHART: MTA V330 E TO V520 W 16000
(Document on V330 FAA Form 8260-16)

#CHART: MTA V465 NE TO V330 W OR V520 W 16000
(Document on V465 FAA Form 8260-16)

When an MTA is required by TERPS, Volume 1, paragraph 1714(c), enter the MTA information in the REMARKS section of FAA Form 8260-2, Radio Fix and Holding Data Record, as specified on the appropriate FAA Form 8260-16, Transmittal of Airways/Route Data, using the following format:

MTA: V330 E TO V520 W 16000

MTA: V465 NE TO V330 W OR V520 W 16000

Minimum Crossing Altitude (MCA)

An MCA is the lowest altitude at certain fixes at which the aircraft must cross when proceeding in the direction of a higher minimum en route IFR altitude. [Figure 2-62] When applicable, MCAs are depicted on the en route chart. [Figure 2-59] MCAs are established in all cases where obstacles

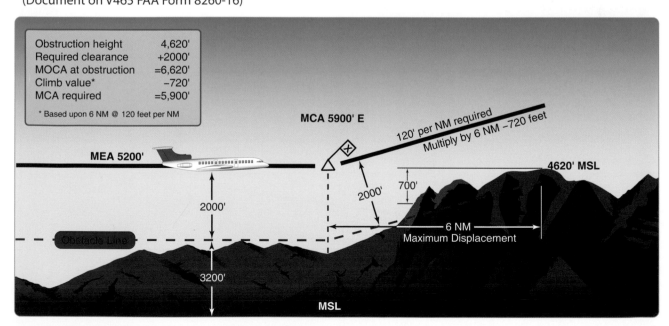

Figure 2-63. Minimum crossing altitude (MCA) determination point.

intervene to prevent pilots from maintaining obstacle clearance during a normal climb to a higher MEA after passing a point beyond which the higher MEA applies. The same protected en route area vertical obstacle clearance requirements for the primary and secondary areas are considered in the determination of the MCA. The standard for determining the MCA is based upon the following climb gradients and is computed from the flight altitude:

- Sea level through 5,000 feet MSL—150 feet per NM

- 5000 feet through 10,000 feet MSL—120 feet per NM

- 10,000 feet MSL and over—100 feet per NM

To determine the MCA seen on an en route chart, the distance from the obstacle to the fix is computed from the point where the centerline of the en route course in the direction of flight intersects the farthest displacement from the fix. [Figure 2-63] When a change of altitude is involved with a course change, course guidance must be provided if the change of altitude is more than 1,500 feet and/or if the course change is more than 45°, although there is an exception to this rule. In some cases, course changes of up to 90° may be approved without course guidance provided that no obstacles penetrate the established MEA requirement of the previous airway or route segment. Outside United States airspace, pilots may encounter different flight procedures regarding MCA and transitioning from one MEA to a higher MEA. In this case, pilots are expected to be at the higher MEA crossing the fix, similar to an MCA. Pilots must thoroughly review flight procedure differences when flying outside United States airspace. On IFR en route low altitude charts, routes and associated data outside the conterminous United States are shown for transitional purposes only and are not part of the high altitude jet route and RNAV route systems. [Figure 2-64]

Minimum IFR Altitude (MIA)

The MIA for operations is prescribed in 14 CFR Part 91. These MIAs are published on aeronautical charts and prescribed in 14 CFR Part 95 for airways and routes, and in 14 CFR Part 97 for standard instrument approach procedures. If no applicable minimum altitude is prescribed in 14 CFR Parts 95 or 97, the following MIA applies: In designated mountainous areas, 2,000 feet above the highest obstacle within a horizontal distance of 4 NM from the course to be flown; or other than mountainous areas, 1,000 feet above the highest obstacle within a horizontal distance of 4 NM from the course to be flown; or as otherwise authorized by the Administrator or assigned by ATC. MIAs are not flight checked for communication.

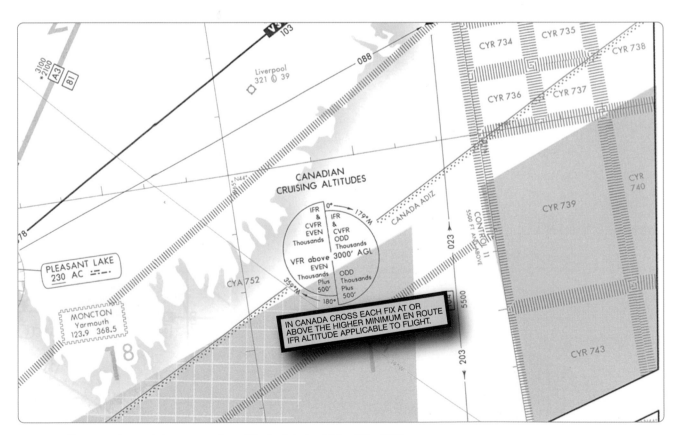

Figure 2-64. En route chart minimum crossing altitude data (outside of the U.S.).

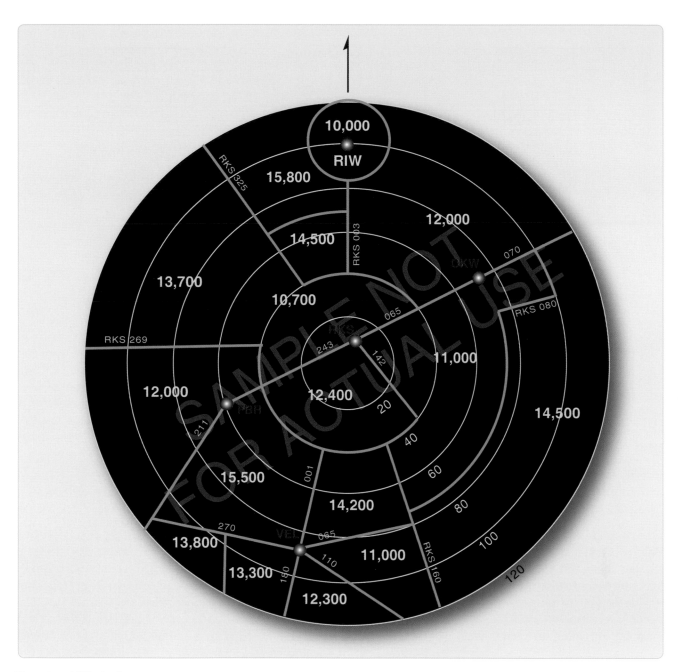

Figure 2-65. MVA chart.

Minimum Vectoring Altitudes (MVA)

MVAs are established for use by ATC when radar ATC is exercised. The MVA provides 1,000 feet of clearance above the highest obstacle in non-mountainous areas and 2,000 feet above the highest obstacle in designated mountainous areas. Because of the ability to isolate specific obstacles, some MVAs may be lower than MEAs, MOCAs, or other minimum altitudes depicted on charts for a given location. While being radar vectored, IFR altitude assignments by ATC are normally at or above the MVA.

Air traffic controllers use MVAs only when they are assured an adequate radar return is being received from the aircraft.

Charts depicting MVAs are available to controllers and have recently become available to pilots. They can be found at http://www.faa.gov/air_traffic/flight_info/aeronav/digital_products/mva_mia/ Situational Awareness is always important, especially when being radar vectored during a climb into an area with progressively higher MVA sectors, similar to the concept of MCA. Except where diverse vector areas have been established, when climbing, pilots should not be vectored into a sector with a higher MVA unless at or above the next sector's MVA. Where lower MVAs are required in designated mountainous areas to achieve compatibility with terminal routes or to permit vectoring to an instrument approach procedure, 1,000 feet

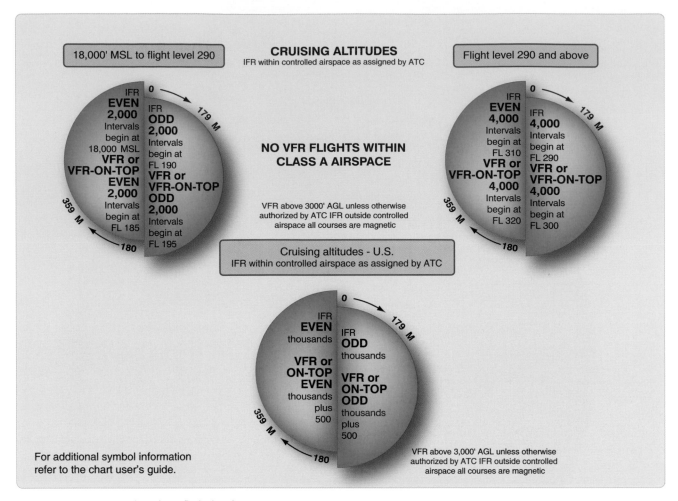

Figure 2-66. IFR cruising altitude or flight level.

of obstacle clearance may be authorized with the use of Airport Surveillance Radar (ASR). The MVA provides at least 300 feet above the floor of controlled airspace. The MVA charts are developed to the maximum radar range. Sectors provide separation from terrain and obstructions. Each MVA chart has sectors large enough to accommodate vectoring of aircraft within the sector at the MVA. [Figure 2-65]

IFR Cruising Altitude or Flight Level

In controlled airspace, pilots must maintain the altitude or flight level assigned by ATC, although if the ATC clearance assigns "VFR conditions on-top," an altitude or flight level as prescribed by 14 CFR Part 91, section 91.159 must be maintained. In uncontrolled airspace (except while in a holding pattern of two minutes or less or while turning) if operating an aircraft under IFR in level cruising flight, an appropriate altitude as depicted in the legend of IFR en route high and low altitude charts must be maintained. [Figure 2-66]

When operating on an IFR flight plan below 18,000 feet MSL in accordance with a VFR-on-top clearance, any VFR cruising altitude appropriate to the direction of flight

between the MEA and 18,000 feet MSL may be selected that allows the flight to remain in VFR conditions. Any change in altitude must be reported to ATC, and pilots must comply with all other IFR reporting procedures. VFR-on-top is not authorized in Class A airspace. When cruising below 18,000 feet MSL, the altimeter must be adjusted to the current setting, as reported by a station within 100 NM of your position. In areas where weather-reporting stations are more than 100 NM from the route, the altimeter setting of a station that is closest may be used.

During IFR flight, ATC advises flights periodically of the current altimeter setting, but it remains the responsibility of the pilot or flight crew to update altimeter settings in a timely manner. Altimeter settings and weather information are available from weather reporting facilities operated or approved by the U.S. National Weather Service, or a source approved by the FAA. Some commercial operators have the authority to act as a government-approved source of weather information, including altimeter settings, through certification under the FAA's Enhanced Weather Information System.

Flight level operations at or above 18,000 feet MSL require

the altimeter to be set to 29.92 inches of mercury (" Hg). A flight level (FL) is defined as a level of constant atmospheric pressure related to a reference datum of 29.92 " Hg. Each flight level is stated in three digits that represent hundreds of feet. For example, FL 250 represents an altimeter indication of 25,000 feet. Conflicts with traffic operating below 18,000 feet MSL may arise when actual altimeter settings along the route of flight are lower than 29.92 " Hg. Therefore, 14 CFR Part 91, section 91.121 specifies the lowest usable flight levels for a given altimeter setting range.

Reduced Vertical Separation Minimums (RSVM)

Reduced vertical separation minimums (RVSM) is a term used to describe the reduction of the standard vertical separation required between aircraft flying at levels between FL 290 (29,000 feet) and FL 410 (41,000 feet) from 2,000 feet to 1,000 feet. The purpose, therefore, increases the number of aircraft that can safely fly in a particular volume of airspace. Historically, standard vertical separation was 1,000 feet from the surface to FL 290, 2,000 feet from FL 290 to FL 410 and 4,000 feet above this. This was because the accuracy of the pressure altimeter (used to determine altitude) decreases with height. Over time, air data computers (ADCs) combined with altimeters have become more accurate and autopilots more adept at maintaining a set level; therefore, it became apparent that for many modern aircraft, the 2,000 foot separation was not required . It was therefore proposed by ICAO that this be reduced to 1,000 feet.

Between 1997 and 2005, RVSM was implemented in all of Europe, North Africa, Southeast Asia, North America, South America, and over the North Atlantic, South Atlantic, and Pacific Oceans. The North Atlantic implemented initially in March 1997, at FL 330 through FL 370. The entire western hemisphere implemented RVSM FL 290–FL 410 on January 20, 2005.

Only aircraft with specially certified altimeters and autopilots may fly in RVSM airspace, otherwise the aircraft must fly lower or higher than the airspace, or seek special exemption from the requirements. Additionally, aircraft operators (airlines or corporate operators) must receive specific approval from the aircraft's state of registry in order to conduct operations in RVSM airspace. Non-RVSM approved aircraft may transit through RVSM airspace provided they are given continuous climb throughout the designated airspace, and 2,000 feet vertical separation is provided at all times between the non-RVSM flight and all others for the duration of the climb/descent.

Critics of the change were concerned that by reducing the space between aircraft, RVSM may increase the number of mid-air collisions and conflicts. In the ten years since RVSM was first implemented, not one collision has been attributed to RVSM. In the United States, this program was known as the Domestic Reduced Vertical Separation Minimum (DRVSM).

Cruise Clearance

The term "cruise" may be used instead of "maintain" to assign a block of airspace to an aircraft. The block extends from the minimum IFR altitude up to and including the altitude that is specified in the cruise clearance. On a cruise clearance, you may level off at any intermediate altitude within this block of airspace. You are allowed to climb or descend within the block at your own discretion. However, once you start descent and verbally report leaving an altitude in the block to ATC, you may not return to that altitude without an additional ATC clearance. A cruise clearance also authorizes you to execute an approach at the destination airport.

Lowest Usable Flight Level

When the barometric pressure is 31.00 " Hg or less and pilots are flying below 18,000 feet MSL, use the current reported altimeter setting. When an aircraft is en route on an instrument flight plan, air traffic controllers furnish this information at least once while the aircraft is in the controller's area of jurisdiction. When the barometric pressure exceeds 31.00 " Hg, the following procedures are placed in effect by NOTAM defining the geographic area affected: Set 31.00 " Hg for en route operations below 18,000 feet MSL and maintain this setting until beyond the affected area. ATC issues actual altimeter settings and advises pilots to set 31.00 " Hg in their altimeter, for en route operations below 18,000 feet MSL in affected areas. If an aircraft has the capability of setting the current altimeter setting and operating into airports with the capability of measuring the current altimeter setting, no additional restrictions apply. At or above 18,000 feet MSL, altimeters should be set to 29.92 " Hg (standard setting). Additional procedures exist beyond the en route phase of flight.

The lowest usable flight level is determined by the atmospheric pressure in the area of operation. As local altimeter settings fall below 29.92 " Hg, pilots operating in Class A airspace must cruise at progressively higher indicated altitudes to ensure separation from aircraft operating in the low altitude structure as follows:

Current Altimeter Setting	Lowest Usable Flight Level
29.92 or higher	180
29.91 to 29.42	185
29.41 to 28.92	190
28.91 to 28.42	195
28.41 to 27.91	200

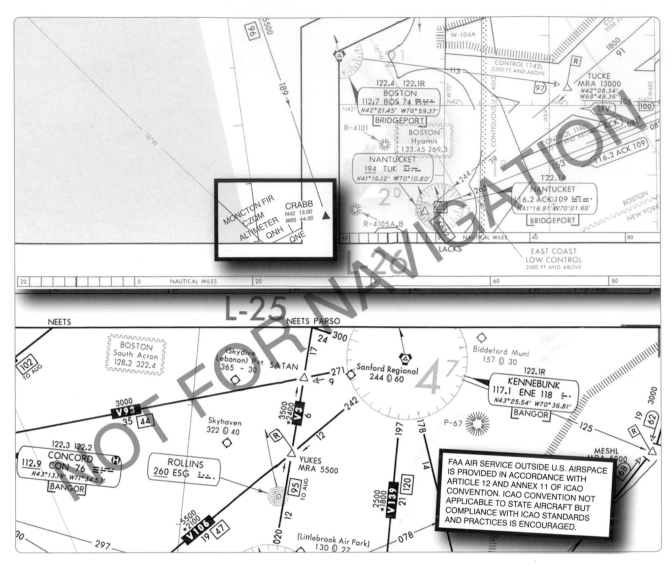

Figure 2-67. Altimeter setting changes.

When the minimum altitude, as prescribed in 14 CFR Part 91, sections 91.159 and 91.177, is above 18,000 feet MSL, the lowest usable flight level is the flight level equivalent of the minimum altitude plus the number of feet specified according to the lowest flight level correction factor as follows:

Altimeter Setting	Correction Factor
29.92 or higher	—
29.91 to 29.42	500 feet
29.41 to 28.92	1,000 feet
28.91 to 28.42	1,500 feet
28.41 to 27.91	2,000 feet
27.91 to 27.42	2,500 feet

Operations in Other Countries

When flight crews transition from the U.S. NAS to another country's airspace, they should be aware of differences not only in procedures but also airspace. For example, when flying into Canada as depicted in Figure 2-67, notice the change from transition level (QNE) to transition altitude (QNH) when flying north-bound into the Moncton flight information region (FIR).

Operations in international airspace demand that pilots are aware of, and understand the use of, the three types of altimeter settings. *Most overseas airports give altimeter settings in hectopascals (hPa) (millibars). Therefore, it is imperative that pilots or on-board equipment are able to accurately convert inches of mercury to hPa, or hPa to inches of mercury.*

Altitude Above Ground (QFE)

A local altimeter setting equivalent to the barometric pressure measured at an airport altimeter datum, usually signifying the approach end of the runway is in use. At the airport altimeter datum, an altimeter set to QFE indicates zero altitude. If required to use QFE altimetry, altimeters are set to QFE while operating at or below the transition

RADAR/NON-RADAR REPORTS
These reports should be made at all times without a specific ATC request.

REPORTS	EXAMPLE
Leaving one assigned flight altitude or flight level for another	"Marathon 564, leaving 8,000, climb to 10,000."
VFR-on-top change in altitude	"Marathon 564, VFR-on-top, climbing to 10,500."
Leaving any assigned holding fix or point	"Marathon 564, leaving FARGO Intersection."
Missed approach	"Marathon 564, missed approach, request clearance to Chicago."
Unable to climb or descend at least 500 feet per minute	"Marathon 564, maximum climb rate 400 feet per minute."
TAS variation from filed speed of 5% or 10 knots, whichever is greater	"Marathon 564, advises TAS decrease to140 knots."
Time and altitude or flight level upon reaching a holding fix or clearance limit	"Marathon 564, FARGO Intersection at 05, 10,000, holding east."
Loss of Nav/Comm capability (required by Part 91.187)	"Marathon 564, ILS receiver inoperative."
Unforecast weather conditions or other information relating to the safety of flight (required by Part 91.183)	"Marathon 564, experiencing moderate turbulence at 10,000."

NON-RADAR REPORTS
When you are not in radar contact, these reports should be made without a specific request from ATC.

REPORTS	EXAMPLE
Leaving FAF or OM inbound on final approach	"Marathon 564, outer marker inbound, leaving 2,000."
Revised ETA of more than three minutes	"Marathon 564, revising SCURRY estimate to 55."

Figure 2-68. ATC reporting procedures.

altitude and below the transition level. On the airport, the altimeter will read "0" feet.

Barometric Pressure for Standard Altimeter Setting (QNE)

Use the altimeter setting (en route) at or above the transition altitude (FL 180 in the United States). The altimeter setting is always 29.92 inches of mercury/1013.2 hPa for a QNE altitude. *Transition levels differ from country to country and pilots should be particularly alert when making a climb or descent in a foreign area.*

Barometric Pressure for Local Altimeter Setting (QNH)

A local altimeter setting equivalent to the barometric pressure measured at an airport altimeter datum and corrected to sea level pressure. At the airport altimeter datum, an altimeter set to QNH indicates airport elevation above mean sea level (MSL). Altimeters are set to QNH while operating at and below the transition altitude and below the transition level.

For flights in the vicinity of airports, express the vertical position of aircraft in terms of QNH or QFE at or below the transition altitude and in terms of QNE at or above the transition level. While passing through the transition layer, express vertical position in terms of FLs when ascending and in terms of altitudes when descending.

When an aircraft that receives a clearance as number one to land completes its approach using QFE, express the vertical position of the aircraft in terms of height above the airport elevation during that portion of its flight for which you may use QFE.

It is important to remember that most pressure altimeters are subject to mechanical, elastic, temperature, and installation errors. In addition, extremely cold temperature differences may also require altimeter correction factors as appropriate.

En Route Reporting Procedures

In addition to acknowledging a handoff to another Center en route controller, there are reports that should be made without a specific request from ATC. Certain reports should be made at all times regardless of whether a flight is in radar contact with ATC, while others are necessary only if radar contact has been lost or terminated. [Figure 2-68]

Non-Radar Position Reports

If radar contact has been lost or radar service terminated, the CFRs require pilots to provide ATC with position reports over designated VORs and intersections along their route of flight. These compulsory reporting points are depicted on IFR en route charts by solid triangles. Position reports over fixes indicated by open triangles are noncompulsory reporting points and are only necessary when requested by ATC. If on a direct course that is not on an established airway, report over the fixes used in the flight plan that define the route, since they automatically become compulsory reporting points. Compulsory reporting points also apply when conducting an IFR flight in accordance with a VFR-on-top clearance.

Whether a route is on an airway or direct, position reports are mandatory in a non-radar environment, and they must include specific information. A typical position report includes information pertaining to aircraft position, expected route, and ETA. When a position report is to be made passing a VOR radio facility, the time reported should be the time at which the first complete reversal of the TO/FROM indicator is accomplished. When a position report is made passing a facility by means of an airborne ADF, the time reported should be the time at which the indicator makes a complete reversal. When an aural or a light panel indication is used to determine the time passing a reporting point, such as a fan marker, Z marker, cone of silence or intersection of range courses, the time should be noted when the signal is first received and again when it ceases. The mean of these two times should then be taken as the actual time over the fix. If a position is given with respect to distance and direction from a reporting point, the distance and direction should be computed as accurately as possible. Except for terminal area transition purposes, position reports or navigation with reference to aids not established for use in the structure in which flight is being conducted are not normally required by ATC.

Flights in a Radar Environment

When informed by ATC that their aircraft are in "Radar Contact," pilots should discontinue position reports over designated reporting points. They should resume normal position reporting when ATC advises "radar contact lost" or "radar service terminated." ATC informs pilots that they are in radar contact:

1. When their aircraft is initially identified in the ATC system; and

2. When radar identification is reestablished after radar service has been terminated or radar contact lost.

Subsequent to being advised that the controller has established radar contact, this fact is not repeated to the pilot when handed off to another controller. At times, the aircraft identity is confirmed by the receiving controller; however, this should not be construed to mean that radar contact has been lost. The identity of transponder equipped aircraft is confirmed by asking the pilot to "ident," "squawk standby," or to change codes. Aircraft without transponders are advised of their position to confirm identity. In this case, the pilot is expected to advise the controller if in disagreement with the position given. Any pilot who cannot confirm the accuracy of the position given because of not being tuned to the NAVAID referenced by the controller should ask for another radar position relative to the tuned in NAVAID.

Position Report Items

Position reports should include the following items:

1. Aircraft identification

2. Position

3. Time

4. Altitude or flight level (include actual altitude or flight level when operating on a clearance specifying VFR-on-top)

5. Type of flight plan (not required in IFR position reports made directly to ARTCCs or approach control)

6. ETA and name of next reporting point

7. The name only of the next succeeding reporting point along the route of flight

8. Pertinent remarks

Additional Reports

The following reports should be made at all times to ATC or FSS facilities without a specific ATC request:

1. When vacating any previously assigned altitude or flight level for a newly assigned altitude or flight level.

2. When an altitude change is made if operating on a clearance specifying VFR-on-top.

3. When unable to climb/descend at a rate of a least 500 feet per minute (fpm).

4. When approach has been missed. (Request clearance for specific action (i.e., to alternative airport, another approach).

5. Change in the average true airspeed (at cruising altitude) when it varies by 5 percent or 10 knots (whichever is greater) from that filed in the flight plan.

6. The time and altitude or flight level upon reaching a holding fix or point to which cleared.

7. When leaving any assigned holding fix or point.

NOTE: The reports stated in subparagraphs 6 and 7 may be omitted by pilots of aircraft involved in instrument training at military terminal area facilities when radar service is being provided.

8. Any loss, in controlled airspace, of VOR, TACAN, ADF, low frequency navigation receiver capability, GPS anomalies while using installed IFR-certified GPS/GNSS receivers, complete or partial loss of ILS receiver capability or impairment of air/ground communications capability. Reports should include aircraft identification, equipment affected, degree to which the capability to operate under IFR in the ATC system is impaired, and the nature and extent of assistance desired from ATC.

9. Any information relating to the safety of flight.

Other equipment installed in an aircraft may effectively impair safety and/or the ability to operate under IFR. If such equipment (e.g., airborne weather radar) malfunctions and in the pilot's judgment either safety or IFR capabilities are affected, reports should be made as stated above. When reporting GPS anomalies, include the location and altitude of the anomaly. Be specific when describing the location and include duration of the anomaly if necessary.

Communication Failure

Two-way radio communication failure procedures for IFR operations are outlined in 14 CFR Part 91, section 91.185. Unless otherwise authorized by ATC, pilots operating under IFR are expected to comply with this regulation. Expanded procedures for communication failures are found in the Aeronautical Information Manual (AIM). Pilots can use the transponder to alert ATC to a radio communication failure by squawking code 7600. [Figure 2-69] If only the transmitter is inoperative, listen for ATC instructions on any operational receiver, including the navigation receivers. It is possible ATC may try to make contact with pilots over a VOR, VORTAC, NDB, or localizer frequency. In addition to monitoring NAVAID receivers, attempt to reestablish communications by contacting ATC on a previously assigned frequency or calling an FSS.

The primary objective of the regulations governing communication failures is to preclude extended IFR no-radio operations within the ATC system since these operations may adversely affect other users of the airspace. If the radio fails while operating on an IFR clearance, but in VFR conditions, or if encountering VFR conditions at any time after the failure, continue the flight under VFR conditions, if possible, and land as soon as practicable. The requirement to land as soon as practicable should not be construed to mean as soon as possible. Pilots retain the prerogative of exercising their best judgment and are not

required to land at an unauthorized airport, at an airport unsuitable for the type of aircraft flown, or to land only minutes short of their intended destination. However, if IFR conditions prevail, pilots must comply with procedures designated in the CFRs to ensure aircraft separation.

If pilots must continue their flight under IFR after experiencing two-way radio communication failure, they should fly one of the following routes:

- The route assigned by ATC in the last clearance received.

- If being radar vectored, the direct route from the point of radio failure to the fix, route, or airway specified in the radar vector clearance.

- In the absence of an assigned route, the route ATC has advised to expect in a further clearance.

- In the absence of an assigned or expected route, the route filed in the flight plan.

It is also important to fly a specific altitude should two-way radio communications be lost. The altitude to fly after a

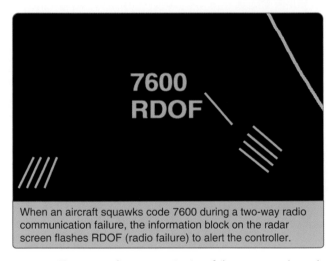

When an aircraft squawks code 7600 during a two-way radio communication failure, the information block on the radar screen flashes RDOF (radio failure) to alert the controller.

Figure 2-69. Two-way radio communications failure transponder code.

communication failure can be found in 14 CFR Part 91, section 91.185 and must be the highest of the following altitudes for each route segment flown.

- The altitude or flight level assigned in the last ATC clearance.

- The minimum altitude or flight level for IFR operations.

- The altitude or flight level ATC has advised to expect in a further clearance.

In some cases, the assigned or expected altitude may not be as high as the MEA on the next route segment. In this situation, pilots normally begin a climb to the higher MEA when they reach the fix where the MEA rises. If the fix also has a published MCA, they start the climb so they are at or above the MCA when reaching the fix. If the next succeeding route segment has a lower MEA, descend to the applicable altitude either the last assigned altitude or the altitude expected in a further clearance—when reaching the fix where the MEA decreases.

ARTCC Radio Frequency Outage

ARTCCs normally have at least one back-up radio receiver and transmitter system for each frequency that can usually be placed into service quickly with little or no disruption of ATC service. Occasionally, technical problems may cause a delay but switchover seldom takes more than 60 seconds. When it appears that the outage is not quickly remedied, the ARTCC usually requests a nearby aircraft, if there is one, to switch to the affected frequency to broadcast communications instructions. It is important that the pilot wait at least 1 minute before deciding that the ARTCC has actually experienced a radio frequency failure. When such an outage does occur, the pilot should, if workload and equipment capability permit, maintain a listening watch on the affected frequency while attempting to comply with the following recommended communications procedures:

1. If two-way communications cannot be established with the ARTCC after changing frequencies, a pilot should attempt to re-contact the transferring controller for the assignment of an alternative frequency or other instructions.

2. When an ARTCC radio frequency failure occurs after two-way communications have been established, the pilot should attempt to reestablish contact with the center on any other known ARTCC frequency, preferably that of the next responsible sector when practicable, and ask for instructions. However, when the next normal frequency change along the route is known to involve another ATC facility, the pilot should contact that facility, if feasible, for instructions. If communications cannot be reestablished by either method, the pilot is expected to request communication instructions from the FSS appropriate to the route of flight.

NOTE: The exchange of information between an aircraft and an ARTCC through an FSS is quicker than relay via company radio because the FSS has direct interphone lines to the responsible ARTCC sector. Accordingly, when circumstances dictate a choice between the two during an ARTCC frequency outage relay via FSS radio is recommended.

Climbing and Descending En Route

When ATC issues a clearance or instruction, pilots are expected to execute its provisions upon receipt. In some cases, ATC includes words that modify their expectation. For example, the word "immediately" in a clearance or instruction is used to impress urgency to avoid an imminent situation, and expeditious compliance is expected and necessary for safety. The addition of a climb point or time restriction, for example, does not authorize pilots to deviate from the route of flight or any other provision of the ATC clearance. If the pilot receives the term "climb at pilot's discretion" in the altitude information of an ATC clearance, it means that the pilot has the option to start a climb when they desire and are authorized to climb at any rate, and to temporarily level off at any intermediate altitude as desired, although once you vacate an altitude, you may not return to that altitude. When ATC has not used the term nor imposed any climb restrictions, pilots should climb promptly on acknowledgment of the clearance. Climb at an optimum rate consistent with the operating characteristics of the aircraft to 1,000 feet below the assigned altitude, and then attempt to climb at a rate of between 500 and 1,500 fpm until the assigned altitude is reached. If at any time the pilot is unable to climb at a rate of at least 500 fpm, advise ATC. If it is necessary to level off at an intermediate altitude during climb, advise ATC.

When ATC issues the instruction, "Expedite climb," this normally indicates that the pilot should use the approximate best rate of climb without an exceptional change in aircraft handling characteristics. Normally controllers inform pilots of the reason for an instruction to expedite. If flying a turbojet airplane equipped with afterburner engines, such as a military aircraft, pilots should advise ATC prior to takeoff if intending to use afterburning during the climb to the en route altitude. Often, the controller may be able to plan traffic to accommodate a high performance climb and allow the pilot to climb to the planned altitude without "expedite" clearance from restriction. If you receive an ATC instruction, and your altitude to maintain is subsequently changed or restated without an expedite instruction, the expedite instruction is canceled.

During en route climb, as in any other phase of flight, it is essential that you clearly communicate with ATC regarding clearances. In the following example, a flight crew experienced an apparent clearance readback/hearback error, that resulted in confusion about the clearance and, ultimately, to inadequate separation from another aircraft. "Departing IFR, clearance was to maintain 5,000 feet, expect 12,000 in 10 minutes." After handoff to Center, the pilot understood and read back, "Leaving 5,000 turn left heading 240° for vector on course." The pilot turned to the assigned heading climbing through 5,000 feet. At 5,300 feet, Center advised assigned altitude was 5,000 feet. The pilot immediately descended to 5,000. Center then informed the pilot that there was traffic at 12 o'clock and a mile at 6,000. After passing traffic, a higher altitude was assigned and climb resumed. The pilot then believed the clearance was probably "reaching" 5,000, etc. Even the readback to the controller with "leaving" did not catch the different wording. "Reaching" and "leaving" are commonly used ATC terms having different usages. They may be used in clearances involving climbs, descents, turns, or speed changes. In the flight deck, the words "reaching" and "leaving" sound much alike.

For altitude awareness during climb, pilots often call out altitudes on the flight deck. The pilot monitoring may call 2,000 and 1,000 feet prior to reaching an assigned altitude. The callout may be, "two" climbing through the transit to go altitude (QNH), both pilots set their altimeters to 29.92 inches of mercury and announce "2992 inches" (or 'standard,' on some airplanes) and the flight level passing. For example, "2992 inches" (standard), flight level one eight zero. The second officer on three pilot crews may ensure that both pilots have inserted the proper altimeter setting. On international flights, pilots must be prepared to differentiate, if necessary, between barometric pressure equivalents with inches of mercury, and millibars or hectopascals, to eliminate any potential for error. For example, 996 millibars erroneously being set as 2996.

For a typical IFR flight, the majority of in-flight time often is flown in level flight at cruising altitude from top of climb (TOC) to top of descent (TOD). Generally, TOD is used in airplanes with a FMS and represents the point at which descent is first initiated from cruise altitude. FMS also assist in level flight by cruising at the most fuel saving speed, providing continuing guidance along the flight plan route including great circle direct routes, and continuous evaluation and prediction of fuel consumption along with changing clearance data.

Aircraft Speed and Altitude

During the en route descent phase of flight, an additional benefit a FMS is that it provides fuel saving idle thrust descent to your destination airport. This allows an uninterrupted profile descent from level cruising altitude to an appropriate MIA, except where level flight is required for speed adjustment.. Controllers anticipate and plan that the pilot may level off at 10,000 feet MSL on descent to comply with the 14 CFR Part 91 indicated airspeed limit of 250 knots. Leveling off at any other time on descent may seriously affect air traffic handling by ATC. It is imperative that pilots make every effort to fulfill ATC expected actions on descent to aid in safely handling and expediting air traffic.

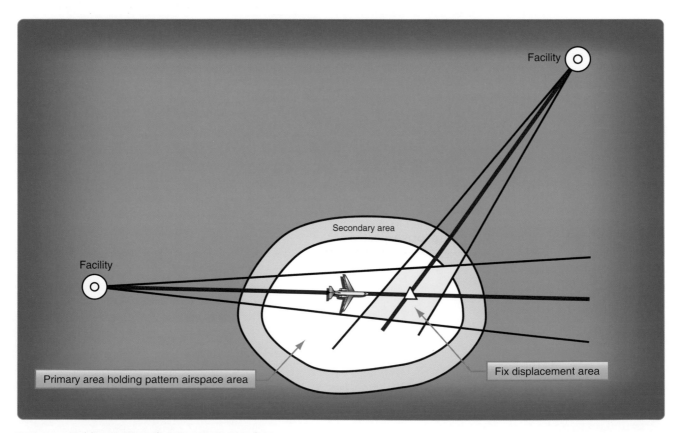

Figure 2-70. Holding pattern design criteria template.

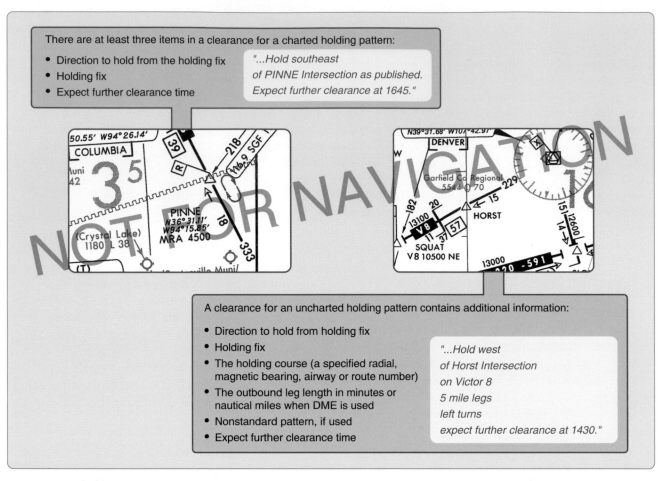

There are at least three items in a clearance for a charted holding pattern:

- Direction to hold from the holding fix
- Holding fix
- Expect further clearance time

"...Hold southeast of PINNE Intersection as published. Expect further clearance at 1645."

A clearance for an uncharted holding pattern contains additional information:

- Direction to hold from holding fix
- Holding fix
- The holding course (a specified radial, magnetic bearing, airway or route number)
- The outbound leg length in minutes or nautical miles when DME is used
- Nonstandard pattern, if used
- Expect further clearance time

"...Hold west of Horst Intersection on Victor 8 5 mile legs left turns expect further clearance at 1430."

Figure 2-71. ATC holding instructions.

ATC issues speed adjustments if the flight is being radar controlled to achieve or maintain required or desired spacing. They express speed adjustments in terms of knots based on indicated airspeed in 10 knot increments except that at or above FL 240 speeds may be expressed in terms of Mach numbers in 0.01 increments. The use of Mach numbers by ATC is restricted to turbojets. If complying with speed adjustments, pilots are expected to maintain that speed within plus or minus 10 knots or 0.02 Mach.

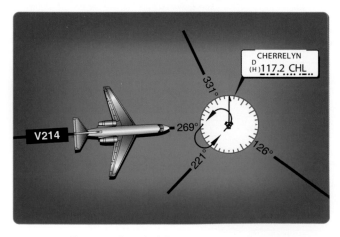

Figure 2-72. Clearance limit holding.

Speed and altitude restrictions in clearances are subject to misinterpretation, as evidenced in this case where a corporate flight crew treated instructions in a published procedure as a clearance. The aircraft was at FL 310 and had already programmed the 'expect-crossing altitude' of 17,000 feet at the VOR. When the altitude alerter sounded, the pilot advised Center that we were leaving FL 310. ATC acknowledged with a "Roger." At FL 270, Center questioned the pilot about the aircrafts descent. The pilot told the controller that the reason for the descent was to cross the VOR at 17,000 feet. ATC advised the pilot that he did not have clearance to descend. What the pilot thought was a clearance was in fact an "expect" clearance. Whenever pilots are in doubt about a clearance it is imperative they request clarity from ATC. Also, the term "Roger" only means that ATC received the transmission, not that they understood the transmission. "Expect" altitudes are published for planning purposes and are not considered crossing restrictions until verbally issued by ATC.

En Route Holding Procedures

The criteria for holding pattern airspace is developed both to provide separation of aircraft, as well as obstacle

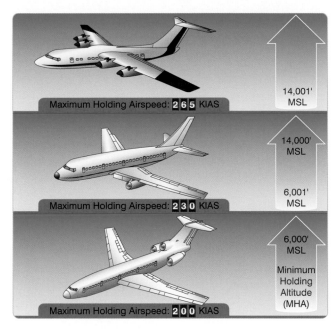

Maximum Holding Airspeed: 2 6 5 KIAS

14,001' MSL

Maximum Holding Airspeed: 2 3 0 KIAS

14,000' MSL

6,001' MSL

6,000' MSL

Minimum Holding Altitude (MHA)

Maximum Holding Airspeed: 2 0 0 KIAS

Figure 2-73. Maximum holding speeds for different altitudes.

clearance. The alignment of holding patterns typically coincides with the flight course you fly after leaving the holding fix. For level holding, a minimum of 1,000 feet obstacle clearance is provided throughout the primary area. In the secondary area, 500 feet of obstacle clearance is provided at the inner edge, tapering to zero feet at the outer edge. Allowance for precipitous terrain is considered, and the altitudes selected for obstacle clearance may be rounded to the nearest 100 feet. When criteria for a climb in hold are applied, no obstacle penetrates the holding surface. [Figure 2-70]

There are many factors that affect aircraft during holding maneuvers, including navigational aid ground and airborne tolerance, effect of wind, flight procedures, application of ATC, outbound leg length, maximum holding airspeeds, fix to NAVAID distance, DME slant range effect, holding airspace size, and altitude holding levels.

ATC Holding Instructions

When controllers anticipate a delay at a clearance limit or fix, pilots are usually issued a holding clearance at least 5 minutes before the ETA at the clearance limit or fix. If the holding pattern assigned by ATC is depicted on the appropriate aeronautical chart, pilots are expected to hold as charted. In the following example, the controller issues a holding clearance that includes the name of the fix, directs the pilot to hold as charted, and includes an expect further clearance (EFC) time.

"Marathon five sixty four, hold east of MIKEY Intersection as published, expect further clearance at 1521."

When ATC issues a clearance requiring you to hold at a fix where a holding pattern is not charted, pilots are issued complete holding instructions. The holding instructions include the direction from the fix, name of the fix, course, leg length, if appropriate, direction of turns (if left turns are required), and the EFC time. Pilots are required to maintain the last assigned altitude unless a new altitude is specifically included in the holding clearance and should fly right turns unless left turns are assigned. Note that all holding instructions should include an EFC time. In the event that two-way radio communication is lost, the EFC allows the pilot to depart the holding fix at a definite time. Pilots should plan the last lap of the holding pattern to leave the fix as close as possible to the exact time. [Figure 2-71]

When approaching the clearance limit and you have not received holding instructions from ATC, pilots are expected to follow certain procedures. First, call ATC and request further clearance before reaching the fix. If further clearance cannot be obtained, pilots are expected to hold at the fix in compliance with the charted holding pattern. If a holding pattern is not charted at the fix, pilots are expected to hold on the inbound course using right turns. This procedure ensures that ATC provides adequate separation. [Figure 2-72] For example, the aircraft is heading eastbound on V214 and the Cherrelyn VORTAC is the clearance limit and the pilot has not been able to obtain further clearance and has not received holding instructions, plan to hold southwest on the 221° radial using left-hand turns, as depicted. If this holding pattern is not charted, hold west of the VOR on V214 using right-hand turns.

Where required for aircraft separation, ATC may request that the pilot hold at any designated reporting point in a standard holding pattern at the MEA or the MRA, whichever altitude is the higher at locations where a minimum holding altitude has not been established. Unplanned holding at en route fixes may be expected on airway or route radials, bearings, or courses. If the fix is a facility, unplanned holding could be on any radial or bearing and there may be holding limitations required if standard holding cannot be accomplished at the MEA or MRA.

Maximum Holding Speed

The size of the holding pattern is directly proportional to the speed of the airplane. In order to limit the amount of airspace that must be protected by ATC, maximum holding speeds in knots indicated airspeed (KIAS) have been designated for specific altitude ranges. [Figure 2-73] Even so, some holding patterns may have additional speed restrictions to keep faster airplanes from flying out of the protected area. If a holding pattern has a nonstandard speed restriction, it is depicted by an icon with the limiting airspeed. If the holding speed limit is less than the pilot feels necessary, advise ATC of the revised holding speed. Also, if the indicated airspeed exceeds the applicable maximum holding speed, ATC expects the pilot to slow to the speed limit within three minutes of the ETA at the holding fix. Often pilots can avoid flying a holding pattern, or reduce the length of time spent in the holding pattern, by slowing down on the way to the holding fix.

High Performance Holding

When operating at higher airspeeds, there are certain limitations that must be adhered to. For example, aircraft do not make standard rate turns in holding patterns if the bank angle exceeds 30°. If your aircraft is using a flight director system, the bank angle is limited to 25°. The aircraft must be traveling over 210 knots true airspeed (TAS) for the bank angle in a standard rate turn to exceed 30°, therefore this limit applies to relatively fast airplanes. An aircraft using a flight director would have to be holding at more than 170 knots TAS to come up against the 25° limit. These true airspeeds correspond to indicated airspeeds of about 183 and 156 knots, respectively, at 6,000 feet in a standard atmosphere.

En Route Safety Considerations

Fuel State Awareness

In order to increase fuel state awareness, pilots are required to monitor the time and fuel remaining during an IFR flight. For example, on a flight scheduled for one hour or less, the flight crew may record the time and fuel remaining at the top of climb (TOC) and at one additional waypoint listed in the flight plan. Generally, TOC is used in airplanes with an FMS, and represents the point at which cruise altitude is first reached. TOC is calculated based on current airplane altitude, climb speed, and cruise altitude. The pilot may elect to delete the additional waypoint recording requirement if the flight is so short that the record will not assist in the management of the flight. For flights scheduled for more than one hour, the pilot may record the time and fuel remaining shortly after TOC and at selected waypoints listed in the flight plan, conveniently spaced approximately one hour apart. The actual fuel burn is then compared to the planned fuel burn. Each fuel tank must be monitored to verify proper burn off and appropriate fuel remaining. For two-pilot airplanes, the pilot monitoring (PM) keeps the flight plan record. On three-pilot airplanes, the second officer and PM coordinate recording and keeping the flight plan record. In all cases, the crew member(s) making the recording communicates the information to the pilot flying.

Diversion Procedures

OpSpecs for commercial operators include provisions for en route emergency diversion airport requirements. Operators are expected to develop a sufficient set of emergency diversion airports, so that one or more can be reasonably expected to be available in varying weather conditions. The flight must be able to make a safe landing, and the airplane maneuvered off of the runway at the selected diversion airport. In the event of a disabled airplane following landing, the capability to move the disabled airplane must exist so as not to block the operation of any recovery airplane. In addition, those airports designated for use must be capable of protecting the safety of all personnel by being able to:

- Offload the passengers and flight crew in a safe manner during possible adverse weather conditions.

- Provide for the physiological needs of the passengers and flight crew for the duration until safe evacuation.

- Be able to safely extract passengers and flight crew as soon as possible. Execution and completion of the recovery is expected within 12 to 48 hours following diversion.

Part 91 operators also need to be prepared for a diversion. Designation of an alternate on the IFR flight plan is a good first step; but changing weather conditions or equipment issues may require pilots to consider other options.

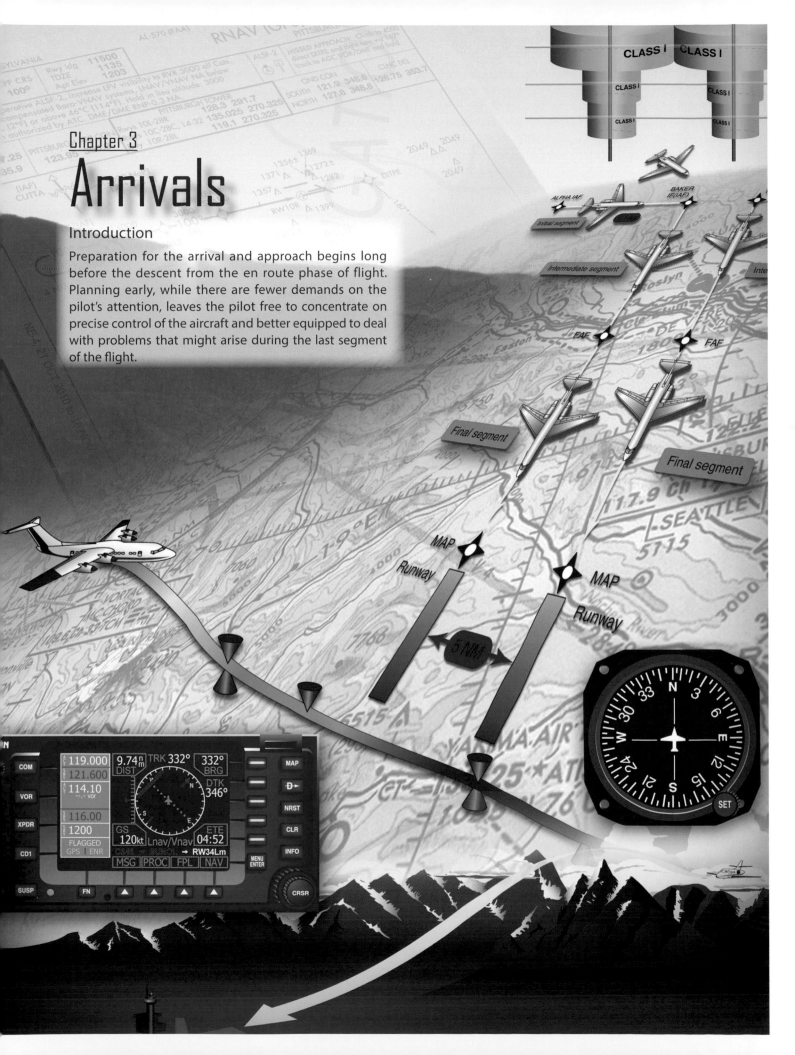

Chapter 3
Arrivals

Introduction

Preparation for the arrival and approach begins long before the descent from the en route phase of flight. Planning early, while there are fewer demands on the pilot's attention, leaves the pilot free to concentrate on precise control of the aircraft and better equipped to deal with problems that might arise during the last segment of the flight.

This chapter focuses on the current procedures pilots and air traffic control (ATC) use for instrument flight rule (IFR) arrivals in the National Airspace System (NAS). The objective is to provide pilots with an understanding of ATC arrival procedures and pilot responsibilities as they relate to the transition between the en route and approach phases of flight. This chapter emphasizes standard terminal arrival routes (STARs), descent clearances, descent planning, and ATC procedures, while the scope of coverage focuses on transitioning from the en route phase of flight, typically the origination point of a STAR to the STAR termination fix.

Optimum IFR arrival options include flying directly from the en route structure to an approach gate or initial approach fix (IAF), a visual arrival, STARs, and radar vectors. Within controlled airspace, ATC routinely uses radar vectors for separation purposes, noise abatement considerations when it is an operational advantage, or when requested by pilots. Vectors outside of controlled airspace are provided only on pilot request. The controller tells the pilot the purpose of the vector when the vector is controller-initiated and takes the aircraft off a previously assigned non-radar route. Typically, when operating on area navigation (RNAV) routes, pilots are allowed to remain on their own navigation.

Navigation in the Arrival Environment

The most significant and demanding navigational requirement is the need to safely separate aircraft. In a non-radar environment, ATC does not have an independent means to separate air traffic and must depend entirely on information relayed from flight crews to determine the actual geographic position and altitude. In this situation, precise navigation is critical to ATC's ability to provide separation.

Even in a radar environment, precise navigation and position reports, when required, are still a primary means of providing separation. In most situations, ATC does not have the capability or the responsibility for navigating an aircraft. Because they rely on precise navigation by the flight crew, flight safety in all IFR operations depends directly on the pilot's ability to achieve and maintain certain levels of navigational performance. ATC uses radar to monitor navigational performance, detect possible navigational errors, and expedite traffic flow. In a non-radar environment, ATC has no independent knowledge of the actual position of the aircraft or its relationship to other aircraft in adjacent airspace. Therefore, ATC's ability to detect a navigational error and resolve collision hazards is seriously degraded when a deviation from a clearance occurs.

The concept of navigation performance, previously discussed in this book, involves the precision that must be maintained for both the assigned route and altitude. Required levels of navigation performance vary from area to area depending on traffic density and complexity of the routes flown. The level of navigation performance must be more precise in domestic airspace than in oceanic and remote land areas since air traffic density in domestic airspace is much greater. For example, there are three million flight operations conducted within Chicago Center's airspace each year. The minimum lateral distance permitted between co-altitude aircraft in Chicago Center's airspace is 8 nautical miles (NM) (3 NM when radar is used). The route ATC assigns an aircraft has protected airspace on both sides of the centerline, equal to one-half of the lateral separation minimum standard. For example, the overall level of lateral navigation performance necessary for flight safety must be better than 4 NM in Center airspace. When STARs are reviewed subsequently in this chapter, it is demonstrated how the navigational requirements become more restrictive in the arrival phase of flight where air traffic density increases and procedural design and obstacle clearance become more limiting.

The concept of navigational performance is fundamental to the code of federal regulations and is best defined in Title 14 of the Code of Federal Regulations (14 CFR) Part 121, sections 121.103 and 121.121, which state that each aircraft must be navigated to the degree of accuracy required for ATC. The requirements of 14 CFR Part 91, section 91.123 related to compliance with ATC clearances and instructions also reflect this fundamental concept. Commercial operators must comply with their Operations Specifications (OpSpecs) and understand the categories of navigational operations and be able to navigate to the degree of accuracy required for the control of air traffic.

In the broad concept of air navigation, there are two major categories of navigational operations consisting of Class I navigation and Class II navigation. Class I navigation is any en route flight operation conducted in controlled or uncontrolled airspace that is entirely within operational service volumes of International Civil Aviation Organization (ICAO) standard navigational aids (NAVAIDs) (very high frequency (VHF) omnidirectional radio range (VOR), VOR/distance measuring equipment (DME), non-directional beacon (NDB), etc.).

Class II navigation is any en route operation that is not categorized as Class I navigation and includes any operation or portion of an operation that takes place outside the operational service volumes of ICAO standard NAVAIDs. For example, aircraft equipped only with VORs conducts

Class II navigation when the flight operates in an area outside the operational service volumes of federal VORs. Class II navigation does not automatically require the use of long-range, specialized navigational systems if special navigational techniques are used to supplement conventional NAVAIDs. Class II navigation includes transoceanic operations and operations in desolate and remote land areas, such as the Arctic. The primary types of specialized navigational systems approved for Class II operations include inertial navigation system (INS), Doppler, and global positioning system (GPS). Figure 3-1 provides several examples of Class I and II navigation.

Descent Planning

Planning the descent from cruise is important because of the need to dissipate altitude and airspeed in order to arrive at the approach gate properly configured. Descending early results in more flight at low altitudes with increased fuel consumption, and starting down late results in problems controlling both airspeed and descent rates on the approach. Prior to flight, pilots need to calculate the fuel, time, and distance required to descend from the cruising altitude to the approach gate altitude for the specific instrument approach at the destination airport. While in flight prior to the descent, it is important for pilots to verify landing weather to include winds at their intended destination. Inclimate weather at the destination airport can cause slower descents and missed approaches that require a sufficient amount of fuel that should be calculated prior to starting the descent. In order to plan the descent, the pilot needs to know the cruise altitude, approach gate altitude or initial approach fix altitude, descent groundspeed, and

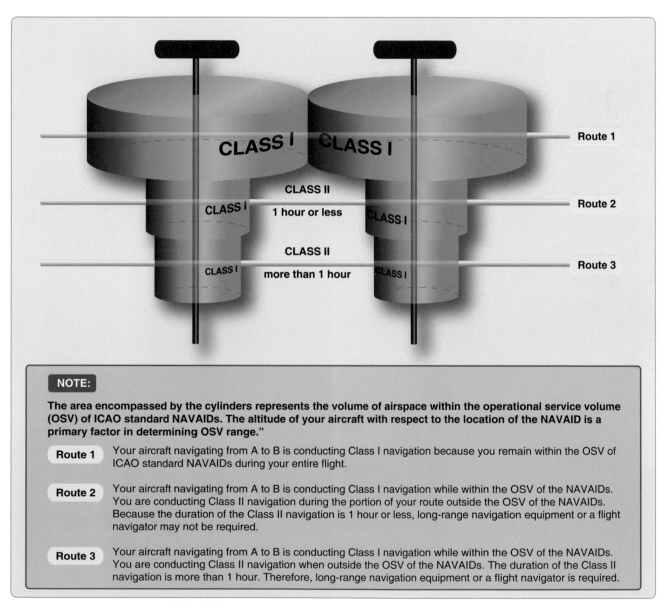

Figure 3-1. Example of Class I and II navigation.

descent rate. This information must be updated while in flight for changes in altitude, weather, and wind. The approach gate is an imaginary point used by ATC to vector aircraft to the final approach course. The approach gate is established along the final approach course 1 NM from the final approach fix (FAF) on the side away from the airport and is located no closer than 5 NM from the landing threshold. Flight manuals or operating handbooks may also contain a fuel, time, and distance to descend chart that contains the same information.

One technique that is often used is the descent rule of thumb, which is used to determine when you need to descend in terms of the number of miles prior to the point at which you desire to arrive at your new altitude. First, divide the altitude needed to be lost by 300. For example, if cruising altitude is 7,000 feet and you want to get down to a pattern altitude of 1,000 feet. The altitude you want to lose is 6,000 feet, which when divided by 300 results in 20. Therefore, you need to start your descent 20 NM out and leave some extra room so that you are at pattern altitude prior to the proper entry. It is also necessary to know what rate-of-descent (ROD) to use.

To determine ROD for a three-degree path, simply multiply your groundspeed by 5. If you are going 120 knots, your ROD to fly the desired path would be 600 feet per minute (120 × 5 = 600). It was determined in the previous example that

a descent should be initiated at 20 NM to lose 6,000 feet. If the groundspeed is 120 knots, that means the aircraft is moving along at 2 NM per minute. So to go 20 NM, it takes 10 minutes. Ten minutes at 600 feet per minute means you will lose 6,000 feet.

The calculations should be made before the flight and rules of thumb updates should be applied in flight. For example, from the charted STAR pilots might plan a descent based on an expected clearance to "cross 40 DME West of Brown VOR at 6,000" and then apply rules of thumb for slowing down from 250 knots. These might include planning airspeed at 25 NM from the runway threshold to be 250 knots, 200 knots at 20 NM, and 150 knots at 15 NM until gear and flap speeds are reached, never to fall below approach speed.

Vertical Navigation (VNAV) Planning

Vertical navigation (VNAV) is the vertical component of the flight plan. This approach path is computed from the top-of- descent (TOD) point down to the end-of-descent waypoint (E/D), which is generally the runway or missed approach point, which is slightly different than to the approach gate for non-flight management system (FMS) equipped aircraft. [Figure 3-2] The VNAV path is computed based upon the aircraft performance, approach constraints, weather data (winds, temperature, icing conditions, etc.) and aircraft weight.

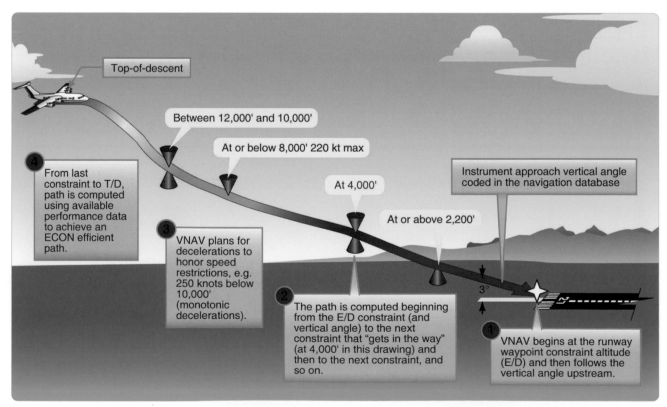

Figure 3-2. VNAV path construction.

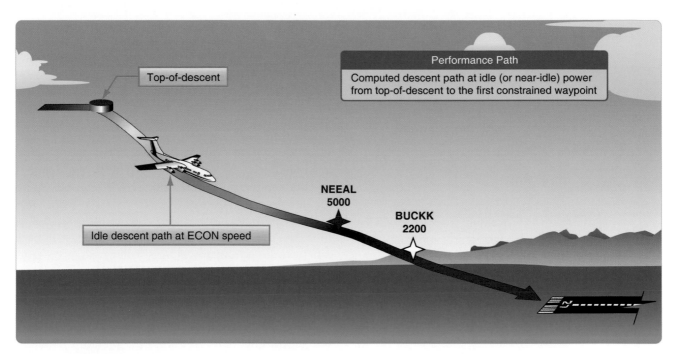

Figure 3-3. VNAV performance path.

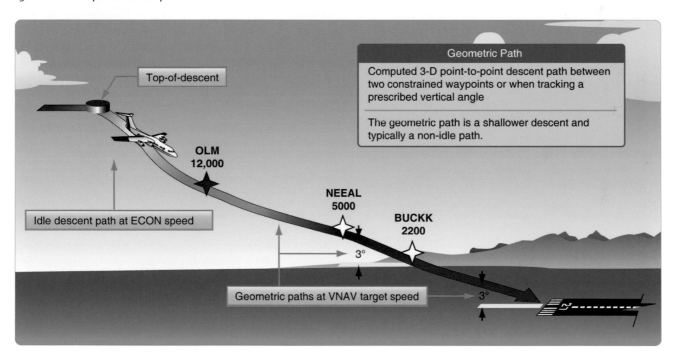

Figure 3-4. VNAV geometric path.

The two types of VNAV paths that the FMS use is either a performance path or a geometric path. The performance path is computed using at idle or near idle power from the TOD to the first constrained waypoint. [Figure 3-3] The geometric path is computed from point to point between two constrained waypoints or when on an assigned vertical angle. The geometric path is shallower than the performance path and is typically a non-idle path. [Figure 3-4]

LNAV/VNAV Equipment

Lateral navigation/vertical navigation (LNAV/VNAV) equipment is similar to an instrument landing system (ILS) in that it provides both lateral and vertical approach course guidance. Since precise vertical position information is beyond the current capabilities of the GPS, approaches with LNAV/VNAV minimums make use of certified barometric VNAV (baro-VNAV) systems for vertical guidance and/or the wide area augmentation system (WAAS) to improve GPS accuracy for this purpose.

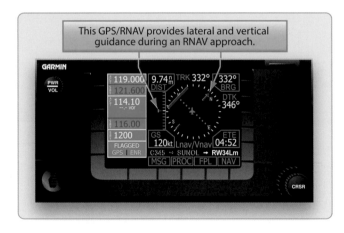

Figure 3-5. WAAS data provide lateral and vertical guidance.

NOTE: WAAS makes use of a collection of ground stations that are used to detect and correct inaccuracies in the position information derived from the GPS. Using WAAS, the accuracy of vertical position information is increased to within three meters.

To make use of WAAS, however, the aircraft must be equipped with an IFR-approved GPS receiver with WAAS signal reception that integrates WAAS error correction signals into its position determining processing. The WAAS enabled GPS receiver [Figure 3-5] allows the pilot to load an RNAV approach and receive guidance along the lateral and vertical profile shown on the approach chart. [Figure 3-6] It is very important to know what kind of equipment is

installed in an aircraft, and what it is approved to do. It is also important to understand that the VNAV function of non-WAAS capable or non-WAAS equipped IFR-approved GPS receivers does not make the aircraft capable of flying approaches to LNAV/VNAV minimums.

FMS are the primary tool for most modern aircraft, air carriers, and any operators requiring performance based navigation. Most of the modern FMS are fully equipped with LNAV/VNAV and WAAS. The FMS provides flight control steering and thrust guidance along the VNAV path. Some less integrated systems may only advise the flight crew of the VNAV path but have no auto-throttle capability. These less integrated systems require an increase in pilot workload during the arrival/approach phase in order to maintain the descent path.

Descent Planning for High Performance Aircraft

The need to plan the IFR descent into the approach gate and airport environment during the preflight planning stage of flight is particularly important for turbojet powered airplanes. TOD from the en route phase of flight for high performance airplanes is often used in this process and is calculated manually or automatically through a FMS based upon the altitude of the approach gate. A general rule of thumb for initial IFR descent planning in jets is the 3 to 1 formula. This means that it takes 3 nautical miles (NM) to descend 1,000 feet. If an airplane is at flight level (FL) 310 and the approach gate or initial approach fix is at 6,000 feet, the initial descent requirement equals 25,000 feet (31,000–6,000). Multiplying 25 times 3 equals 75; therefore begin descent 75 NM from the approach gate, based on a normal jet airplane, idle thrust, speed Mach 0.74 to 0.78, and vertical speed of 1,800–2,200 feet per minute (fpm). For a tailwind adjustment, add 2 NM for each 10 knots of tailwind. For a headwind adjustment, subtract 2 NM for each 10 knots of headwind. During the descent planning stage, try to determine which runway is in use at the destination airport, either by reading the latest aviation routine weather report (METAR) or checking the automatic terminal information service (ATIS) information. There can be big differences in distances depending on the active runway and STAR. The objective is to determine the most economical point for descent.

An example of a typical jet descent-planning chart is depicted in Figure 3-7. Item 1 is the pressure altitude from which the descent begins; item 2 is the time required for the descent in minutes; item 3 is the amount of fuel consumed in pounds during descent to sea level; and item 4 is the distance covered in NM. Item 5 shows that the chart is based on a Mach .80 airspeed until 280 knots

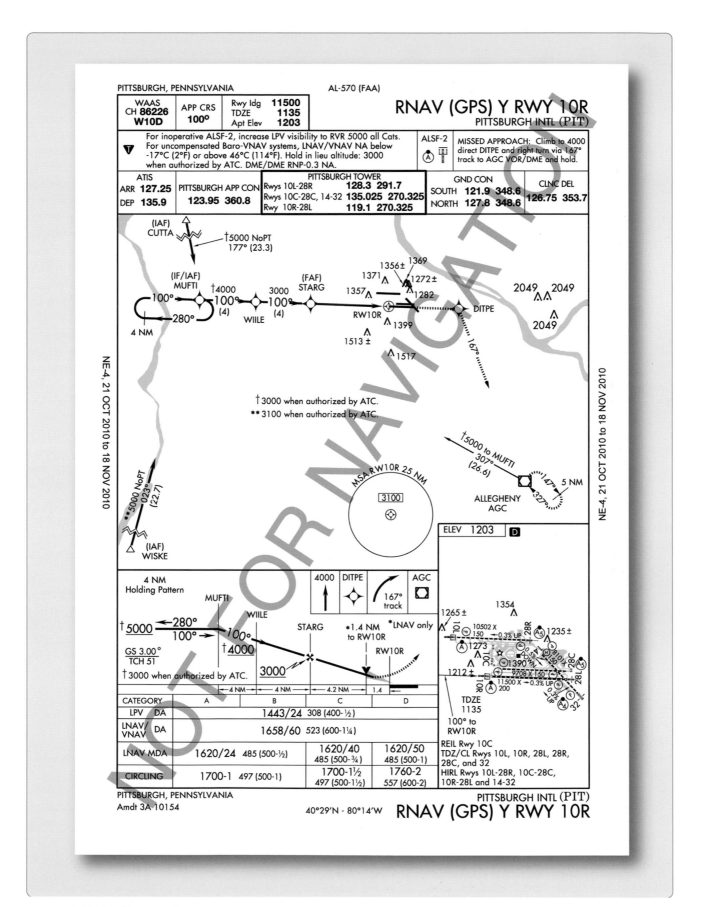

Figure 3-6. RNAV (GPS) approach.

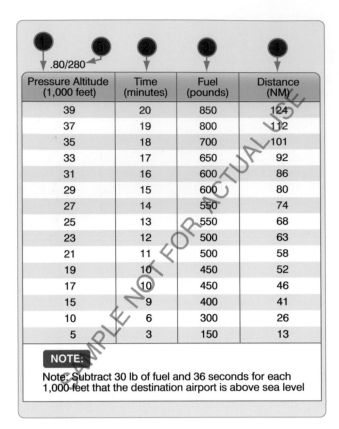

Pressure Altitude (1,000 feet)	Time (minutes)	Fuel (pounds)	Distance (NM)
39	20	850	124
37	19	800	112
35	18	700	101
33	17	650	92
31	16	600	86
29	15	600	80
27	14	550	74
25	13	550	68
23	12	500	63
21	11	500	58
19	10	450	52
17	10	450	46
15	9	400	41
10	6	300	26
5	3	150	13

NOTE:
Note: Subtract 30 lb of fuel and 36 seconds for each 1,000 feet that the destination airport is above sea level

Figure 3-7. Jet descent task.

indicated airspeed (KIAS) is obtained. The 250 knot airspeed limitation below 10,000 feet mean sea level (MSL) is not included on the chart, since its effect is minimal. Also, the effect of temperature or weight variation is negligible and is therefore omitted.

Due to the increased flight deck workload, pilots should get as much done ahead of time as possible. As with the climb and cruise phases of flight, aircrews should consult the proper performance charts to compute their fuel requirements, as well as the time and distance needed for their descent.

During the cruise and descent phases of flight, pilots need to monitor and manage the airplane according to the appropriate manufacturer's recommendations. Flight manuals and operating handbooks contain cruise and descent checklists, performance charts for specific cruise configurations, and descent charts that provide information regarding the fuel, time, and distance required to descend. Aircrews should review this information prior to the departure of every flight so they have an understanding of how the airplane is supposed to perform at cruise and during descent. A stabilized descent constitutes a preplanned maneuver in which the power is properly set, and minimum control input is required to maintain the appropriate descent path. Excessive corrections or control inputs indicate the descent was improperly planned. Plan the IFR descent from cruising altitude so that the aircraft arrives at the approach gate altitude or initial approach fix altitude prior to beginning the instrument approach. For example, suppose you are asked to descend from 11,000 feet to meet a crossing restriction at 3,000 feet. [Figure 3-8] Since there is a 200 knot speed restriction while approaching the destination airport, you choose a descent speed of 190 knots and a descent rate of 1,000 fpm. Assuming a 10 knot headwind component, groundspeed in the descent is 180 knots.

Descending From the En Route Altitude

Making the transition from cruise flight to the beginning of an instrument approach procedure sometimes requires arriving at a given waypoint at an assigned altitude. When this requirement is prescribed by a published arrival

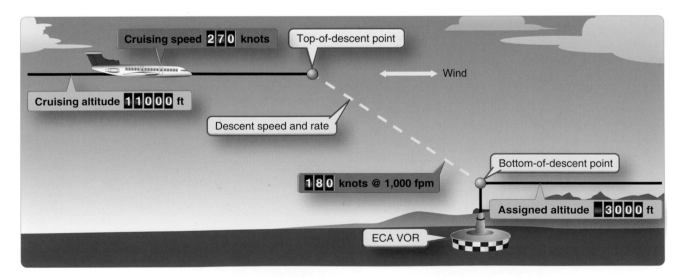

Figure 3-8. The descent planning task.

procedure or issued by ATC, it is called a crossing restriction. Even when ATC allows a descent at the pilot's discretion, aircrews need to choose a waypoint and altitude for positioning convenient to start the approach. In either case, descending from a cruising altitude to a given waypoint or altitude requires both planning and precise flying.

ATC may ask the pilot to descend to and maintain a specific altitude. Generally, this clearance is for en route traffic separation purposes, and pilots need to respond to it promptly. Descend at the optimum rate for the aircraft being flown until 1,000 feet above the assigned altitude, then descend at a rate between 500 and 1,500 fpm to the assigned altitude. If at any time, other than when slowing to 250 KIAS at 10,000 feet MSL, the pilot cannot descend at a rate of at least 500 fpm, advise ATC.

The second type of clearance allows the pilot to descend "… at pilot's discretion." When ATC issues a clearance to descend at pilot's discretion, pilots may begin the descent whenever they choose and at any rate of their choosing. Pilots are also authorized to level off, temporarily, at any intermediate altitude during the descent. However, once the aircraft leaves an altitude, it may not return to that altitude.

A descent clearance may also include a segment where the descent is at the pilots' discretion—such as "cross the Joliet VOR at or above 12,000, descend and maintain 5,000." This clearance authorizes pilots to descend from their current altitude whenever they choose, as long as they cross the Joliet VOR at or above 12,000 feet MSL. After that, they are expected to descend at a normal rate until they reach the assigned altitude of 5,000 feet MSL.

Clearances to descend at pilots' discretion are not just an option for ATC. Pilots may also request this type of clearance so that they can operate more efficiently. For example, if a pilot was en route above an overcast layer, he or she might ask for a descent at his or her discretion to allow the aircraft to remain above the clouds for as long as possible. This might be particularly important if the atmosphere is conducive to icing and the aircraft's icing protection is limited. The pilot's request permits the aircraft to stay at its cruising altitude longer to conserve fuel or to avoid prolonged IFR flight in icing conditions. This type of descent can also help to minimize the time spent in turbulence by allowing pilots to level off at an altitude where the air is smoother.

Controlled Flight Into Terrain (CFIT)

Inappropriate descent planning and execution during arrivals has been a contributing factor to many fatal aircraft accidents. Since the beginning of commercial jet operations, more than 9,000 people have died worldwide because of controlled flight into terrain (CFIT). CFIT is described as an event in which a normally functioning aircraft is inadvertently flown into the ground, water, or an obstacle. Of all CFIT accidents, 7.2 percent occurred during the descent phase of flight.

The basic causes of CFIT accidents involve poor flight crew situational awareness, or SA. One definition of SA is an accurate perception by pilots of the factors and conditions currently affecting the safe operation of the aircraft and the crew. The causes of CFIT are the flight crews' lack of vertical position awareness or their lack of horizontal position awareness in relation to the ground, water, or an obstacle. More than two-thirds of all CFIT accidents are the result of an altitude error or lack of vertical SA. CFIT accidents most often occur during reduced visibility associated with instrument meteorological conditions (IMC), darkness, or a combination of both.

The inability of controllers and pilots to properly communicate has been a factor in many CFIT accidents. Heavy workloads can lead to hurried communication and the use of abbreviated or non-standard phraseology. The importance of good communication during the arrival phase of flight was made evident in a report by an air traffic controller and the flight crew of an MD-80.

The controller reported that he was scanning his radarscope for traffic and noticed that the MD-80 was descending through 6,400 feet. He immediately instructed a climb to at least 6,500 feet. The pilot returned to 6,500 feet, but responded to ATC that he had been cleared to 5,000 feet. When he had read back 5,000 feet to the controller, he received no correction from the controller. After almost simultaneous ground proximity warning system (GPWS) and controller warnings, the pilot climbed and avoided the terrain. The recording of the radio transmissions confirmed that the airplane was cleared to 7,000 feet and the pilot mistakenly read back 5,000 feet then attempted to descend to 5,000 feet. The pilot stated in the report: "I don't know how much clearance from the mountains we had, but it certainly makes clear the importance of good communications between the controller and pilot."

ATC is not always responsible for safe terrain clearance for the aircraft under its jurisdiction. Many times ATC issue en route clearances for pilots to proceed off airway direct to a point. Pilots who accept this type of clearance also are accepting the shared responsibility for maintaining safe terrain clearance. Know the height of the highest terrain and obstacles in the operating area and your position in relation to the surrounding high terrain.

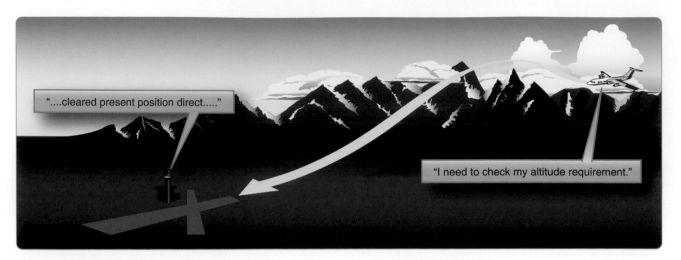

Figure 3-9. Altitude management when cleared direct.

The following are excerpts from CFIT accidents related to descending on arrival: "…delayed the initiation of the descent…"; "Aircraft prematurely descended too early…"; "…late getting down…"; "During a descent…incorrectly cleared down…"; "…aircraft prematurely let down…"; "…lost situational awareness…"; "Premature descent clearance…"; "Prematurely descended…"; "Premature descent clearance while on vector…"; "During initial descent…" [Figure 3-9]

Practicing good communication skills is not limited to just pilots and controllers. In its findings from a 1974 air carrier accident, the National Transportation Safety Board (NTSB) wrote, "…the extraneous conversation conducted by the flight crew during the descent was symptomatic of a lax atmosphere in the flight deck that continued throughout the approach." The NTSB listed the probable cause as "… the flight crew's lack of altitude awareness at critical points during the approach due to poor flight deck discipline in that the crew did not follow prescribed procedures."

In 1981, the FAA issued 14 CFR Part 121, section 121.542 and Part 135, section 135.100, Flight Crewmember Duties, commonly referred to as "sterile flight deck rules." The provisions in this rule can help pilots, operating under any regulations, to avoid altitude and course deviations during arrival. In part, it states: (a) No certificate holder should require, nor may any flight crewmember perform, any duties during a critical phase of flight except those duties required for the safe operation of the aircraft. Duties such as company required calls made for such purposes as ordering galley supplies and confirming passenger connections, announcements made to passengers promoting the air carrier or pointing out sights of interest, and filling out company payroll and related records are not required for the safe operation of the aircraft. (b) No flight crewmember

may engage in, nor may any pilot in command permit, any activity during a critical phase of flight that could distract any flight crewmember from the performance of his or her duties or which could interfere in any way with the proper conduct of those duties. Activities such as eating meals, engaging in nonessential conversations within the flight deck and nonessential communications between the cabin and flight deck crews, and reading publications not related to the proper conduct of the flight are not required for the safe operation of the aircraft. (c) Critical phases of flight include all ground operations involving taxi, takeoff and landing, and all other flight operations conducted below 10,000 feet, except cruise flight.

Standard Terminal Arrival Routes (STARs)

A STAR is an ATC-coded IFR route established for application to arriving IFR aircraft destined for certain airports. A STAR provides a critical form of communication between pilots and ATC. Once a flight crew has accepted a clearance for a STAR, they have communicated with the controller what route, and in some cases what altitude and airspeed, they fly during the arrival, depending on the type of clearance. The STAR provides a common method for leaving the en route structure and navigating to your destination. It is a preplanned instrument flight rule ATC arrival procedure published for pilot use in graphic and textual form that simplifies clearance delivery procedures.

The principal difference between standard instrument departure (SID) or departure procedures (DPs) and STARs is that the DPs start at the airport pavement and connect to the en route structure. STARs on the other hand, start at the en route structure but do not make it down to the pavement; they end at a fix or NAVAID designated by ATC, where radar vectors commonly take over. This is primarily

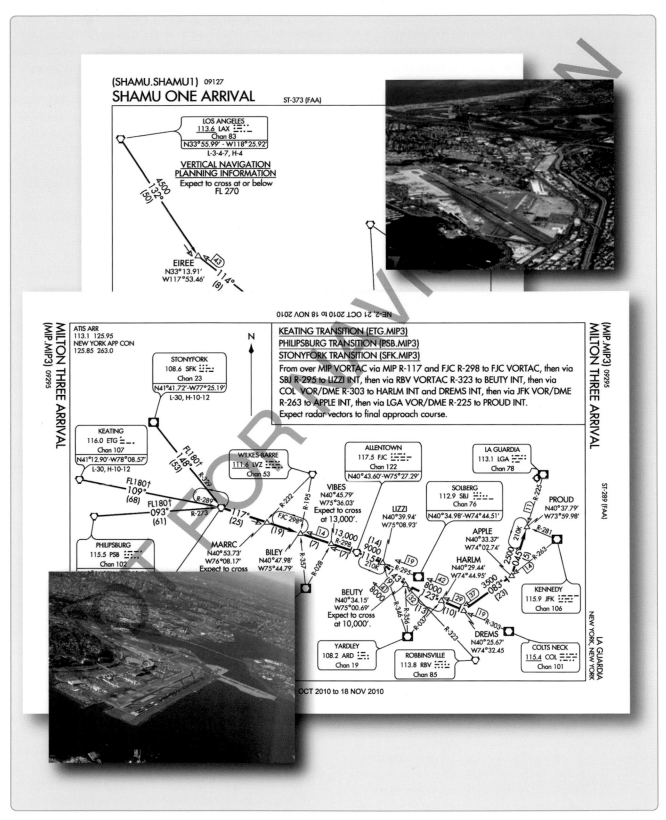

Figure 3-10. Arrival charts.

because STARs serve multiple airports. STARs greatly help to facilitate the transition between the en route and approach phases of flight. The objective when connecting a STAR to an instrument approach procedure is to ensure a seamless lateral and vertical transition. The STAR and approach procedure should connect to one another in such a way as to maintain the overall descent and deceleration profiles. This often results in a seamless transition between the en route, arrival, and approach phases of flight, and serves as a preferred route into high volume terminal areas. [Figure 3-10]

STARs provide a transition from the en route structure to an approach gate, outer fix, instrument approach fix, or arrival waypoint in the terminal area, and they usually terminate with an instrument or visual approach procedure. STARs are included at the front of each Terminal Procedures Publication (TPP) regional booklet.

For STARs based on conventional NAVAIDs, the procedure design and obstacle clearance criteria are essentially the same as that for en route criteria, covered in Chapter 2, En Route Operations. STAR procedures typically include a standardized descent gradient at and above 10,000 feet MSL of 318 feet per nautical mile (FPNM), or 3 degrees. Below 10,000 feet MSL, the maximum descent rate is 330 FPNM, or approximately 3.1 degrees. In addition to standardized descent gradients, STARs allow for deceleration segments at any waypoint that has a speed restriction. As a general guideline, deceleration considerations typically add 1 NM of distance for each 10 knots of speed reduction required.

RNAV STARs or STAR Transitions

STARs designated RNAV serve the same purpose as conventional STARs, but are only used by aircraft equipped with FMS or GPS. An RNAV STAR or STAR transition typically includes flyby waypoints, with fly over waypoints used only when operationally required. These waypoints may be assigned crossing altitudes and speeds to optimize the descent and deceleration profiles. RNAV STARs often are designed, coordinated, and approved by a joint effort between air carriers, commercial operators, and the ATC facilities that have jurisdiction for the affected airspace.

RNAV STAR procedure design, such as minimum leg length, maximum turn angles, obstacle assessment criteria, including widths of the primary and secondary areas, use the same design criteria as RNAV DPs. Likewise, RNAV STAR procedures are designated as either RNAV 1 or RNAV 2, based on the aircraft navigation equipment required, flight crew procedures, and the process and criteria used to develop the STAR. The RNAV 1 or RNAV 2 designation appears in the notes on the chart. RNAV 1

STARs have higher equipment requirements and, often, tighter required navigation performance (RNP) tolerances than RNAV 2. For RNAV 1 STARS, pilots are required to use a course deviation indicator (CDI)/flight director, and/or autopilot in LNAV mode while operating on RNAV courses. (These requirements are detailed in Chapter 1 of this book, under RNAV Departures.) RNAV 1 STARs are generally designated for high-traffic areas. Controllers may clear a pilot to use an RNAV STAR in various ways.

If the pilots clearance simply states, "cleared Hadly One arrival," the pilot is to use the arrival for lateral routing only.

- A clearance such as "cleared Hadly One arrival, descend and maintain flight level two four zero," clears the pilot to descend only to the assigned altitude, and then should maintain that altitude until cleared for further VNAV.

- If the pilot is cleared using the phrase "descend via," the controller expects the pilot to use the equipment for both lateral and VNAV, as published on the chart.

- The controller may also clear the pilot to use the arrival with specific exceptions—for example, "Descend via the Haris One arrival, except cross Bruno at one three thousand then maintain one zero thousand." In this case, the pilot should track the arrival both laterally and vertically, descending so as to comply with all altitude and airspeed restrictions until reaching Bruno, and then maintain 10,000 feet until cleared by ATC to continue to descend.

- Pilots might also be given direct routing to intercept a STAR and then use it for VNAV. For example, "Proceed direct Mahem, descend via the Mahem Two arrival."

Interpreting the STAR

STARs use much of the same symbology as departure and approach charts. In fact, a STAR may at first appear identical to a similar graphic DP, except the direction of flight is reversed and the procedure ends at an approach fix. The STAR officially begins at the common NAVAID, intersection, or fix where all the various transitions to the arrival come together. A STAR transition is a published segment used to connect one or more en route airways, jet routes, or RNAV routes to the basic STAR procedure. It is one of several routes that bring traffic from different directions into one STAR. This way, arrivals from several directions can be accommodated on the same chart, and traffic flow is routed appropriately within the congested airspace.

To illustrate how STARs can be used to simplify a complex clearance and reduce frequency congestion, consider

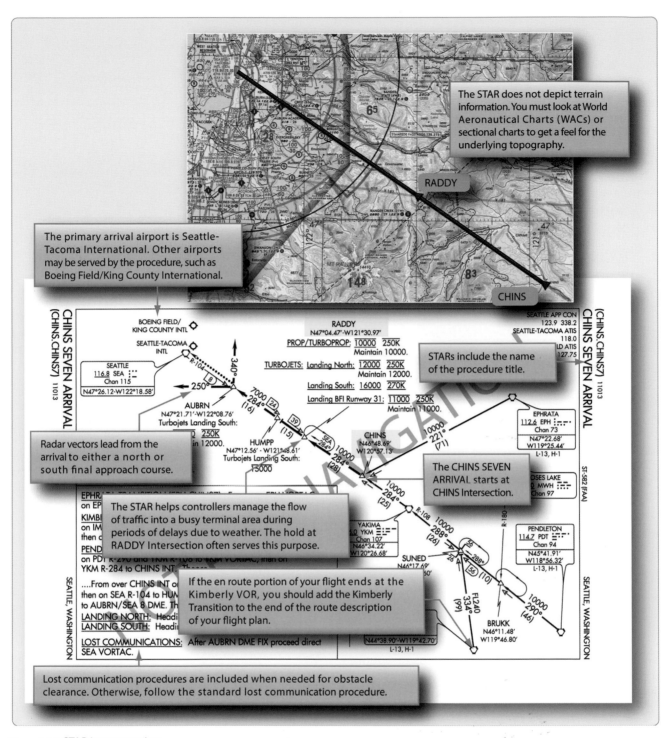

Figure 3-11. STAR interpretation.

the following arrival clearance issued to a pilot flying to Seattle, Washington, depicted in Figure 3-11: "Cessna 32G, cleared to the Seattle/Tacoma International Airport as filed. Maintain 12,000. At the Ephrata VOR, intercept the 221° radial to CHINS Intersection. Intercept the 284° radial of the Yakima VOR to RADDY Intersection. Cross RADDY at 10,000. Continue via the Yakima 284° radial to AUBRN Intersection. Expect radar vectors to the final approach course."

Now consider how this same clearance is issued when a STAR exists for this terminal area. "Cessna 32G, cleared to Seattle/Tacoma International Airport as filed, then CHINS EIGHT ARRIVAL, Ephrata Transition. Maintain 10,000 feet." A shorter transmission conveys the same information.

Safety is enhanced when both pilots and controllers know what to expect. Effective communication increases with the

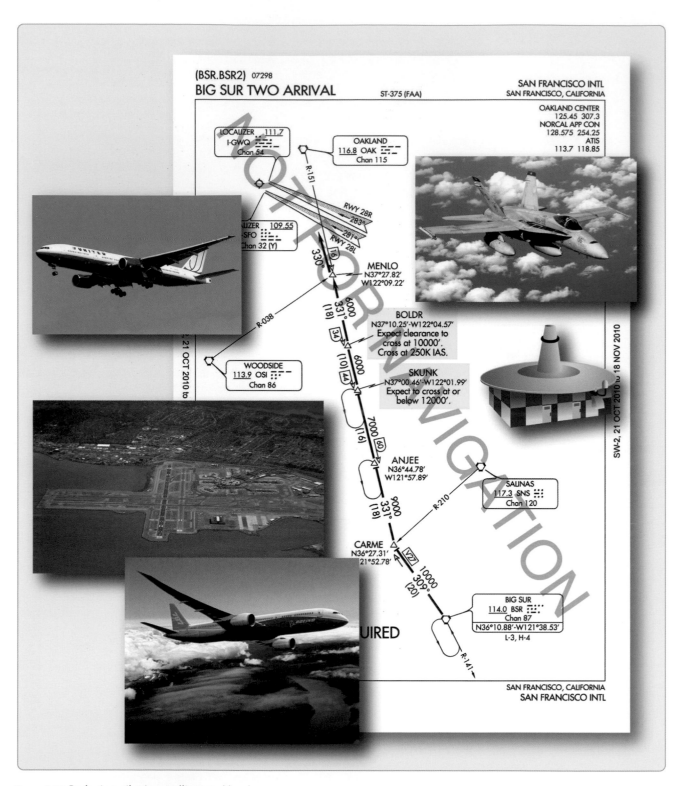

Figure 3-12. Reducing pilot/controlling workload.

reduction of repetitive clearances, decreasing congestion on control frequencies. To accomplish this, STARs are developed according to the following criteria:

- STARs must be simple, easily understood and, if possible, limited to one page.

- A STAR transition should be able to accommodate as many different types of aircraft as possible.

- VHF Omnidirectional Range/Tactical Aircraft Control (VORTACs) are used wherever possible, with some exceptions on RNAV STARs, so that military and civilian aircraft can use the same arrival.

- DME arcs within a STAR should be avoided since

not all aircraft operating under IFR are equipped to navigate them.

- Altitude crossing and airspeed restrictions are included when they are assigned by ATC a majority of the time. [Figure 3-12]

STARs usually are named according to the point at which the procedure begins. In the United States, typically there are en route transitions before the STAR itself. So the STAR name is usually the same as the last fix on the en route transitions where they come together to begin the basic STAR procedure. A STAR that commences at the CHINS Intersection becomes the CHINS SEVEN ARRIVAL. When a

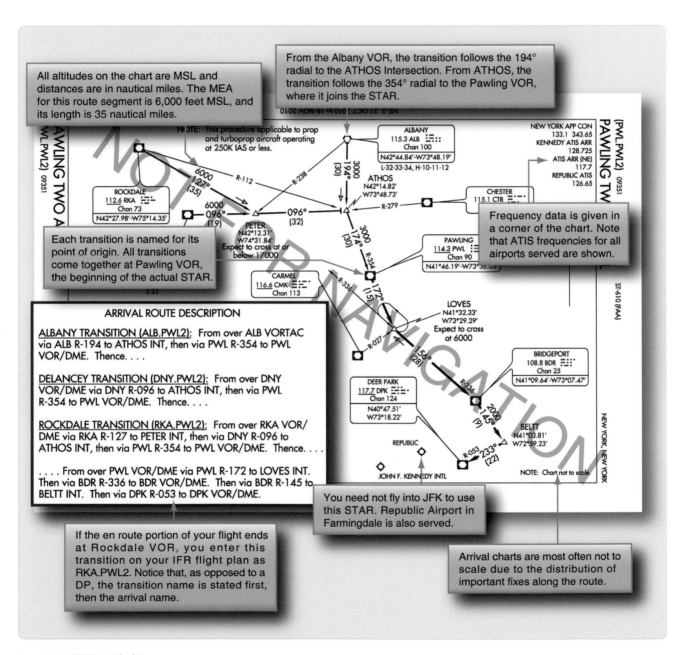

Figure 3-13. STAR symbology.

3-15

significant portion of the arrival is revised, such as an altitude, a route, or data concerning the NAVAID, the number of the arrival changes. For example, the CHINS SEVEN ARRIVAL is now the CHINS EIGHT ARRIVAL due to modifications in the procedure.

Studying the STARs for an airport may allow pilots to perceive the specific topography of the area. Note the initial fixes and where they correspond to fixes on the Aeronautical Information Systems (AIS) en route or area chart. Arrivals may incorporate step-down fixes when necessary to keep aircraft within airspace boundaries or for obstacle clearance. Routes between fixes contain courses, distances, and minimum altitudes, alerting aircrews to possible obstructions or terrain under their arrival path. Airspeed restrictions also appear where they aid in managing the traffic flow. In addition, some STARs require that pilots use DME and/or ATC radar. Aircrews can decode the symbology on the PAWLING TWO ARRIVAL by referring to the legend at the beginning of the TPP. [Figure 3-13]

STAR Procedures

Pilots may accept a STAR within a clearance or they may file for one in their flight plan. As the aircraft nears its destination airport, ATC may add a STAR procedure to its original clearance. Keep in mind that ATC can assign a STAR even if the aircrew has not requested one. Use of a STAR requires pilot possession of at least the approved chart. RNAV STARs must be retrievable by the procedure name from the aircraft database and conform to charted procedure. If an aircrew does not want to use a STAR, they must specify "No STAR" in the remarks section of their flight plan. Pilots may also refuse the STAR when it is given to them verbally by ATC, but the system works better if the aircrew advises ATC ahead of time.

Preparing for the Arrival

As mentioned before, STARs include navigation fixes that are used to provide transition and arrival routes from the en route structure to the final approach course. They also may lead to a fix where radar vectors are provided to intercept the final approach course. Pilots may have noticed that minimum crossing altitudes and airspeed restrictions appear on some STARs. These expected altitudes and airspeeds are not part of the clearance until ATC includes them verbally. A STAR is simply a published routing; it does not have the force of a clearance until issued specifically by ATC. For example, minimum en route altitude (MEAs) printed on STARs are not valid unless stated within an ATC clearance or in cases of lost communication. After receiving the arrival clearance, the aircrew should review the assigned STAR procedure and ensure the FMS has the appropriate procedure loaded (if so equipped). Obtain the airport and weather information as early

as practical. It is recommended that pilots have this information prior to flying the STAR. If you are landing at an airport with approach control services that has two or more published instrument approach procedures, you will receive advance notice of which instrument approaches to expect. This information is broadcast either by ATIS or by a controller. [Figure 3-14] It may not be provided when the visibility is 3 statute miles (SM) or better and the ceiling is at or above the highest initial approach altitude established for any instrument approach procedure for the airport.

For STAR procedures charted with radar vectors to the final approach, look for routes from the STAR terminating fixes to the IAF. If no route is depicted, you should have a predetermined plan of action to fly from the STAR terminating fix to the IAF in the event of a communication failure.

Reviewing the Approach

Once the aircrew has determined which approach to expect, review the approach chart thoroughly before entering the terminal area. Aircrews should check fuel level and make sure a prolonged hold or increased headwinds have not cut into the aircraft's fuel reserves because there is always a chance the pilot has to make a missed approach or go to an alternate. By completing landing checklists early, aircrews can concentrate on the approach.

In setting up for the expected approach procedure when using an RNAV, GPS, or FMS system, it is important to understand how multiple approaches to the same runway are coded in the database. When more than one RNAV procedure is issued for the same runway, there must be a way to differentiate between them within the equipment's database, as well as to select which procedure is to be used. (Multiple procedures may exist to accommodate GPS receivers and FMS, both with and without VNAV capability.) Each procedure name incorporates a letter of the alphabet, starting with Z and working backward through Y, X, W, and so on. (Naming conventions for approaches are covered in more depth in the next chapter). [Figure 3-15]

Altitude

Upon arrival in the terminal area, ATC either clears the aircraft to a specific altitude, or they give it a "descend via" clearance that instructs the pilot to follow the altitudes published on the STAR. [Figure 3-16] Pilots are not authorized to leave their last assigned altitude unless specifically cleared to do so. If ATC amends the altitude or route to one that is different from the published procedure, the rest of the charted descent procedure is cancelled. ATC assigns any further route, altitude, or airspeed clearances, as necessary. Notice the JANESVILLE FOUR ARRIVAL depicts only one published arrival route, with no named transition

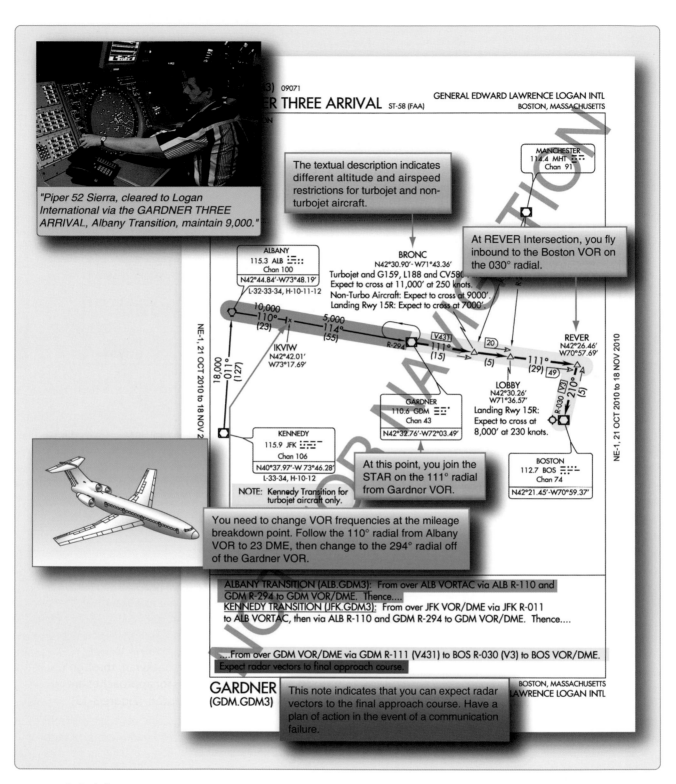

"Piper 52 Sierra, cleared to Logan International via the GARDNER THREE ARRIVAL, Albany Transition, maintain 9,000."

(3) 09071

...ER THREE ARRIVAL ST-58 (FAA)

GENERAL EDWARD LAWRENCE LOGAN INTL
BOSTON, MASSACHUSETTS

The textual description indicates different altitude and airspeed restrictions for turbojet and non-turbojet aircraft.

At REVER Intersection, you fly inbound to the Boston VOR on the 030° radial.

MANCHESTER
114.4 MHT
Chan 91

ALBANY
115.3 ALB
Chan 100
N42°44.84'-W73°48.19'
L-32-33-34, H-10-11-12

BRONC
N42°30.90'- W71°43.36'
Turbojet and G159, L188 and CV580...
Expect to cross at 11,000' at 250 knots.
Non-Turbo Aircraft: Expect to cross at 9000'.
Landing Rwy 15R: Expect to cross at 7000'.

REVER
N42°26.46'
W70°57.69'

10,000
110°
(23)
5,000
114°
(55)

IKVIW
N42°42.01'
W73°17.69'

R-294

111°
(15)

V431

20

(5)

111°
(29)

49

210°
(5)

LOBBY
N42°30.26'
W71°36.57'
Landing Rwy 15R:
Expect to cross at
8,000' at 230 knots.

R-030 V3

GARDNER
110.6 GDM
Chan 43
N42°32.76'-W72°03.49'

At this point, you join the STAR on the 111° radial from Gardner VOR.

BOSTON
112.7 BOS
Chan 74
N42°21.45'-W70°59.37'

KENNEDY
115.9 JFK
Chan 106
N40°37.97'-W 73°46.28'
L-33-34, H-10-12

NOTE: Kennedy Transition for turbojet aircraft only.

You need to change VOR frequencies at the mileage breakdown point. Follow the 110° radial from Albany VOR to 23 DME, then change to the 294° radial off of the Gardner VOR.

ALBANY TRANSITION (ALB.GDM3): From over ALB VORTAC via ALB R-110 and GDM R-294 to GDM VOR/DME. Thence....
KENNEDY TRANSITION (JFK.GDM3): From over JFK VOR/DME via JFK R-011 to ALB VORTAC, then via ALB R-110 and GDM R-294 to GDM VOR/DME. Thence....

....From over GDM VOR/DME via GDM R-111 (V431) to BOS R-030 (V3) to BOS VOR/DME. Expect radar vectors to final approach course.

GARDNER
(GDM.GDM3)

This note indicates that you can expect radar vectors to the final approach course. Have a plan of action in the event of a communication failure.

BOSTON, MASSACHUSETTS
...LAWRENCE LOGAN INTL

NE-1, 21 OCT 2010 to 18 NOV 2...   NE-1, 21 OCT 2010 to 18 NOV 2010

Figure 3-14. Arrival clearance.

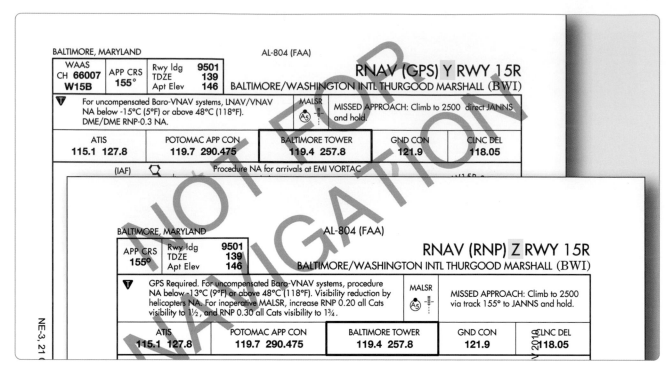

Figure 3-15. Two RNAV (GPS) approaches to Runway 15R at Baltimore. A controller issuing a clearance for one of these approaches would speak the identifying letter—for example, "...cleared for the RNAV (GPS) Yankee approach, Runway 15R..."

routes leading to the basic STAR procedure beginning at the Janesville VOR/DME. VNAV planning information is included for turbojet and turboprop airplanes at the bottom of the chart. Additionally, note that there are several ways to identify the BRIBE reporting point using alternate formation radials, some of which are from off-chart NAVAIDs. ATC may issue a descent clearance that includes a crossing altitude restriction. In the PENNS ONE ARRIVAL, the ATC clearance authorizes aircraft to descend at the pilots' discretion, as long as the pilot crosses the PENNS Intersection at 6,000 feet MSL. [Figure 3-17]

In the United States, Canada, and many other countries, the common altitude for changing to the standard altimeter setting of 29.92 inches of mercury ("Hg) (or 1013.2 hectopascals or millibars) when climbing to the high altitude structure is 18,000 feet. When descending from high altitude, the altimeter should be changed to the local altimeter setting when passing through FL 180, although in most countries throughout the world the change to or from the standard altimeter setting is not done at the same altitude for each instance.

For example, the flight level where aircrews change their altimeter setting to the local altimeter setting is specified by ATC each time they arrive at a specific airport. This information is shown on STAR charts outside the United States with the words: TRANS LEVEL: BY ATC. When departing from that same airport (also depicted typically on

the STAR chart), the altimeter should be set to the standard setting when passing through 5,000 feet, as an example. This means that altimeter readings when flying above 5,000 feet are actual flight levels, not feet. This is common for Europe, but very different for pilots experienced with flying in the United States and Canada.

Although standardization of these procedures for terminal locations is subject to local considerations, specific criteria apply in developing new or revised arrival procedures. Normally, high performance airplanes enter the terminal area at or above 10,000 feet above the airport elevation and begin their descent 30 to 40 NM from touchdown on the landing runway. Unless pilots indicate an operational need for a lower altitude, descent below 5,000 feet above the airport elevation is typically limited to an altitude where final descent and glideslope/glidepath intercept can be made without exceeding specific obstacle clearance and other related arrival, approach, and landing criteria.

Arrival delays typically are absorbed at a metering fix. This fix is established on a route prior to the terminal airspace, 10,000 feet or more above the airport elevation. The metering fix facilitates profile descents, rather than controllers using delaying vectors or a holding pattern at low altitudes. Descent restrictions normally are applied prior to reaching the final approach phase to preclude relatively high descent rates close in to the destination airport. At least 10 NM from initial descent from 10,000

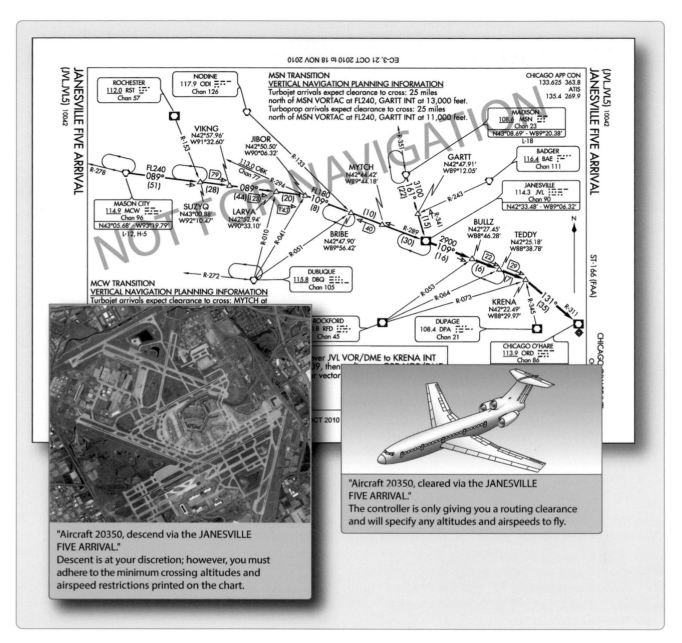

Figure 3-16. Assigned altitudes.

feet above the airport elevation, the controller issues an advisory that details when to expect to commence the descent. ATC typically uses the phraseology, "Expect descent in (number) miles." Standard ATC phraseology is, "Maintain (altitude) until specified point (e.g., abeam landing runway end), cleared for visual approach or expect visual or contact approach clearance in (number of miles, minutes, or specified point)."

Once the determination is made regarding the instrument approach and landing runway pilots use, ATC will not permit a change to another NAVAID that is not aligned with the landing runway. When altitude restrictions are required

for separation purposes, ATC avoids assigning an altitude below 5,000 feet above the airport elevation.

There are numerous exceptions to the high performance airplane arrival procedures previously outlined. For example, in a non-radar environment, the controller may clear the flight to use an approach based on a NAVAID other than the one aligned with the landing runway, such as a circling approach. In this case, the descent to a lower altitude usually is limited to the circling approach area with the circle-to-land maneuver confined to the traffic pattern.

IFR en route descent procedures should include a review

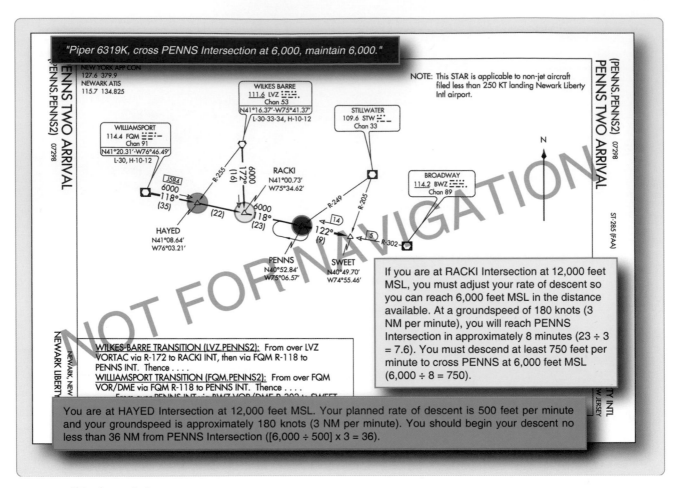

Figure 3-17. Altitude restrictions.

of minimum, maximum, mandatory, and recommended altitudes that normally precede the fix or NAVAID facility to which they apply. The initial descent gradient for a low altitude instrument approach procedure does not exceed 500 FPNM (approximately 5°), and for a high altitude approach, the maximum allowable initial gradient is 1,000 FPNM (approximately 10°).

Remember during arrivals, when cleared for an instrument approach, maintain the last assigned altitude until established on a published segment of the approach or on a segment of a published route. If no altitude is assigned with the approach clearance and the aircraft is already on a published segment, the pilot can descend to its minimum altitude for that segment of the approach.

Airspeed

During the arrival, expect to make adjustments in speed at the controller's request. When pilots fly a high-performance aircraft on an IFR flight plan, ATC may ask them to adjust their airspeed to achieve proper traffic sequencing and separation. This also reduces the amount of radar vectoring required in the terminal area. When operating

a reciprocating engine or turboprop airplane within 20 NM from the destination airport, 150 knots is usually the slowest airspeed that is assigned. If the aircraft cannot maintain the assigned airspeed, the pilot must advise ATC. Controllers may ask pilots to maintain the same speed as the aircraft ahead of or behind them on the approach. Pilots are expected to maintain the specified airspeed ±10 knots. At other times, ATC may ask pilots to increase or decrease airspeed by 10 knots, or multiples thereof. When the speed adjustment is no longer needed, ATC advises the pilot to "… resume normal speed."

Keep in mind that the maximum speeds specified in 14 CFR Part 91, section 91.117 still apply during speed adjustments. It is the pilot's responsibility to advise ATC if an assigned speed adjustment would cause an exceedence of these limits. For operations in Class C or D airspace at or below 2,500 feet above ground level (AGL), within 4 NM of the primary airport, ATC has the authority to approve a faster speed than those prescribed in 14 CFR Part 91, section 91.117.

Pilots operating at or above 10,000 feet MSL on an assigned speed adjustment that is greater than 250 KIAS are expected to reduce speed to 250 KIAS to comply with 14 CFR Part 91, section 91.117(a) when cleared below 10,000 feet MSL, within domestic airspace. This speed adjustment is made without notifying ATC. Pilots are expected to comply with the other provisions of 14 CFR Part 91, section 91.117 without notifying ATC. For example, it is normal for faster aircraft to level off at 10,000 feet MSL while slowing to the 250 KIAS limit that applies below that altitude, and to level off at 2,500 feet above airport elevation to slow to the 200 KIAS limit that applies within the surface limits of Class C or D airspace. Controllers anticipate this action and plan accordingly.

Speed restrictions of 250 knots do not apply to aircraft operating beyond 12 NM from the coastline within the United States Flight Information Region in offshore Class E airspace below 10,000 feet MSL. In airspace underlying a Class B airspace area designated for an airport, pilots are expected to comply with the 200 KIAS limit specified in 14 CFR Part 91, section 91.117(c). (See 14 CFR Part 91, sections 91.117(c) and 91.703.) Approach clearances cancel any previously assigned speed adjustment.

Holding Patterns

If aircraft reach a clearance limit before receiving a further clearance from ATC, a holding pattern is required at the last assigned altitude. Controllers assign holds for a variety of reasons, including deteriorating weather or high traffic volume. Holding might also be required following a missed approach. Since flying outside the area set aside for a holding pattern could lead to an encounter with terrain or other aircraft, aircrews need to understand the size of the protected airspace that a holding pattern provides.

Each holding pattern has a fix, a direction to hold from the fix, and an airway, bearing, course, radial, or route on which the aircraft is to hold. These elements, along with the direction of the turns, define the holding pattern.

Since the speed of the aircraft affects the size of a holding pattern, maximum holding airspeeds have been designated to limit the amount of airspace that must be protected. The three airspeed limits are shown in Figure 2-73 in Chapter 2, En Route Operations, of this book. Some holding patterns have additional airspeed restrictions to keep faster airplanes from flying out of the protected area. These are depicted on charts by using an icon and the limiting airspeed.

DME and IFR-certified GPS equipment offer some additional options for holding. Rather than being based on time, the leg lengths for DME/GPS holding patterns are based on distances in nautical miles. These patterns use the same entry and holding procedures as conventional holding patterns. The controller or the instrument approach procedure chart specifies the length of the outbound leg. The end of the outbound leg is determined by the DME or the along track distance (ATD) readout. The holding fix on conventional procedures, or controller-defined holding based on a conventional navigation aid with DME, is a specified course or radial and distances are from the DME station for both the inbound and outbound ends of the holding pattern. When flying published GPS overlay or standalone procedures with distance specified, the holding fix is a waypoint in the database and the end of the outbound leg is determined by the ATD. Instead of using the end of the outbound leg, some FMS are programmed

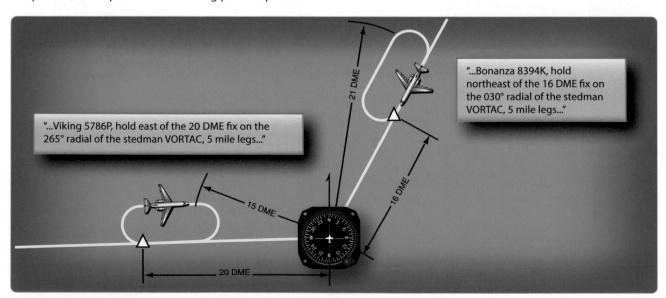

Figure 3-18. Instead of flying for a specific time after passing the holding fix, these holding patterns use distances to mark where the turns are made. The distances come from DME or IFR-certified GPS equipment.

to cue the inbound turn so that the inbound leg length matches the charted outbound leg length.

Normally, the difference is negligible, but in high winds, this can enlarge the size of the holding pattern. Aircrews need to understand their aircraft's FMS holding program to ensure that the holding entry procedures and leg lengths match the holding pattern. Some situations may require pilot intervention in order to stay within protected airspace. [Figure 3-18]

Approach Clearance

The approach clearance provides guidance to a position from where the pilot can execute the approach. It is also a clearance to fly that approach. If only one approach procedure exists, or if ATC authorizes the aircrew to execute the approach procedure of their choice, the clearance may be worded as simply as "… cleared for approach." If ATC wants to restrict the pilot to a specific approach,

the controller names the approach in the clearance. For example, "…cleared ILS Runway 35 Right approach."

When the landing is to be made on a runway that is not aligned with the approach being flown, the controller may issue a circling approach clearance, such as "…cleared for VOR Runway 17 approach, circle to land Runway 23." Approaches whose final approach segment is more than 30 degrees different from the landing runway alignment are always designated as circling approaches. Unless a specific landing runway is specified in the approach clearance, the pilot may land on any runway. Pilots landing at non-towered airports are reminded of the importance of making radio calls as set forth in the AIM.

When cleared for an approach prior to reaching a holding fix, ATC expects the pilot to continue to the holding fix, along the feeder route associated with the fix, and then to the IAF. If a feeder route to an IAF begins at a fix located

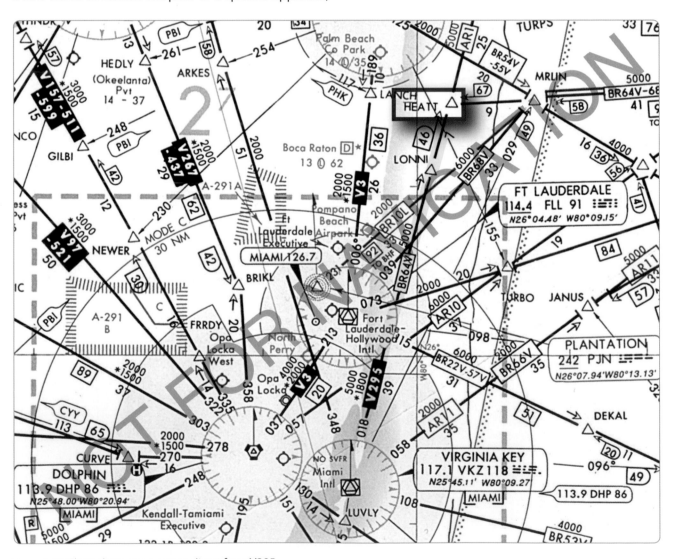

Figure 3-19. Cleared present position direct from V295.

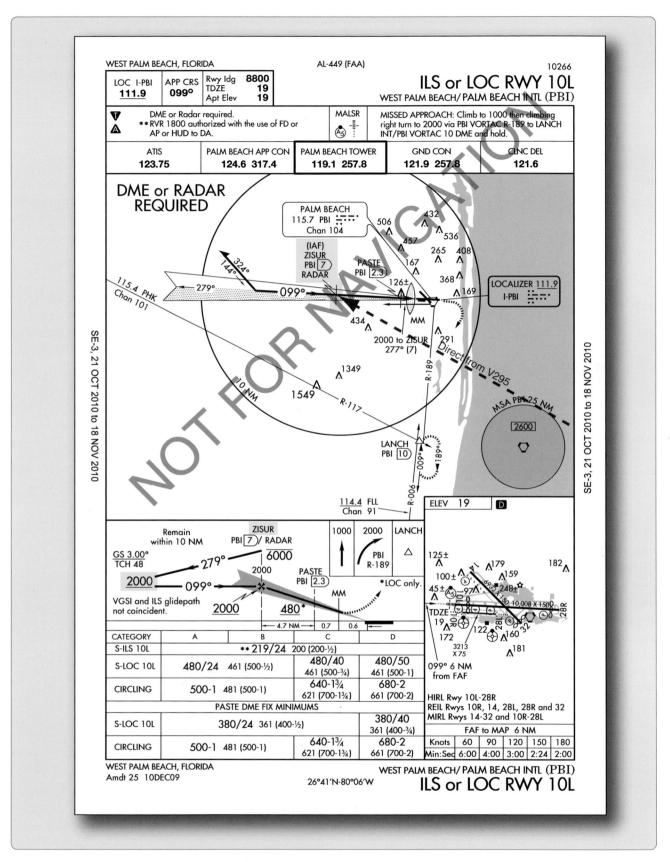

Figure 3-20. Cleared for the Palm Beach ILS approach.

along the route of flight prior to reaching the holding fix, and clearance for an approach is issued, the pilot should commence the approach via the published feeder route. The pilot is expected to commence the approach in a similar manner at the IAF, if the IAF is located along the route to the holding fix.

ATC also may clear an aircraft directly to the IAF by using language such as "direct" or "proceed direct." Controllers normally identify an approach by its published name, even if some component of the approach aid (such as the glideslope of an ILS) is inoperative or unreliable. The controller uses the name of the approach as published but advises the aircraft when issuing the approach clearance that the component is unusable.

Present Position Direct

In addition to using high and low altitude en route charts as resources for arrivals, area charts can be helpful as a planning aid for SA. Many pilots find the area chart helpful in locating a depicted fix after ATC clears them to proceed to a fix and hold, especially at unfamiliar airports.

Looking at Figures 3-19 and 3-20 assume the pilot is on V295 northbound en route to Palm Beach International Airport. The pilot is en route on the airway when the controller clears him present position direct to the ZISUR (IAF) and for the ILS approach. There is no transition authorized or charted between his present position and the approach facility. There is no minimum altitude published for the route the pilot is about to travel.

In Figure 3-20, the pilot is just north of HEATT Intersection at 5,000 feet when the approach controller states, "Citation 9724J, 2 miles from HEATT, cleared present position direct ZISUR, cleared for the Palm Beach ILS Runway 10L Approach, contact Palm Beach Tower on 119.1 established inbound." With no minimum altitude published from that point to the ZISUR intersection, the pilot should maintain the last assigned altitude until he reaches the IAF. In Figure 3-19, after passing ZISUR intersection outbound, commence the descent to 2,000 feet for the course reversal.

The ILS procedure relies heavily on the controller's recognition of the restriction upon the pilot to maintain the last assigned altitude until "established" on a published segment of the approach. Prior to issuing a clearance for the approach, the controller usually assigns the pilot an altitude to maintain until established on the final approach course , compatible with glideslope intercept.

Radar Vectors to Final Approach Course

Arriving aircraft usually are vectored to intercept the final approach course, except with vectors for a visual approach, at least 2 NM outside the approach gate unless one of the following exists:

1. When the reported ceiling is at least 500 feet above the minimum vectoring altitude or minimum IFR altitude and the visibility is at least 3 NM (report may be a pilot report if no weather is reported for the airport), aircraft may be vectored to intercept the final approach course closer than 2 NM outside the approach gate but no closer than the approach gate.

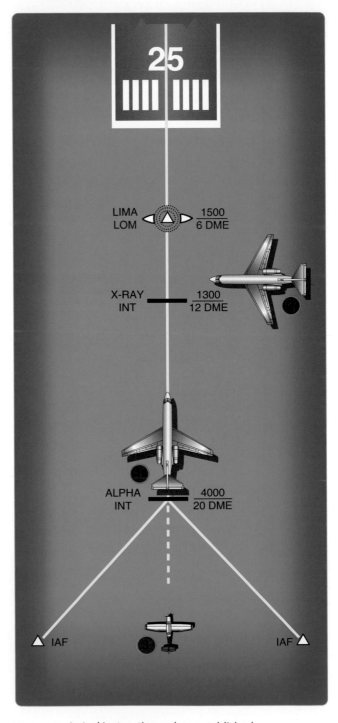

Figure 3-21. Arrival instructions when established.

2. If specifically requested by a pilot, ATC may vector aircraft to intercept the final approach course inside the approach gate but no closer than the final approach fix (FAF).

For a precision approach, aircraft are vectored at an altitude that is not above the glideslope/glidepath or below the minimum glideslope/glidepath intercept altitude specified on the approach procedure chart. For a non-precision approach, aircraft are vectored at an altitude that allows descent in accordance with the published procedure.

When a vector takes the aircraft across the final approach course, pilots are informed by ATC and the reason for the action is stated. In the event that ATC is not able to inform the aircraft, the pilot is not expected to turn inbound on the final approach course unless an approach clearance has been issued. An example of ATC phraseology in this case is, "… expect vectors across final for spacing."

The following ATC arrival instructions are issued to an IFR aircraft before it reaches the approach gate:

1. Position relative to a fix on the final approach course. If none is portrayed on the controller's radar display or if none is prescribed in the instrument approach procedure, ATC issues position information relative to the airport or relative to the NAVAID that provides final approach guidance.

2. Vector to intercept the final approach course if required.

3. Approach clearance except when conducting a radar approach. ATC issues the approach clearance only after the aircraft is established on a segment of a published route or instrument approach procedure, or in the following examples as depicted in Figure 3-21.

Aircraft 1 was vectored to the final approach course but clearance was withheld. It is now at 4,000 feet and established on a segment of the instrument approach procedure. "Seven miles from X-RAY. Cleared ILS runway three six approach."

Aircraft 2 is being vectored to a published segment of the final approach course, 4 NM from LIMA at 2,000 feet. The minimum vectoring altitude for this area is 2,000 feet. "Four miles from LIMA. Turn right heading three four zero. Maintain two thousand until established on the localizer. Cleared ILS runway three six approach."

Aircraft 3: There are many times when it is desirable to position an aircraft onto the final approach course prior to a published, charted segment of an instrument approach procedure (IAP).

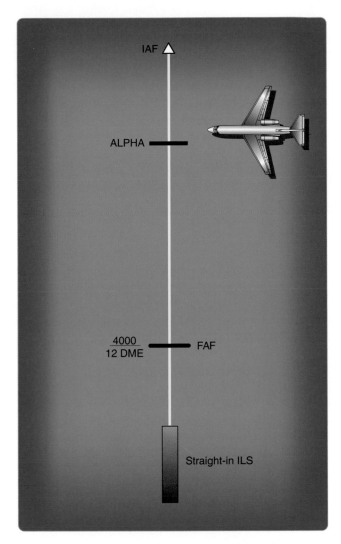

Figure 3-22. Arrival instructions when not established.

Sometimes IAPs have no initial segment and require vectors. "RADAR REQUIRED" is charted in the plan view. Sometimes a route intersects an extended final approach course making a long intercept desirable.

When ATC issues a vector or clearance to the final approach course beyond the published segment, controllers assign an altitude to maintain until the aircraft is established on a segment of a published route or IAP. This ensures that both the pilot and controller know precisely what altitude is to be flown and precisely where descent to appropriate minimum altitudes or step-down altitudes can begin.

Most aircraft are vectored onto a localizer or final approach course between an intermediate fix and the approach gate. These aircraft normally are told to maintain an altitude until established on a segment of the approach.

When an aircraft is assigned a route that is not a published segment of an approach, the controller must issue an altitude to maintain until the aircraft is established on a published segment of the approach. [Figure 3-21] Assume the aircraft is established on the final approach course beyond the approach segments, 8 NM from Alpha at 6,000 feet. The minimum vectoring altitude for this area is 4,000 feet. "Eight miles from Alpha. Cross Alpha at or above four thousand. Cleared ILS runway three six approach."

If an aircraft is not established on a segment of a published approach and is not conducting a radar approach, ATC assigns an altitude to maintain until the aircraft is established on a segment of a published route or instrument approach procedure. [Figure 3-22]

The aircraft is being vectored to a published segment of the ILS final approach course, 3 NM from Alpha at 4,000 feet. The minimum vectoring altitude for this area is 4,000 feet. "Three miles from Alpha. Turn left heading two one zero. Maintain four thousand until established on the localizer. Cleared ILS runway one eight approach."

The ATC assigned altitude ensures IFR obstruction clearance from the point at which the approach clearance is issued until established on a segment of a published route or instrument approach procedure.

ATC tries to make frequency changes prior to passing the FAF, although when radar is used to establish the FAF, ATC informs the pilot to contact the tower on the local control frequency after being advised that the aircraft is over the fix. For example, "Three miles from final approach fix. Turn left heading zero one zero. Maintain two thousand until established on the localizer. Cleared ILS runway three six approach. I will advise when over the fix."

"Over final approach fix. Contact tower one-one eight point one."

Special Airport Qualification

It is important to note an example of additional resources that are helpful for arrivals, especially into unfamiliar airports requiring special pilot or navigation qualifications. The operating rules governing domestic and flag air carriers require pilots in command to be qualified over the routes and into airports where scheduled operations are conducted, including areas, routes, and airports in which special pilot qualifications or special navigation qualifications are needed. For Part 119 certificate holders who conduct operations under 14 CFR Part 121, section 121.443, there are provisions in OpSpecs under which operators can comply with this regulation. Figure 3-27 provides some examples of special airports in the United States along with associated comments.

Special Airports	Comments
Kodiak, AK	Airport is surrounded by mountainous terrain. Any go-around beyond ILS or GCA MAP will not provide obstruction clearance.
Petersburg, AK	Mountainous terrain in immediate vicinity of airport, all quadrants.
Cape Newenham AFS, AK	Runway located on mountain slope with high gradient factor; nonstandard instrument approach.
Washington, DC (National)	Special arrival/departure procedures.
Shenandoah Valley, VA (Stanton-Waynesboro-Harrisonburg)	Mountainous terrain.
Aspen, CO	High terrain; special procedures.
Gunnison, CO	VOR only; uncontrolled; numerous obstructions in airport area; complete departure procedures.
Missoula, MT	Mountainous terrain; special procedures.
Jackson Hole, WY	Mountainous terrain; all quadrants; complex departure procedures.
Hailey, ID (Friedman Memorial)	Mountainous terrain; special arrival/departure procedures.
Hayden, Yampa Valley, CO	Mountainous terrain; no control tower; special engine-out procedures for certain large airplanes.
Lihue, Kauai, HI	High terrain; mountainous to 2,300 feet within 3 miles of the localizer.
Ontario, CA	Mountainous terrain and extremely limited visibility in haze conditions.

Figure 3-27. Special airports and comments.

Chapter 4
Approaches

Introduction

This chapter discusses general planning and conduct of instrument approaches by pilots operating under Title 14 of the Code of Federal Regulations (14 CFR) Parts 91,121, 125, and 135. The operations specifications (OpSpecs), standard operating procedures (SOPs), and any other Federal Aviation Administration (FAA) approved documents for each commercial operator are the final authorities for individual authorizations and limitations as they relate to instrument approaches. While coverage of the various authorizations and approach limitations for all operators is beyond the scope of this chapter, an attempt is made to give examples from generic manuals where it is appropriate.

Approach Planning

Depending on speed of the aircraft, availability of weather information, and the complexity of the approach procedure or special terrain avoidance procedures for the airport of intended landing, the in-flight planning phase of an instrument approach can begin as far as 100-200 nautical miles (NM) from the destination. Some of the approach planning should be accomplished during preflight. In general, there are five steps that most operators incorporate into their flight standards manuals for the in-flight planning phase of an instrument approach:

- Gathering weather information, field conditions, and Notices to Airmen (NOTAMs) for the airport of intended landing.

- Calculation of performance data, approach speeds, and thrust/power settings.

- Flight deck navigation/communication and automation setup.

- Instrument approach procedure (IAP) review and, for flight crews, IAP briefing.

- Operational review and, for flight crews, operational briefing.

Although often modified to suit each individual operator, these five steps form the basic framework for the in-flight planning phase of an instrument approach. The extent of detail that a given operator includes in their SOPs varies from one operator to another; some may designate which pilot performs each of the above actions, the sequence, and the manner in which each action is performed. Others may leave much of the detail up to individual flight crews and only designate which tasks should be performed prior to commencing an approach. Flight crews of all levels, from single-pilot to multi-crewmember Part 91 operators, can benefit from the experience of commercial operators in developing techniques to fly standard instrument approach procedures (SIAPs).

Determining the suitability of a specific IAP can be a very complex task, since there are many factors that can limit the usability of a particular approach. There are several questions that pilots need to answer during preflight planning and prior to commencing an approach. Is the approach procedure authorized for the company, if Part 91, subpart K, 121, 125, or 135? Is the weather appropriate for the approach? Is the aircraft currently at a weight that will allow it the necessary performance for the approach and landing or go around/ missed approach? Is the aircraft properly equipped for the approach? Is the flight crew qualified and current for the approach? Many of these types of issues must be considered during preflight planning and

within the framework of each specific air carrier's OpSpecs, or Part 91.

Weather Considerations

Weather conditions at the field of intended landing dictate whether flight crews need to plan for an instrument approach and, in many cases, determine which approaches can be used, or if an approach can even be attempted. The gathering of weather information should be one of the first steps taken during the approach-planning phase. Although there are many possible types of weather information, the primary concerns for approach decision-making are windspeed, wind direction, ceiling, visibility, altimeter setting, temperature, and field conditions. It is also a good idea to check NOTAMs at this time, in case there were any changes since preflight planning.

Windspeed and direction are factors because they often limit the type of approach that can be flown at a specific location. This typically is not a factor at airports with multiple precision approaches, but at airports with only a few or one approach procedure, the wrong combination of wind and visibility can make all instrument approaches at an airport unavailable. Pilots must be prepared to execute other available approaches, not just the one that they may have planned for. As an example, consider the available approaches at the Chippewa Valley Regional Airport (KEAU) in Eau Claire, Wisconsin. [Figure 4-1] In the event that the visibility is reported as less than one mile, the only useable approaches for Category C airplanes is the Instrument Landing System (ILS) and Lateral navigation (LNAV)/vertical navigation (VNAV) to Runway 22. This leaves very few options for flight crews if the wind does not favor Runway 22; and, in cases where the wind restricts a landing on that runway altogether, even a circling approach cannot be flown because of the visibility.

Weather Sources

Most of the weather information that flight crews receive is issued to them prior to the start of each flight segment, but the weather used for in-flight planning and execution of an instrument approach is normally obtained en route via government sources, company frequency, or Aircraft Communications Addressing and Reporting System (ACARS).

Air carriers and operators certificated under the provisions of Part 119 (Certification: Air Carriers and Commercial Operators) are required to use the aeronautical weather information systems defined in the OpSpecs issued to that certificate holder by the FAA. These systems may use basic FAA/National Weather Service (NWS) weather services, contractor or operator-proprietary weather services, and/

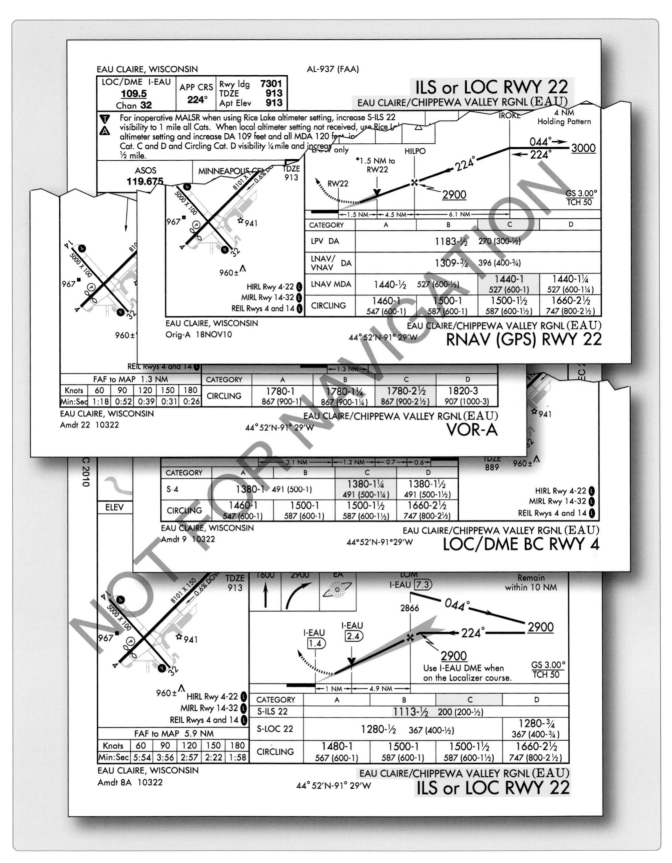

Figure 4-1. Chippewa Regional Airport (KEAU), Eau Claire, Wisconsin.

or Enhanced Weather Information System (EWINS) when approved in the OpSpecs. As an integral part of EWINS approval, the procedures for collecting, producing, and disseminating aeronautical weather information, as well as the crewmember and dispatcher training to support the use of system weather products, must be accepted or approved.

Operators not certified under the provisions of 14 CFR Part 119 are encouraged to use FAA/NWS products through Flight Service Station (FSS)/Automated Flight Service Station (AFSS). FSS and AFSS provide pilot weather briefings, en route weather, receive and process instrument flight rule (IFR) and visual flight rule (VFR) flight plans, relay air traffic control (ATC) clearances, and issue NOTAMs. They also provide assistance to lost aircraft and aircraft in emergency situations and conduct VFR search and rescue services.

Direct User Access Terminal System (DUATS), funded by the FAA, allows any pilot with a current medical certificate to access weather information and file a flight plan via computer. Two contract vendors currently provide information services within the DUATS system, and can be accessed via the Internet at www.duats.com or www. duat.com. The current vendors of DUATS service and the associated phone numbers are listed in Chapter 7 of the Aeronautical Information Manual (AIM).

Flight Information Service—Broadcast (FIS-B) provides certain aviation weather and other aeronautical information to aircraft equipped with an appropriate flight deck display. Reception of FIS-B services can be expected within a ground station coverage volume when line-of-sight geometry is maintained between the aircraft and ground station. National Airspace System (NAS) wide service availability was targeted for 2013 and is currently available within certain regions. FIS-B provides the following textual and graphical aviation weather and aeronautical products free-of-charge. A detailed description of these products can be found in the AIM.

- Aviation Digital Data Services (ADDS) provides the aviation community with text, digital and graphical forecasts, analyses, and observations of aviation related weather variables. ADDS is a joint effort of National Oceanic and Atmospheric Administration's (NOAA) Earth System Research Laboratory, National Center for Atmospheric Research (NCAR) Research Applications Laboratory (RAL), and the Aviation Weather Center (AWC).

- Hazardous In-flight Weather Advisory Service (HIWAS) is a national program for broadcasting hazardous weather information continuously over selected navigation aids (NAVAIDs). The broadcasts include advisories such as Airman's Meteorological Information (AIRMETs), Significant Meteorological Information (SIGMETs), convective SIGMETs, and urgent pilot weather reports (PIREPs/UUA). These broadcasts are only a summary of the information, and pilots should contact an FSS/AFSS or En Route Flight Advisory Service (EFAS) for detailed information.

- Telephone Information Briefing Service (TIBS) is a service prepared and disseminated by selected AFSS. It provides continuous telephone recordings of meteorological and aeronautical information. Specifically, TIBS provides area and route briefings, as well as airspace procedures and special announcements, if applicable. It is designed to be a preliminary briefing tool and is not intended to replace a standard briefing from a flight service specialist. The TIBS service is available 24 hours a day and is updated when conditions change, but it can only be accessed by a touch tone phone. The phone numbers for the TIBS service are listed in the Airport/ Facility Directory (A/FD). TIBS should also contain, but is not limited to: surface observations, terminal aerodrome forecast (TAFs), and winds/temperatures aloft forecasts.

The suite of available aviation weather product types is expanding with the development of new sensor systems, algorithms, and forecast models. The FAA and NWS, supported by the NCAR and the NOAA Forecast Systems Laboratory (FSL), develop and implement new aviation weather product types through a comprehensive process known as the Aviation Weather Technology Transfer process. This process ensures that user needs and technical and operational readiness requirements are met as experimental product types mature to operational application.

The development of enhanced communications capabilities, most notably the internet, has allowed pilots access to an increasing range of weather service providers and proprietary products. It is not the intent of the FAA to limit operator use of this weather information. However, pilots and operators should be aware that weather services provided by entities other than the FAA, NWS, or their contractors (such as the DUATS and flight information services data link (FISDL) providers) may not meet FAA/ NWS quality control standards.

Broadcast Weather

The most common method used by flight crews to obtain

specific in-flight weather information is to use a source that broadcasts weather for the specific airport. Information about ceilings, visibility, wind, temperature, barometric pressure, and field conditions can be obtained from most types of broadcast weather services. Broadcast weather can be transmitted to the aircraft in radio voice format or digital format, if it is available, via an ACARS system.

Automated Terminal Information Service (ATIS)

Automatic terminal information service (ATIS) is the continuous broadcast of recorded non-control information in selected high activity terminal areas. Its purpose is to improve controller effectiveness and to relieve frequency congestion by automating the repetitive transmission of essential but routine information. The information is continuously broadcast over a discrete very high frequency (VHF) radio frequency or the voice portion of a local NAVAID. ATIS transmissions on a discrete VHF radio frequency are engineered to be receivable to a maximum of 60 NM from the ATIS site and a maximum altitude of 25,000 feet above ground level (AGL). At most locations, ATIS signals may be received on the surface of the airport, but local conditions may limit the maximum ATIS reception distance and/or altitude. Pilots are urged to cooperate in the ATIS program as it relieves frequency congestion on approach control, ground control, and local control frequencies. The A/FD indicates airports for which ATIS is provided.

ATIS information includes the time of the latest weather sequence, ceiling, visibility, obstructions to visibility, temperature, dew point (if available), wind direction (magnetic), velocity, altimeter, other pertinent remarks, instrument approach and runway in use. The ceiling/sky condition, visibility, and obstructions to vision may be omitted from the ATIS broadcast if the ceiling is above 5,000 feet and the visibility is more than five miles. The departure runway will only be given if different from the landing runway except at locations having a separate ATIS for departure. The broadcast may include the appropriate frequency and instructions for VFR arrivals to make initial contact with approach control. Pilots of aircraft arriving or departing the terminal area can receive the continuous ATIS broadcast at times when flight deck duties are least pressing and listen to as many repeats as desired. ATIS broadcast will be updated upon the receipt of any official hourly and special weather. A new recording will also be made when there is a change in other pertinent data, such as runway change and instrument approach in use.

Automated Weather Observing Programs

Automated weather reporting systems are increasingly being installed at airports. These systems consist of various sensors, a processor, a computer-generated voice subsystem, and a transmitter to broadcast local, minute-by-minute weather data directly to the pilot.

Automated Weather Observing System

The automated weather observing system (AWOS) observations include the prefix "AUTO" to indicate that the data are derived from an automated system. Some AWOS locations are augmented by certified observers who provide weather and obstruction to vision information in the remarks of the report when the reported visibility is less than 7 miles. These sites, along with the hours of augmentation, are published in the A/FD. Augmentation is identified in the observation as "OBSERVER WEATHER." The AWOS wind speed, direction and gusts, temperature, dew point, and altimeter setting are exactly the same as for manual observations. The AWOS also reports density altitude when it exceeds the field elevation by more than 1,000 feet. The reported visibility is derived from a sensor near the touchdown of the primary instrument runway. The visibility sensor output is converted to a visibility value using a 10-minute harmonic average. The reported sky condition/ ceiling is derived from the ceilometer located next to the visibility sensor. The AWOS algorithm integrates the last 30 minutes of ceilometer data to derive cloud layers and heights. This output may also differ from the observer sky condition in that the AWOS is totally dependent upon the cloud advection over the sensor site.

Automated Surface Observing System (ASOS)/ Automated Weather Sensor System (AWSS)

The automated surface observing system (ASOS)/ automated weather sensor system (AWSS) is the primary surface weather observing system of the United States. The program to install and operate these systems throughout the United States is a joint effort of the NWS, the FAA, and the Department of Defense (DOD). AWSS is a follow-on program that provides identical data as ASOS. ASOS/AWSS is designed to support aviation operations and weather forecast activities. The ASOS/ AWSS provides continuous minute-by-minute observations and performs the basic observing functions necessary to generate a aviation routine weather report (METAR) and other aviation weather information. The information may be transmitted over a discrete VHF radio frequency or the voice portion of a local NAVAID. ASOS/AWSS transmissions on a discrete VHF radio frequency are engineered to be receivable to a maximum of 25 NM from the ASOS/AWSS site and a maximum altitude of 10,000 feet AGL.

At many locations, ASOS/AWSS signals may be received on the surface of the airport, but local conditions may limit the maximum reception distance and/or altitude.

While the automated system and the human may differ in their methods of data collection and interpretation, both produce an observation quite similar in form and content. For the objective elements, such as pressure, ambient temperature, dew point temperature, wind, and precipitation accumulation, both the automated system and the observer use a fixed location and time-averaging technique. The quantitative differences between the observer and the automated observation of these elements are negligible. For the subjective elements, however, observers use a fixed time (spatial averaging technique) to describe the visual elements (sky condition, visibility, and present weather, etc.), while the automated systems use a fixed location and time averaging technique. Although this is a fundamental change, the manual and automated techniques yield remarkably similar results within the limits of their respective capabilities.

The use of the aforementioned visibility reports and weather services are not limited for Part 91 operators. Part 121 and 135 operators are bound by their individual OpSpecs documents and are required to use weather reports that come from the NWS or other approved sources. While all OpSpecs are individually tailored, most operators are required to use ATIS information, runway visual range (RVR) reports, and selected reports from automated weather stations. All reports coming from an AWOS-3 station are usable for Part 121 and 135 operators. Each type of automated station has different levels of approval as outlined in individual OpSpecs. Ceiling and visibility reports given by the tower with the departure information are always considered official weather, and RVR reports are typically the controlling visibility reference. Refer to Chapter 1, Departures, of this manual, as well as the AIM section 7-1-12 for further description of automated weather systems.

Center Weather Advisories (CWA)

Center weather advisories (CWAs) are unscheduled inflight, flow control, air traffic, and aircrew advisories. By nature of its short lead time, the CWA is not a flight planning product. It is generally a nowcast for conditions beginning in the next 2 hours. CWAs will be issued:

1. As a supplement to an existing SIGMET, convective SIGMET, or AIRMET.

2. When an in-flight advisory has not been issued but observed or expected weather conditions meet SIGMET/AIRMET criteria based on current pilot reports and reinforced by other sources of information about existing meteorological conditions.

3. When observed or developing weather conditions do not meet SIGMET, convective SIGMET, or

AIRMET criteria (e.g., in terms of intensity or area coverage), but current pilot reports or other weather information sources indicate that existing or anticipated meteorological phenomena will adversely affect the safe and efficient flow of air traffic within the ARTCC area of responsibility.

Weather Regulatory Requirements

There are many practical reasons for reviewing weather information prior to initiating an instrument approach. Pilots must familiarize themselves with the condition of individual airports and runways so that they may make informed decisions regarding fuel management, diversions, and alternate planning. Because this information is critical, 14 CFR requires pilots to comply with specific weather minimums for planning and execution of instrument flights and approaches..

Weather Requirements and Part 91 Operators

According to 14 CFR Part 91, section 91.103, the pilot in command (PIC) must become familiar with all available information concerning a flight prior to departure. Included in this directive is the fundamental basis for pilots to review NOTAMs and pertinent weather reports and forecasts for the intended route of flight. This review should include current weather reports and terminal forecasts for all intended points of landing and alternate airports. In addition, a thorough review of an airport's current weather conditions should always be conducted prior to initiating an instrument approach. Pilots should also consider weather information as a planning tool for fuel management.

For flight planning purposes, weather information must be reviewed in order to determine the necessity and suitability of alternate airports. For Part 91 operations, the 600-2 and 800-2 rule applies to airports with precision and non-precision approaches, respectively. Approaches with vertical guidance (APV) are non-precision approaches because they do not meet the International Civil Aviation Organization (ICAO) Annex 10 standards for a precision approach. (See Final Approach Segment section later in this chapter for more information regarding APV approaches.) Exceptions to the 600-2 and 800-2 alternate minimums are listed in the front of the National Aeronautical Information Systems (AIS) in the Terminal Procedures Publication (TPP) and are indicated by a symbol **A** on the approach charts for the airport. This does not preclude flight crews from initiating instrument approaches at alternate airports when the weather conditions are below these minimums. The 600-2 and 800-2 rules, or any exceptions, only apply to flight planning purposes, while published landing minimums apply to the actual approach at the alternate.

Weather Requirements and Part 135 Operators

Unlike Part 91 operators, Part 135 operators may not depart for a destination unless the forecast weather there will allow an instrument approach and landing. According to 14 CFR Part 135, section 135.219, flight crews and dispatchers may only designate an airport as a destination if the latest weather reports or forecasts, or any combination of them, indicate that the weather conditions will be at or above IFR landing minimums at the estimated time of arrival (ETA). This ensures that Part 135 flight crews consider weather forecasts when determining the suitability of destinations. Departures for airports can be made when the forecast weather shows the airport will be at or above IFR minimums at the ETA, even if current conditions indicate the airport to be below minimums. Conversely, 14 CFR Part 135, section 135.219 prevents departures when the first airport of intended landing is currently above IFR landing minimums, but the forecast weather is below those minimums at the ETA.

Another very important difference between Part 91 and Part 135 operations is the Part 135 requirement for airports of intended landing to meet specific weather criteria once the flight has been initiated. For Part 135, not only is the weather required to be forecast at or above instrument flight rules (IFR) landing minimums for planning a departure, but it also must be above minimums for initiation of an instrument approach and, once the approach is initiated, to begin the final approach segment of an approach. 14 CFR Part 135, section 135.225 states that pilots may not begin an instrument approach unless the latest weather report indicates that the weather conditions are at or above the authorized IFR landing minimums for that procedure. 14 CFR Part 135, section 135.225 provides relief from this rule if the aircraft has already passed the final approach fix (FAF) when the weather report is received. It should be noted that the controlling factor for determining whether or not the aircraft can proceed is reported visibility. RVR, if available, is the controlling visibility report for determining that the requirements of this section are met. The runway visibility value (RVV), reported in statute miles (SM), takes precedent over prevailing visibility. There is no required timeframe for receiving current weather prior to initiating the approach.

Weather Requirements and Part 121 Operators

Like Part 135 operators, flight crews and dispatchers operating under Part 121 must ensure that the appropriate weather reports or forecasts, or any combination thereof, indicate that the weather will be at or above the authorized minimums at the ETA at the airport to which the flight is dispatched (14 CFR Part 121, section 121.613). This regulation attempts to ensure that flight crews will

always be able to execute an instrument approach at the destination airport. Of course, weather forecasts are occasionally inaccurate; therefore, a thorough review of current weather is required prior to conducting an approach. Like Part 135 operators, Part 121 operators are restricted from proceeding past the FAF of an instrument approach unless the appropriate IFR landing minimums exist for the procedure. In addition, descent below the minimum descent altitude (MDA), decision altitude (DA), or decision height (DH) is governed, with one exception, by the same rules that apply to Part 91 operators. The exception is that during Part 121 and 135 operations, the airplane is also required to land within the touchdown zone (TDZ). Refer to the section titled Minimum Descent Altitude, Decision Altitude, and Decision Height later in this chapter for more information regarding MDA, DA, and DH.

Aircraft Performance Considerations

All operators are required to comply with specific airplane performance limitations that govern approach and landing. Many of these requirements must be considered prior to the origination of flight. The primary goal of these performance considerations is to ensure that the aircraft can remain clear of obstructions throughout the approach, landing, and go-around phase of flight, as well as land within the distance required by the FAA. Although the majority of in-depth performance planning for an instrument flight is normally done prior to the aircraft's departure, a general review of performance considerations is usually conducted prior to commencing an instrument approach.

Airplane Performance Operating Limitations

Generally speaking, air carriers must have in place an approved method of complying with Subpart I of 14 CFR Parts 121 and 135 (Airplane Performance Operating Limitations), thereby proving the airplane's performance capability for every flight that it intends to make. Flight crews must have an approved method of complying with the approach and landing performance criteria in the applicable regulations prior to departing for their intended destination. The primary source of information for performance calculations for all operators, including Part 91, is the approved Aircraft Flight Manual (AFM) or Pilot's Operating Handbook (POH) for the make and model of aircraft that is being operated. It is required to contain the manufacturer determined performance capabilities of the aircraft at each weight, altitude, and ambient temperature that are within the airplane's listed limitations. Typically, the AFM for a large turbine powered airplane should contain information that allows flight crews to determine that the airplane will be capable of performing the following actions, considering the airplane's landing weight and other pertinent environmental factor:

- Land within the distance required by the regulations.
- Climb from the missed approach point (MAP) and maintain a specified climb gradient with one engine inoperative.
- Perform a go-around from the final stage of landing and maintain a specified climb gradient with all engines operating and the airplane in the landing configuration.

Many airplanes have more than one allowable flap configuration for normal landing. Often, a reduced flap setting for landing allows the airplane to operate at a higher landing weight into a field that has restrictive obstacles in the missed approach or rejected landing climb path. On these occasions, the full-flap landing speed may not allow the airplane enough energy to successfully complete a go-around and avoid any high terrain and/or obstacles that might exist on the climb out. Therefore, all- engine and engine-out missed approaches, as well as rejected landings, must be taken into consideration in compliance with the regulations.

Aircraft Approach Categories

Aircraft approach category means a grouping of aircraft based on a reference landing speed (V_{REF}), if specified, or if V_{REF} is not specified, 1.3 V_{SO} at the maximum certified landing weight. V_{REF}, V_{SO}, and the maximum certified landing weight are those values as established for the aircraft by the certification authority of the country of registry. A pilot must use the minima corresponding to the category determined during certification or higher. Helicopters may use Category A minima. If it is necessary to operate at a speed in excess of the upper limit of the speed range for an aircraft's category, the minimums for the higher category must be used. For example, an airplane that fits into Category B, but is circling to land at a speed of 145 knots, must use the approach Category D minimums. As an additional example, a Category A airplane (or helicopter) that is operating at 130 knots on a straight-in approach must use the approach Category C minimums. See the following category limits noting that the airspeeds depicted are indicated airspeeds (IAS):

- Category A: Speed less than 91 knots.
- Category B: Speed 91 knots or more but less than 121 knots.
- Category C: Speed 121 knots or more but less than 141 knots.
- Category D: Speed 141 knots or more but less than 166 knots.
- Category E: Speed 166 knots or more.

NOTE: Helicopter pilots may use the Category A line of minimums provided the helicopter is operated at Category A airspeeds.

An airplane is certified in only one approach category, and although a faster approach may require higher category minimums to be used, an airplane cannot be flown to the minimums of a slower approach category. The certified approach category is permanent and independent of the changing conditions of day-to-day operations. From a terminal instrument procedures (TERPS) viewpoint, the importance of a pilot not operating an airplane at a category line of minimums lower than the airplane is certified for is primarily the margin of protection provided for containment of the airplane within the procedure design for a slower airplane. This includes height loss at the decision altitude, missed approach climb surface, and turn containment in the missed approach at the higher category speeds.

Pilots are responsible for determining if a higher approach category applies. If a faster approach speed is used that places the aircraft in a higher approach category, the minimums for the appropriate higher category must be used. Emergency returns at weights in excess of maximum certificated landing weight, approaches made with inoperative flaps, and approaches made in icing conditions for some airplanes are examples of situations that can necessitate the use of higher approach category minima.

Circling approaches are one of the most challenging flight maneuvers conducted in the NAS, especially for pilots of CAT C and CAT D turbine-powered, transport category airplanes. These maneuvers are conducted at low altitude, day and night, and often with precipitation present affecting visibility, depth perception, and the ability to adequately assess the descent profile to the landing runway. Most often, circling approaches are conducted to runways without the benefit of electronic navigation aids to support the descent from the Circling Minimums Decision Altitude (CMDA) to the runway.

Circling approaches conducted at faster-than-normal, straight-in approach speeds also require a pilot to consider the larger circling approach area, since published circling minimums provide obstacle clearance only within the appropriate area of protection and is based on the approach category speed. [Figure 4-2] The circling approach area is the obstacle clearance area for airplanes maneuvering to land on a runway that does not meet the criteria for a straight- in approach. The size of the circling area varies with the approach category of the airplane, as shown in Figure 4-2.

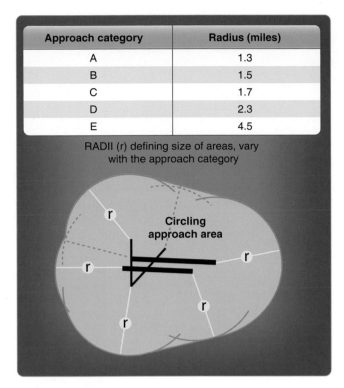

Approach category	Radius (miles)
A	1.3
B	1.5
C	1.7
D	2.3
E	4.5

RADII (r) defining size of areas, vary with the approach category

Circling approach area

Figure 4-2. Construction of circling approach area.

A minimum of 300 feet of obstacle clearance is provided in the circling segment. Pilots should remain at or above the circling altitude until the airplane is continuously in a position from which a descent to a landing on the intended runway can be made at a normal rate of descent and using normal maneuvers. Since an approach category can make a difference in the approach and weather minimums and, in some cases, prohibit flight crews from initiating an approach, the approach speed should be calculated and the effects on the approach determined and briefed in the preflight planning phase, as well as reviewed prior to commencing an approach.

Prior to TERPS Change 21, pilots were often faced with the challenge of descending using a stabilized approach concept if the CMDA height above airport (HAA) exceeded 1,200 feet. Once the HAA approached 1,200 feet, pilots were often forced to increase their rates of descent in order to arrive at the appropriate "in-slot" position. "In-slot" being defined as at a minimum, a CAT C or CAT D turbine-powered airplane should be wings level on a 3 degree - 318'/NM descent path not less than 1 NM from the touchdown point (1,000 feet beyond runway threshold). This was due to the small size of the circling protected airspace that the aircrews must remain within to ensure obstacle clearance.

The TERPS Change 21 to the circling protected airspace afforded much greater obstacle protection. However, it also afforded the pilot the opportunity to use the extra

protected airspace to mitigate the need to conduct a high descent rate, unstabilized approach that was often necessary as a result of the previous criteria for the Circling Approach Radius (CAR). For example, under TERPS Change 21, a sea level airport with a 1,500 ft HAA will have CAT C CAR of 2.86 NM, a 1.16 NM (68.5%) increase over pre-TERPS Change 21 CAR for CAT C. This extra protected airspace can be used by the pilot to maneuver the airplane instead of being forced to use high descent rates which are often necessary for high HAA circling approaches.

Most commercial operators dictate standard procedures for conducting instrument approaches in their FAA-approved manuals. These standards designate company callouts, flight profiles, configurations, and other specific duties for each flight deck crewmember during the conduct of an instrument approach.

Instrument Approach Charts

Beginning in February 2000, the FAA began issuing the current format for instrument approach plates (IAPs). This chart was developed by the Department of Transportation (DOT), Volpe National Transportation Systems Center and is commonly referred to as the Pilot Briefing Information format. The FAA chart format is presented in a logical order, facilitating pilot briefing of the procedures. [Figure 4-3]

Approach Chart Naming Conventions

Individual FAA charts are identified on both the top and bottom of the page by their procedure name (based on the NAVAIDs required for the final approach), runway served, and airport location. The identifier for the airport is also listed immediately after the airport name. [Figure 4-4]

There are several types of approach procedures that may cause some confusion for flight crews unfamiliar with the naming conventions. Although specific information about each type of approach is covered later in this chapter, listed below are a few procedure names that can cause confusion.

Straight-In Procedures

When two or more straight-in approaches with the same type of guidance exist for a runway, a letter suffix is added to the title of the approach so that it can be more easily identified. These approach charts start with the letter Z and continue in reverse alphabetical order. For example, consider the (RNAV) (GPS) Z RWY 13C and RNAV (RNP) Y RWY 13C approaches at Chicago Midway International Airport. [Figure 4-5] Although these two approaches can be flown with a global positioning system (GPS) to the same runway, they are significantly different (e.g., one is a Required Navigation Performance (RNP) Authorization

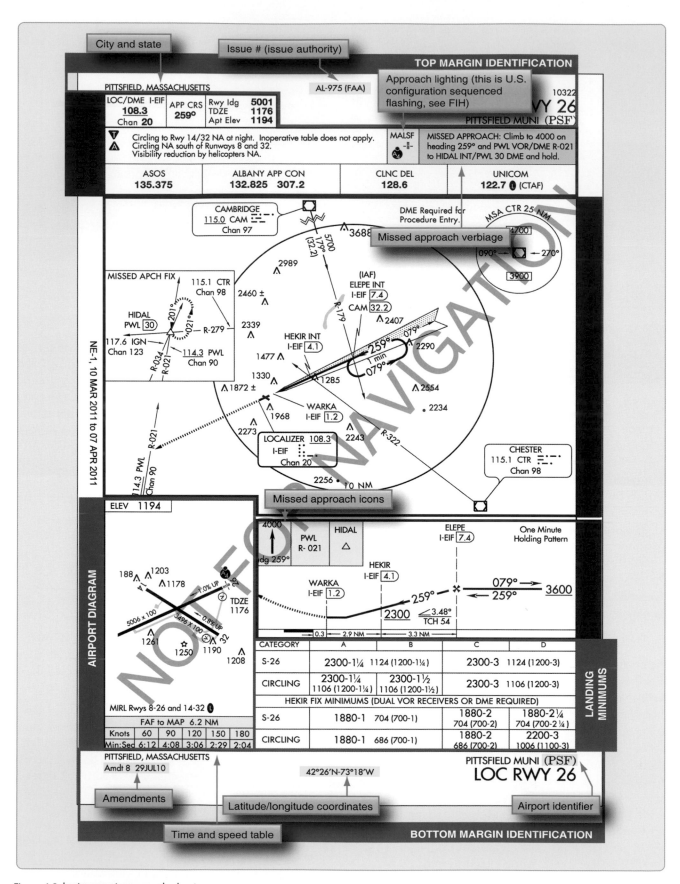

Figure 4-3. Instrument approach chart.

Figure 4-4. Procedure identification.

Required (AR) formally known as SPECIAL AIRCRAFT & AIRCREW AUTHORIZATION REQUIRED (SAAAR);" one has circling minimums and the other does not; the minimums are different; and the missed approaches are not the same). The approach procedure labeled Z has lower landing minimums than Y (some older charts may not reflect this).

In this example, the LNAV MDA for the RNAV (GPS) Z RWY 13C has the lowest minimums of either approach due to the differences in the final approach required obstacle clearance (ROC) evaluation. This convention also eliminates any confusion with approach procedures labeled A and B, where only circling minimums are published. The designation of two area navigation (RNAV) procedures to the same runway can occur when it is desirable to accommodate panel mounted GPS receivers and flight management systems (FMSs), both with and without vertical navigation (VNAV). It is also important to note that only one of each type of approach for a runway, including ILS, VHF omnidirectional range (VOR), and non-directional beacon (NDB) can be coded into a database.

Circling-Only Procedures

Approaches that do not have straight-in landing minimums are identified by the type of approach followed by a letter. Examples in Figure 4-6 show four procedure titles at the same airport that have only circling minimums.

As can be seen from the example, the first approach of this type created at the airport is labeled with the letter A, and the lettering continues in alphabetical order. Typically, circling only approaches are designed for one of the following reasons:

- The final approach course alignment with the runway centerline exceeds 30°.

- The descent gradient is greater than 400 feet per nautical mile (FPNM) from the FAF to the threshold crossing height (TCH). When this maximum gradient is exceeded, the circling only approach procedure may be designed to meet the gradient criteria limits. This does not preclude a straight-in landing if a normal descent and landing can be made in accordance with the applicable CFRs.

- A runway is not clearly defined on the airfield.

Communications

The communication strip provided near the top of FAA approach charts gives flight crews the frequencies that they can expect to be assigned during the approach. The frequencies are listed in the logical order of use from arrival to touchdown. Having this information immediately available during the approach reduces the chances of a loss of contact between ATC and flight crews during this critical phase of flight.

It is important for flight crews to understand their responsibilities with regard to communications in the various approach environments. There are numerous differences in communication responsibilities when operating into and out of airports without ATC towers as compared to airports with control towers. Today's pilots face an increasing range of ATC environments and conflicting traffic dangers, making approach briefing and preplanning more critical. Individual company operating manuals and SOPs dictate the duties for each crewmember.

Advisory Circular (AC) 120-71, Standard Operating Procedures for Flight Deck Crewmembers, contains the following concerning ATC communications: SOPs should state who (Pilot Flying (PF), Pilot Monitoring (PM), Flight Engineer (FE/SO)) handles the radios for each phase of flight, as follows:

- PF makes input to aircraft/autopilot and/or verbally states clearances while PM confirms input is what he or she read back to ATC.

- Any confusion in the flight deck is immediately cleared up by requesting ATC confirmation.

- If any crewmember is off the flight deck, all ATC instructions are briefed upon his or her return. Or,

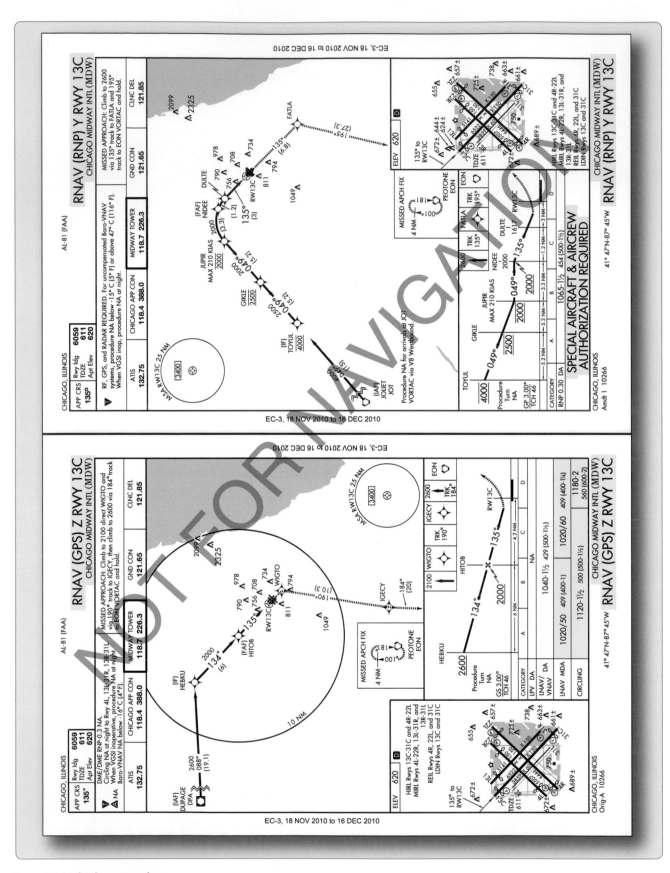

Figure 4-5. Multiple approaches.

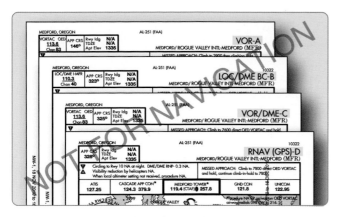

Figure 4-6. Procedures with circling landing minima.

if any crewmember is off the flight deck, all ATC instructions are written down until his or her return and then passed to that crewmember upon return. Similarly, if a crewmember is off ATC frequency when making a precision approach (PA) announcement or when talking on company frequency, all ATC instructions are briefed upon his or her return.

- Company policy should address use of speakers, headsets, boom microphone, and/or hand-held microphone.

- SOPs should state the altitude awareness company policy on confirming assigned altitude.

Example: The PM acknowledges ATC altitude clearance. If the aircraft is on the autopilot, then the PF makes input into the autopilot/altitude alerter. PF points to the input while stating the assigned altitude as he or she understands it. The PM then points to the input stating aloud what he or she understands the ATC clearance to be confirming that the input and clearance match. If the aircraft is being hand-flown, then the PM makes the input into the altitude alerter/autopilot, then points to the input and states clearance. PF then points to the alerter stating aloud what he or she understands the ATC clearance to be confirming that the alerter and clearance match.

Example: If there is no altitude alerter in the aircraft, then both pilots write down the clearance, confirm that they have the same altitude, and then cross off the previously assigned altitude.

Approach Control

Approach control is responsible for controlling all instrument flights operating within its area of responsibility. Approach control may serve one or more airports. Control is exercised primarily through direct pilot and controller communication and airport surveillance radar (ASR). Prior to arriving at the initial approach fix (IAF), instructions will be received from the air route traffic control center (ARTCC) to contact approach control on a specified frequency. Where radar is approved for approach control service, it is used not only for radar approaches, but also for vectors in conjunction with published non-radar approaches using conventional NAVAIDs or RNAV/GPS.

When radar handoffs are initiated between the ARTCC and approach control, or between two approach control facilities, aircraft are cleared (with vertical separation) to an outer fix most appropriate to the route being flown and, if required, given holding instructions. Or, aircraft are cleared to the airport or to a fix so located that the handoff is completed prior to the time the aircraft reaches the fix. When radar handoffs are used, successive arriving flights may be handed off to approach control with radar separation in lieu of vertical separation.

After release to approach control, aircraft are vectored to the final approach course. ATC occasionally vectors the aircraft across the final approach course for spacing requirements. The pilot is not expected to turn inbound on the final approach course unless an approach clearance has been issued. This clearance is normally issued with the final vector for interception of the final approach course, and the vector enables the pilot to establish the aircraft on the final approach course prior to reaching the FAF.

Air Route Traffic Control Center (ARTCC)

ARTCCs are approved for and may provide approach control services to specific airports. The radar systems used by these centers do not provide the same precision as an ASR or precision approach radar (PAR) used by approach control facilities and control towers, and the update rate is not as fast. Therefore, pilots may be requested to report established on the final approach course. Whether aircraft are vectored to the appropriate final approach course or provide their own navigation on published routes to it, radar service is automatically terminated when the landing is completed; or when instructed to change to advisory frequency at airports without an operating ATC tower, whichever occurs first. When arriving on an IFR flight plan at an airport with an operating control tower, the flight plan is closed automatically upon landing.

The extent of services provided by approach control varies greatly from location to location. The majority of Part 121 operations in the NAS use airports that have radar service and approach control facilities to assist in the safe arrival and departure of large numbers of aircraft. Many airports do not have approach control facilities. It is important for pilots to understand the differences between approaches with and without an approach control facility. For example,

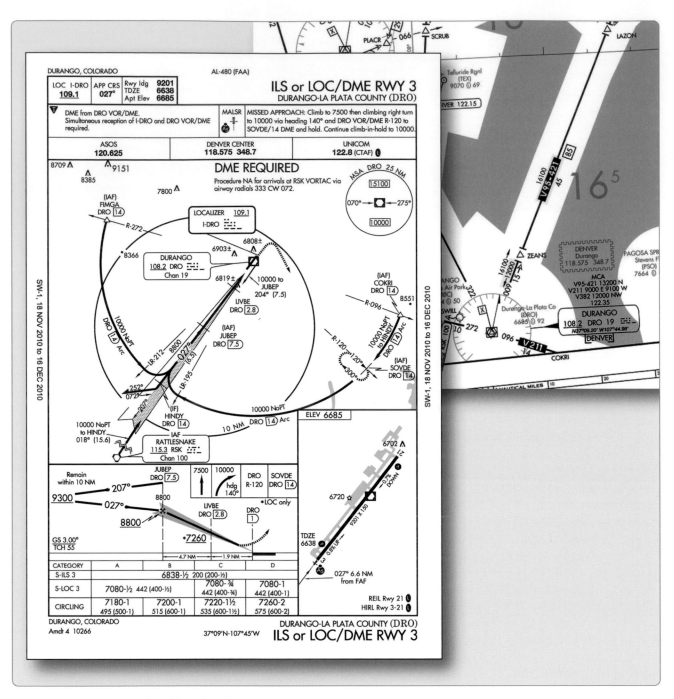

Figure 4-7. Durango approach and low altitude en route excerpt.

consider the Durango, Colorado, ILS DME RWY 2 and low altitude en route chart excerpt shown in Figure 4-7.

High or Lack of Minimum Vectoring Altitudes (MVAs)

Considering the fact that most modern commercial and corporate aircraft are capable of direct, point-to-point flight, it is increasingly important for pilots to understand the limitations of ARTCC capabilities with regard to minimum altitudes. There are many airports that are below the coverage area of Center radar, and, therefore, off-route transitions into the approach environment may require that the aircraft be flown at a higher altitude than would be required for an on-route transition. In the Durango example, an airplane approaching from the northeast on a direct route to the Durango VOR may be restricted to a minimum IFR altitude (MIA) of 17,000 feet mean sea level (MSL) due to unavailability of Center radar coverage in that area at lower altitudes. An arrival on V95 from the northeast would be able to descend to a minimum en route altitude (MEA) of 12,000 feet, allowing a shallower transition to the approach environment. An off-route arrival may necessitate a descent into holding in order to avoid an unstable approach to Durango.

Lack of Approach Control Terrain Advisories

Flight crews must understand that terrain clearance cannot be assured by ATC when aircraft are operating at altitudes that are not served by Center or approach radar. Recent National Transportation Safety Board (NTSB) investigations have identified several accidents that involved controlled flight into terrain (CFIT) by IFR rated and VFR pilots operating under visual flight conditions at night in remote areas. In many of these cases, the pilots were in contact with ATC at the time of the accident and receiving radar service. The pilots and controllers involved all appear to have been unaware that the aircraft were in danger. Increased altitude awareness and better preflight planning would likely have prevented all of these accidents. How can pilots avoid becoming involved in a CFIT accident?

CFIT accidents are best avoided through proper preflight planning.

- Terrain familiarization is critical to safe visual operations at night. Use sectional charts or other topographic references to ensure that your altitude safely clears terrain and obstructions all along your route.

- In remote areas, especially in overcast or moonless conditions, be aware that darkness may render visual avoidance of high terrain nearly impossible and that the absence of ground lights may result in loss of horizon reference.

- When planning a nighttime VFR flight, follow IFR practices, such as climbing on a known safe course, until well above surrounding terrain. Choose a cruising altitude that provides terrain separation similar to IFR flights (2,000 feet AGL in mountainous areas and 1,000 feet above the ground in other areas.)

- When receiving radar services, do not depend on ATC to warn you of terrain hazards. Although controllers try to warn pilots if they notice a hazardous situation, they may not always be able to recognize that a particular VFR aircraft is dangerously close to terrain.

- When issued a heading along with an instruction to "maintain VFR," be aware that the heading may not provide adequate terrain clearance. If you have any doubt about your ability to visually avoid terrain and obstacles, advise ATC immediately and take action to reach a safe altitude if necessary.

- ATC radar software can provide limited prediction and warning of terrain hazards, but the warning system is configured to protect IFR flights and is normally suppressed for VFR aircraft. Controllers can activate the warning system for VFR flights upon pilot request, but it may produce numerous false alarms for aircraft operating below the MIA, especially in en route center airspace.

- If you fly at night, especially in remote or unlit areas, consider whether a GPS-based terrain awareness unit would improve your safety of flight.

- Lack of approach control traffic advisories—if radar service is not available for the approach, the ability of ATC to give flight crews accurate traffic advisories is greatly diminished. In some cases, the common traffic advisory frequency (CTAF) may be the only tool available to enhance an IFR flight's awareness of traffic at the destination airport. Additionally, ATC will not clear an IFR flight for an approach until the preceding aircraft on the approach has cancelled IFR, either on the ground, or airborne once in visual meteorological conditions (VMC).

Airports With an ATC Tower

Control towers are responsible for the safe, orderly, and expeditious flow of all traffic that is landing, taking off, operating on and in the vicinity of an airport and, when the responsibility has been delegated, towers also provide for the separation of IFR aircraft in terminal areas. Aircraft that are departing IFR are integrated into the departure sequence by the tower. Prior to takeoff, the tower controller coordinates with departure control to assure adequate aircraft spacing.

Airports Without A Control Tower

From a communications standpoint, executing an instrument approach to an airport that is not served by an ATC tower requires more attention and care than making a visual approach to that airport. Pilots are expected to self-announce their arrival into the vicinity of the airport no later than 10 NM from the field. Depending on the weather, as well as the amount and type of conflicting traffic that exists in the area, an approach to an airport without an operating ATC tower increases the difficulty of the transition to visual flight.

In many cases, a flight arriving via an instrument approach needs to mix in with VFR traffic operating in the vicinity of the field. For this reason, many companies require that flight crews make contact with the arrival airport CTAF or company operations personnel via a secondary radio over 25 NM from the field in order to receive traffic advisories. In addition, pilots should attempt to listen to the CTAF well in advance of their arrival in order to determine the VFR traffic situation.

Since separation cannot be provided by ATC between IFR and VFR traffic when operating in areas where there is no radar coverage, pilots are expected to make radio announcements on the CTAF. These announcements allow other aircraft operating in the vicinity to plan their departures and arrivals with a minimum of conflicts. In addition, it is very important for crews to maintain a listening watch on the CTAF to increase their awareness of the current traffic situation. Flights inbound on an instrument approach to a field without a control tower should make several self-announced radio calls during the approach:

- Initial call within 4-10 minutes of the aircraft's arrival at the IAF. This call should give the aircraft's location as well as the crew's approach intentions.

- Departing the IAF, stating the approach that is being initiated.

- Procedure turn (or equivalent) inbound.

- FAF inbound, stating intended landing runway and maneuvering direction if circling.

- Short final, giving traffic on the surface notification of imminent landing.

When operating on an IFR flight plan at an airport without a functioning control tower, pilots must initiate cancellation of the IFR flight plan with ATC or an AFSS. Remote communications outlets (RCOs) or ground communications outlets (GCOs), if available, can be used to contact an ARTCC or an AFSS after landing. If a frequency is not available on

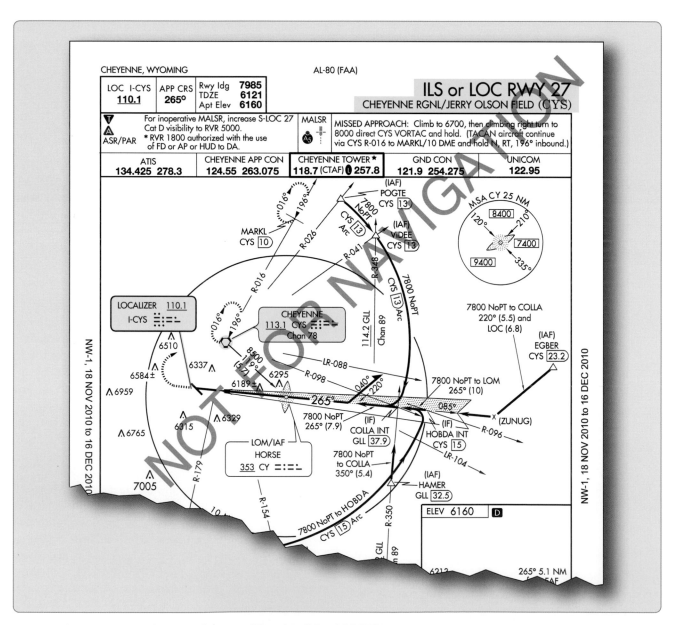

Figure 4-8. Cheyenne Regional (KCYS), Cheyenne, Wyoming, ILS or LOC RWY 27.

the ground, the pilot has the option to cancel IFR while in flight if VFR conditions can be maintained while in contact with ARTCC, as long as those conditions can be maintained until landing. Additionally, pilots can relay a message through another aircraft or contact flight service via telephone.

Primary NAVAID

Most conventional approach procedures are built around a primary final approach NAVAID; others, such as RNAV (GPS) approaches, are not. If a primary NAVAID exists for an approach, it should be included in the IAP briefing, set into the appropriate backup or active navigation radio, and positively identified at some point prior to being used for course guidance. Adequate thought should be given to the appropriate transition point for changing from FMS or other en route navigation over to the conventional navigation to be used on the approach. Specific company standards and procedures normally dictate when this changeover occurs; some carriers are authorized to use FMS course guidance throughout the approach, provided that an indication of the conventional navigation guidance is available and displayed. Many carriers, or specific carrier fleets, are required to change over from RNAV to conventional navigation prior to the FAF of an instrument approach.

Depending on the complexity of the approach procedure, pilots may have to brief the transition from an initial NAVAID to the primary and missed approach NAVAIDs. Figure 4-8 shows the Cheyenne, Wyoming, ILS Runway 27 approach procedure, which requires additional consideration during an IAP briefing.

If the 15 DME arc of the CYS VOR is to be used as the transition to this ILS approach procedure, caution must be paid to the transition from en route navigation to the initial NAVAID and then to the primary NAVAID for the ILS approach. Planning when the transition to each of these NAVAIDs occurs may prevent the use of the incorrect NAVAID for course guidance during approaches where high pilot workloads already exist.

Equipment Requirements

The navigation equipment that is required to join and fly an IAP is indicated by the title of the procedure and notes on the chart. Straight-in IAPs are identified by the navigation system by providing the final approach guidance and the runway with which the approach is aligned (for example, VOR RWY 13). Circling-only approaches are identified by the navigation system by providing final approach guidance and a letter (for example, VOR A). More than one navigation system separated by a slant indicates that more than one type of equipment must be used to execute the final approach (for example, VOR/DME RWY 31). More than one navigation system separated by the word "or" indicates either type of equipment can be used to execute the final approach (for example, VOR or GPS RWY 15).

In some cases, other types of navigation systems, including radar, are required to execute other portions of the approach or to navigate to the IAF (for example, an NDB procedure turn to an ILS, or an NDB in the missed approach, or radar required to join the procedure or identify a fix). When ATC radar or other equipment is required for procedure entry from the en route environment, a note is charted in the plan view of the approach procedure chart (for example, RADAR REQUIRED or AUTOMATIC DIRECTION FINDER (ADF) REQUIRED). When radar or other equipment is required on portions of the procedure outside the final approach segment, including the missed approach, a note is charted in the notes box of the pilot briefing portion of the approach chart (for example, RADAR REQUIRED or DISTANCE MEASURING EQUIPMENT (DME) REQUIRED). Notes are not charted when VOR is required outside the final approach segment. Pilots should ensure that the aircraft is equipped with the required NAVAIDs to execute the approach, including the missed approach. Refer to the AIM paragraph 5-4-5 for additional options with regards to equipment requirements for IAPs.

RNAV systems may be used as a Substitute Means of Navigation when a very high frequency (VHF) Omni-directional Range (VOR), Distance Measuring Equipment (DME), Tactical Air Navigation (TACAN), VOR/TACAN (VORTAC), VOR/DME, non-directional radio beacon (NDB), or compass locator facility including locator outer marker and locator middle marker is out-of-service, i.e., the Navigation Aid (NAVAID) information is not available; an aircraft is not equipped with an automatic direction finder (ADF) or DME; or the installed ADF or DME on an aircraft is not operational. For example, if equipped with a suitable RNAV system, a pilot may hold over an out-of-service NDB. Refer to Advisory Circular 90-108, Use of Suitable RNAV System on Conventional Routes and Procedures, dated March 3, 2011 for additional guidance on the proper times and procedures for substituting a RNAV system for means of navigation.

Courses

Traditional Courses

An aircraft that has been cleared to a holding fix and subsequently "cleared...approach," normally does not receive new routing. Even though clearance for the approach may have been issued prior to the aircraft reaching the holding fix, ATC would expect the pilot to proceed via the holding fix that was the last assigned route, and the feeder route associated with that fix, if a feeder route is published on the approach chart, to the IAF to commence the approach. When cleared for the approach, the published off-airway (feeder) routes that lead from the en route structure to the IAF are part of the approach clearance.

If a feeder route to an IAF begins at a fix located along the route of flight prior to reaching the holding fix, and clearance for an approach is issued, a pilot should commence the approach via the published feeder route. For example, the aircraft would not be expected to overfly the feeder route and return to it. The pilot is expected to commence the approach in a similar manner at the IAF, if the IAF for the procedure is located along the route of flight to the holding fix.

If a route of flight directly to the IAF is desired, it should be so stated by the controller with phraseology to include the words "direct," "proceed direct," or a similar phrase that the pilot can interpret without question. When a pilot is uncertain of the clearance, ATC should be queried immediately as to what route of flight is preferred.

The name of an instrument approach, as published, is used to identify the approach, even if a component of the approach aid is inoperative or unreliable. The controller will use the name of the approach as published, but must advise the aircraft at the time an approach clearance is issued that the inoperative or unreliable approach aid component is unusable. (Example: "Cleared ILS RWY 4, glideslope unusable.")

Area Navigation Courses

RNAV (GPS) approach procedures introduce their own tracking issues because they are flown using an onboard navigation database. They may be flown as coupled approaches or flown manually. In either case, navigation system coding is based on procedure design, including waypoint (WP) sequencing for an approach and missed approach. The procedure design indicates whether the WP is a fly-over (FO) or fly-by (FB), and provides appropriate guidance for each. A FB WP requires the use of turn anticipation to avoid overshooting the next flight segment. A FO WP precludes any turn until the WP is over flown and is followed by either an intercept maneuver of the next flight segment or direct flight to the next WP.

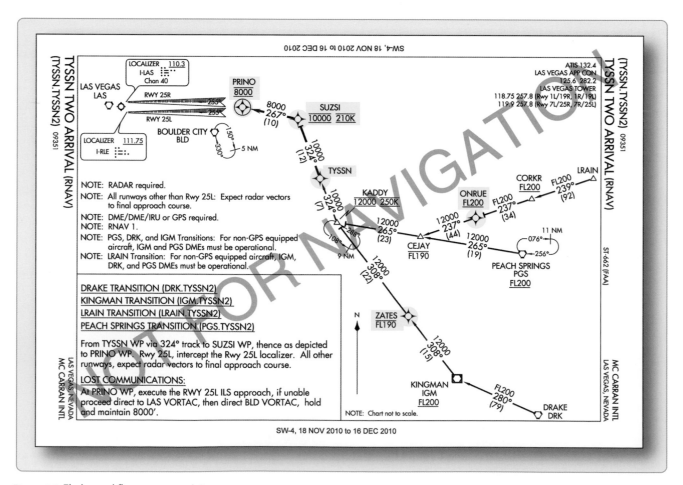

Figure 4-9. Fly-by and fly-over waypoints.

Approach waypoints, except for the missed approach waypoint (MAWP) and the missed approach holding waypoint (MAHWP), are normally FB WPs. Notice that in the plan view in Figure 4-9, there are four FB WPs, but only the circled WP symbol at PRINO is a FO WP. If flying manually to a selected RNAV WP, pilots should anticipate the turn at a FB WP to ensure a smooth transition and avoid overshooting the next flight segment. Alternatively, for a FO WP, no turn is accomplished until the aircraft passes the WP.

There are circumstances when a WP may be coded into the database as both a FB WP and a FO WP, depending on how the WPs are sequenced during the approach procedure. For example, a WP that serves as an IAF may be coded as a FB WP for the approach and as a FO WP when it also serves as the MAWP for the missed approach procedure (MAP). This is just one reason why instrument approaches should be loaded in their entirety from the FMS and not manually built or modified.

Altitudes

Prescribed altitudes may be depicted in four different configurations: minimum, maximum, recommended, and mandatory. The U.S. Government distributes approach charts produced by the FAA. Altitudes are depicted on these charts in the profile view with an underscore or overscore, or both to identify them as minimum, maximum, or mandatory, respectively.

- Minimum altitudes are depicted with the altitude value underscored. Aircraft are required to maintain altitude at or above the depicted value (e.g., 3000).

- Maximum altitudes are depicted with the altitude value overscored. Aircraft are required to maintain altitude at or below the depicted value (e.g., 4800).

- Mandatory altitudes are depicted with the altitude value both underscored and overscored. Aircraft are required to maintain altitude at the depicted value (e.g., 5500).

- Recommended altitudes are depicted without an underscore or overscore.

NOTE: Pilots are cautioned to adhere to altitudes as prescribed because, in certain instances, they may be used as the basis for vertical separation of aircraft by ATC. If a depicted altitude is specified in the ATC clearance, that altitude becomes mandatory as defined above.

Minimum Safe/Sector Altitude

Minimum Safe Altitudes are published for emergency use on IAP charts. MSAs provide 1,000 feet of clearance over all obstacles but do not necessarily assure acceptable navigation signal coverage. The MSA depiction on the plan view of an approach chart contains the identifier of the center point of the MSA, the applicable radius of the MSA, a depiction of the sector(s), and the minimum altitudes above mean sea level which provide obstacle clearance. For conventional navigation systems, the MSA is normally based on the primary omnidirectional facility on which the IAP is predicated, but may be based on the airport reference point (ARP) if no suitable facility is available. For RNAV approaches, the MSA is based on an RNAV waypoint. MSAs normally have a 25 NM radius; however, for conventional navigation systems, this radius may be expanded to 30 NM if necessary to encompass the airport landing surfaces.

Depicted on the Plan View of approach charts, a single sector altitude is normally established. However when it is necessary to obtain obstacle clearance, an MSA area may be further divided with up to four sectors.

Final Approach Fix Altitude

Another important altitude that should be briefed during an IAP briefing is the FAF altitude, designated by the cross on a non-precision approach, and the lightning bolt symbol designating the glideslope/glidepath intercept altitude on a precision approach. Adherence and cross-check of this altitude can have a direct effect on the success and safety of an approach.

Proper airspeed, altitude, and configuration, when crossing the FAF of a non-precision approach, are extremely important no matter what type of aircraft is being flown. The stabilized approach concept, implemented by the FAA within the SOPs of each air carrier, suggests that crossing the FAF at the published altitude is often a critical component of a successful non-precision approach, especially in a large turbojet aircraft.

The glideslope intercept altitude of a precision approach should also be included in the IAP briefing. Awareness of this altitude when intercepting the glideslope can ensure the flight crew that a "false glideslope" or other erroneous indication is not inadvertently followed. Many air carriers include a standard callout when the aircraft passes over the FAF of the non-precision approach underlying the ILS. The PM states the name of the fix and the charted glideslope altitude, thus allowing both pilots to cross-check their respective altimeters and verify the correct indications.

Minimum Descent Altitude (MDA), Decision Altitude (DA), And Decision Height (DH)

MDA—the lowest altitude, expressed in feet MSL, to which descent is authorized on final approach or during circle-to-land maneuvering in execution of a standard instrument approach procedure (SIAP) where no electronic glideslope is provided.

DA—a specified altitude in the precision approach at which a missed approach must be initiated if the required visual reference to continue the approach has not been established.

DH—with respect to the operation of aircraft, means the height at which a decision must be made during an ILS, MLS, or PAR IAP to either continue the approach or to execute a missed approach.

CAT II and III approach DHs are referenced to AGL and measured with a radio altimeter.

The height above touchdown (HAT) for a CAT I precision approach is normally 200 feet above touchdown zone elevation (TDZE). When a HAT of 250 feet or higher is published, it may be the result of the signal-in-space coverage, or there may be penetrations of either the final or missed approach obstacle clearance surfaces (OCSs). If there are OCS penetrations, the pilot has no indication on the approach chart where the obstacles are located. It is important for pilots to brief the MDA, DA, or DH so that there is no ambiguity as to what minimums are being used. These altitudes can be restricted by many factors. Approach category, inoperative equipment in the aircraft or on the ground, crew qualifications, and company authorizations are all examples of issues that may limit or change the height of a published MDA, DA, or DH.

For many air carriers, OpSpecs may be the limiting factor for some types of approaches. NDB and circling approaches are two common examples where the OpSpecs minimum listed altitudes may be more restrictive than the published minimums. Many Part 121 and 135 operators are restricted from conducting circling approaches below 1,000 feet MDA and 3 SM visibility by Part C of their OpSpecs, and many have specific visibility criteria listed for NDB approaches that exceed visibilities published for the approach (commonly 2 SM). In these cases, flight crews must determine which is the more restrictive of the two and comply with those minimums.

In some cases, flight crew qualifications can be the limiting factor for the MDA, DA, or DH for an instrument approach. There are many CAT II and III approach procedures authorized at airports throughout the United States, but RNP AR restricts their use to pilots who have received specific training, and aircraft that are equipped and authorized to conduct those approaches. Other rules pertaining to flight crew qualifications can also determine the lowest usable MDA, DA, or DH for a specific approach. 14 CFR Part 121, section 121.652, 14 CFR Part 125, section 125.379, and 14 CFR Part 135, section 135.225 require

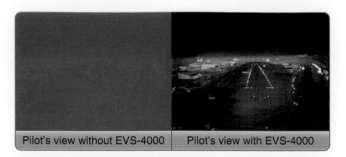

| Pilot's view without EVS-4000 | Pilot's view with EVS-4000 |

Figure 4-10. Enhanced flight vision system.

that some PICs, with limited experience in the aircraft they are operating, increase the approach minimums and visibility by 100 feet and one- half mile respectively. Rules for these "high-minimums" pilots are usually derived from a combination of federal regulations and the company's OpSpecs. There are many factors that can determine the actual minimums that can be used for a specific approach. All of them must be considered by pilots during the preflight and approach planning phases, discussed, and briefed appropriately.

Pilots are cautioned to fully understand and abide by the guidelines set forth in 91.175(c) regarding proper identification of the runway and runway environment when electing to continue any approach beyond the published DA/DH or MDA.

It is imperative to recognize that any delay in making a decision to execute the Missed Approach Procedure at the DA/DH or MDA/Missed Approach Point will put the aircrew at risk of impacting any obstructions that may be penetrating the visual obstacle clearance surface

The visual segment of an IAP begins at DA or MDA and continues to the runway. There are two means of operating in the visual segment, one is by using natural vision under 14 CFR Part 91, section 91.175 (c) and the other is by using an Enhanced Flight Vision System under 14 CFR Part 91, section 91.175 (l).

Enhanced Flight Vision Systems (EFVS) and Instrument Approaches

An Enhanced Flight Vision System (EFVS) is an installed airborne system that uses an electronic means to provide a display of the forward external scene topography (the applicable natural or manmade features of a place or region especially in a way to show their relative positions and elevation) through the use of imaging sensors, such as forward looking infrared, millimeter wave radiometry, millimeter wave radar, and/or low light level image intensifying. The EFVS imagery is displayed along with the additional flight information and aircraft flight symbology required by 14 CFR Part 91, section 91.175(m) on a head-up display (HUD), or an equivalent display, in the same scale and alignment as the external view and includes the display element, sensors, computers and power supplies, indications, and controls. [Figure 4-10]

When the runway environment cannot be visually acquired at the DA or MDA using natural vision, a pilot may use an EFVS to descend below DA or MDA down to 100 feet above the TDZE, provided the pilot determines that the enhanced flight visibility (EFV) observed by using the EFVS is not less than the minimum visibility prescribed in the IAP being flown, the pilot acquires the required visual references prescribed in 14 CFR Part 91, section 91.175 (l)(3), and all of the other requirements of 14 CFR Part 91, section 91.175 (l) and (m) are met. The primary reference for maneuvering the aircraft is based on what the pilot sees through the EFVS. At 100 feet above the TDZE, a pilot can continue to descend only when the visual reference requirements for descent below 100 feet can be seen using natural vision (without the aid of the EFVS). In other words, a pilot may not continue to rely on the EFVS sensor image to identify the required visual references below 100 feet above the TDZE. Supporting information is provided by the flight path vector (FPV) cue, flight path angle (FPA) reference cue, onboard navigation system, and other imagery and flight symbology displayed on the HUD. The FPV and FPA reference cues, along with the EFVS imagery of the TDZ, provide the primary vertical path reference for the pilot when vertical guidance from a precision approach or approach with vertical guidance is not available.

An EFVS may be used to descend below DA or MDA from any straight-in IAP, other than Category II or Category III approaches, provided all of the requirements of 14 CFR Part 91, section 91.175 (l) are met. This includes straight-in precision approaches, approaches with vertical guidance (localizer performance with vertical guidance (LPV) or lateral navigation (LNAV)/vertical navigation (VNAV)), and non-precision approaches (VOR, NDB, localizer (LOC), RNAV, GPS, localizer type directional aid (LDA), simplified directional facility (SDF)). An instrument approach with a circle-to-land maneuver or circle-to-land minimums does not meet criteria for straight-in landing minimums. While the regulations do not prohibit EFVS from being used during any phase of flight, they do prohibit it from being used for operational credit on anything but a straight-in IAP with straight-in landing minima. EFVS may only be used during a circle-to-land maneuver provided the visual references required throughout the circling maneuver are distinctly visible using natural vision. An EFVS cannot be used to satisfy the requirement that an identifiable part of the airport be distinctly visible to the pilot during a circling maneuver at or above MDA or while descending below MDA from a circling maneuver.

The EFVS visual reference requirements of 14 CFR Part 91, section 91.175 (l)(3) comprise a more stringent standard than the visual reference requirements prescribed under 14 CFR Part 91, section 91.175 (c)(3) when using natural vision. The more stringent standard is needed because an EFVS might not display the color of the lights used to identify specific portions of the runway or might not be able to consistently display the runway markings. The main differences for EFVS operations are that the visual glideslope indicator (VGSI) lights cannot be used as a visual reference, and specific visual references from both the threshold and TDZ must be distinctly visible and identifiable. However, when using natural vision, only one of the specified visual references must be visible and identifiable.

Pilots must be especially knowledgeable of the approach conditions and approach course alignment when considering whether to rely on EFVS during a non-precision approach with an offset final approach course. Depending upon the combination of crosswind correction and the lateral field of view provided by a particular EFVS, the required visual references may or may not be within the pilot's view looking through the EFVS display. Pilots conducting any non-precision approach must verify lateral alignment with the runway centerline when determining when to descend from the MDA.

Any pilot operating an aircraft with an EFVS installed should be aware that the requirements of 14 CFR Part 91, section 91.175 (c) for using natural vision, and the requirements of 14 CFR Part 91, section 91.175 (l) for using EFVS are different. A pilot would, therefore, first have to determine whether an approach is commenced using natural vision or using EFVS. While these two sets of requirements provide a parallel decision making process, the requirements for when a missed approach must be executed differ. Using EFVS, a missed approach must be initiated at or below DA or MDA down to 100 feet above TDZE whenever the pilot determines that:

1. The enhanced flight visibility is less than the visibility minima prescribed for the IAP being used;

2. The required visual references for the runway of intended landing are no longer distinctly visible and identifiable to the pilot using the EFVS imagery;

3. The aircraft is not continuously in a position from which a descent to a landing can be made on the intended runway, at a normal rate of descent, using normal maneuvers; or

4. For operations under 14 CFR Part 121 and 135, the descent rate of the aircraft would not allow touchdown to occur within the TDZ of the runway of intended landing.

It should be noted that a missed approach after passing the DA, or beyond the MAP, involves additional risk until established on the published missed approach segment. Initiating a go-around after passing the published MAP may result in loss of obstacle clearance. As with any approach, pilot planning should include contingencies between the published MAP and touchdown with reference to obstacle clearance, aircraft performance, and alternate escape plans.

At and below 100 feet above the TDZE, the regulations do not require the EFVS to be turned off or the display to be stowed in order to continue to a landing. A pilot may continue the approach below this altitude using an EFVS as long as the required visual references can be seen through the display using natural vision. An operator may not continue to descend beyond this point by relying solely on the sensor image displayed on the EFVS. In order to descend below 100 feet above the TDZE, the flight visibility assessed using natural vision must be sufficient

for the following visual references to be distinctly visible and identifiable to the pilot without reliance on the EFVS to continue to a landing:

1. The lights or markings of the threshold, or

2. The lights or markings of the TDZ.

It is important to note that from 100 feet above the TDZE and below, the flight visibility does not have to be equal to or greater than the visibility prescribed for the IAP in order to continue descending. It only has to be sufficient for the visual references required by 14 CFR Part 91, section 91.175 (l)(4) to be distinctly visible and identifiable to the pilot without reliance on the EFVS.

A missed approach must be initiated when the pilot determines that:

1. The flight visibility is no longer sufficient to distinctly see and identify the required visual references listed in 14 CFR Part 91, section 91.175 (l)(4) using natural vision;

2. The aircraft is not continuously in a position from which a descent to a landing can be made on the

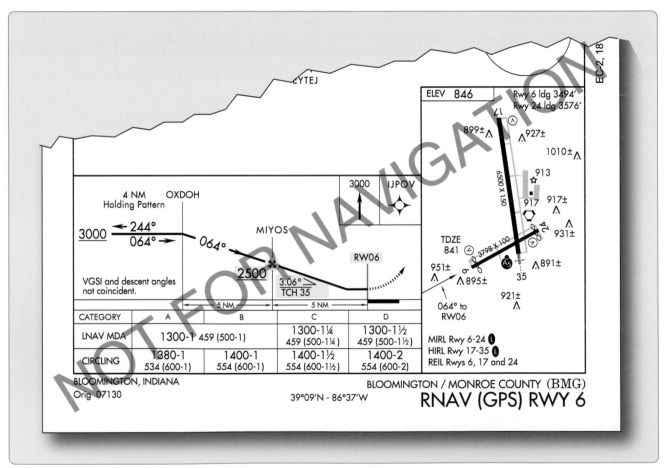

Figure 4-11a. VNAV information.

intended runway, at a normal rate of descent, using normal maneuvers; or

3. For operations under 14 CFR Part 121 and 135, the descent rate of the aircraft would not allow touchdown to occur within the TDZ of the runway of intended landing.

While touchdown within the TDZ is not specifically addressed in the regulations for operators other than Part 121 and 135 operators, continued operations below DA or MDA where touchdown in the TDZ is not assured, where a high sink rate occurs, or where the decision to conduct a MAP is not executed in a timely manner, all create a significant risk to the operation. A missed approach initiated after the DA or MAP involves additional risk. At 100 feet or less above the runway, it is likely that an aircraft is significantly below the TERPS missed approach obstacle clearance surface. Prior planning is recommended and should include contingencies between the published MAP and touchdown with reference to obstacle clearance, aircraft performance, and alternate escape plans.

Vertical Navigation

One of the advantages of some GPS and multi-sensor FMS RNAV avionics is the advisory VNAV capability. Traditionally, the only way to get vertical path information during an approach was to use a ground-based precision NAVAID. Modern RNAV avionics can display an electronic vertical path that provides a constant-rate descent to minimums.

Since these systems are advisory and not primary guidance, the pilot must continuously ensure the aircraft remains at or above any published altitude constraint, including step-down fix altitudes, using the primary barometric altimeter. The pilots, airplane, and operator must be approved to use advisory VNAV inside the FAF on an instrument approach.

VNAV information appears on selected conventional nonprecision, GPS, and RNAV approaches (see "Types of Approaches" later in this chapter). It normally consists of two fixes (the FAF and the landing runway threshold), a FAF crossing altitude, a vertical descent angle (VDA), and may provide a visual descent point (VDP) [Figure 4-11a].

The VDA provides the pilot with advisory information not previously available on nonprecision approaches. It provides a means for the pilot to establish a stabilized descent from the FAF or step-down fix to the MDA. Stabilized descent is a key factor in the reduction of controlled flight into terrain (CFIT) incidents. However, pilots should be aware that the published angle is for information only – it is strictly advisory in nature. There is

no implicit additional obstacle protection below the MDA. Pilots must still respect any published stepdown fixes and the published MDA unless the visual cues stated 14 CFR Section 91.175 are present, and they can visually acquire and avoid both lit and unlit obstacles once below the MDA. The presence of a VDA does not guarantee obstacle protection in the visual segment and does not change any of the requirements for flying a nonprecision approach.

Pilots may use the published angle and estimated/actual groundspeed to find a target rate of descent from the rate of descent table published in the back of the U.S. Terminal Procedures Publication. This rate of descent can be flown with the Vertical Velocity Indicator (VVI) in order to use the VDA as an aid to flying a stabilized descent. No special equipment is required.

In rare cases, the LNAV minima may have a lower HAT than minima with a glide path, due to the location of the obstacles and the nonprecision MAP. This should serve as a clear indication to the pilot that obstacles exist below the MDA, which must be seen in order to ensure adequate clearance. In those cases, the glide path may be treated as a VDA and used to descend to the LNAV MDA, as long as all of the rules for a nonprecision approach are applied at the MDA.

When there are obstacles in the visual area that could cause an aircraft to destabilize the approach between the MDA and touchdown, the IAP will not show a vertical descent angle in the profile view. The charts currently include the following statement: "Descent Angle NA" or "Descent Angle NA-Obstacles" [Figure 4-11b].

Like flying any other IAP, the pilot must see and avoid any obstacles in the visual segment during transition to landing.

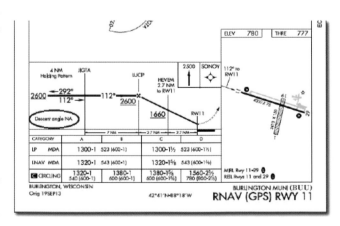

Figure 4-11b. Descent Angle N/A..

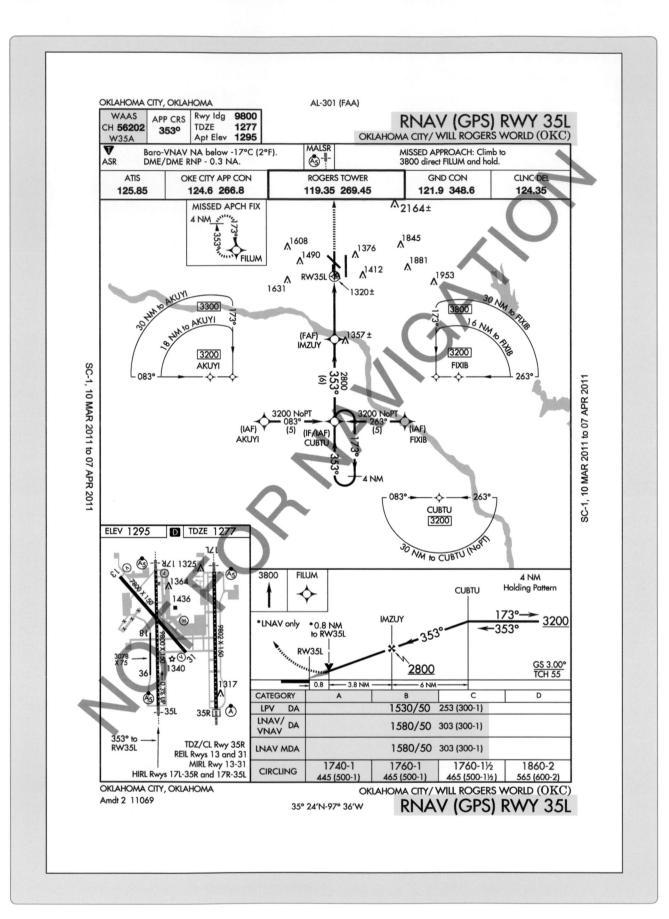

Figure 4-12. RNAV GPS approach minima.

A constant-rate descent has many safety advantages over non-precision approaches that require multiple level-offs at stepdown fixes or manually calculating rates of descent. A stabilized approach can be maintained from the FAF to the landing when a constant-rate descent is used. Additionally, the use of an electronic vertical path produced by onboard avionics can serve to reduce CFIT, and minimize the effects of visual illusions on approach and landing. Some countries even mandate the use of continuous descent final approaches (CDFAs) on non-precision approaches.

Wide Area Augmentation System

The Wide Area Augmentation System (WAAS) offers an opportunity for airports to gain ILS like approach capability without the purchase or installation of any ground-based navigation equipment at the airport. Today, WAAS is already being used at more than 900 runways across the United States to achieve minimums as low as 200 feet height above HAT/one-half mile visibility.

Benefits Of WAAS In The Airport Environment

WAAS is a navigation service using a combination of GPS satellites and the WAAS geostationary satellites to improve the navigational service provided by GPS. WAAS achieved initial operating capability (IOC) in 2003. The system is owned and operated by the FAA and provided free of direct user charges to users across the United States and most of Canada and Mexico.

WAAS improves the navigational system accuracy for en route, terminal, and approach operations over all the continental United States and significant portions of Alaska, Canada, and Mexico. This new navigational technology supports vertically-guided instrument approaches to all qualifying runways in the United States. Vertically-guided approaches reduce pilot workload and provide safety benefits compared to non-precision approaches. The WAAS enabled vertically guided approach procedures are called LPV, which stands for "localizer performance with vertical guidance," and provide ILS equivalent approach minimums as low as 200 feet at qualifying airports. Actual minimums are based on an airport's current infrastructure, as well as an evaluation of any existing obstructions. The FAA plans to publish 300 WAAS approach procedures per year to provide service to all qualifying instrument runways within the NAS.

Advantages Of WAAS Enabled LPV Approaches

The advantages of WAAS enabled LPV approaches include:

- LPV procedures have no requirement for ground-based transmitters at the airport.

- No consideration needs to be given to the placement of navigation facility, maintenance of clear zones around the facility, or access to the facility for maintenance.

- LPV approaches eliminate the need for critical area limitations associated with an ILS.

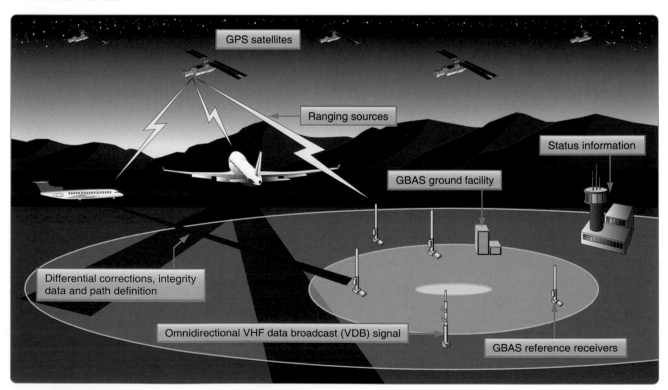

Figure 4-13. GBAS architecture.

- From a pilot's viewpoint, an LPV approach looks and flies like an ILS, but the WAAS approach is more stable than that of an ILS.

- WAAS equipped users can fly RNAV and basic required navigation performance (RNP) procedures, as well as LPV procedures, and the avionics costs are relatively inexpensive considering the total navigation solution provided.

RNAV (GPS) approach charts presently can have up to four lines of approach minimums: LPV, LNAV/VNAV, LNAV, and Circling. Figure 4-12 shows how these minimums might be presented on an approach chart, with the exception of Ground Based Augmentation System (GBAS) Landing System (GLS). This enables as many GPS equipped aircraft to use the procedure as possible and provides operational flexibility if WAAS becomes unavailable. Some aircraft may only be equipped with GPS receivers so they can fly to the LNAV MDA. Some aircraft equipped with GPS and FMS (with approach-certified barometric vertical navigation, or Baro-VNAV) can fly to the LNAV/VNAV MDA. Flying a WAAS LPV approach requires an aircraft with WAAS-LPV avionics. If for some reason the WAAS service becomes unavailable, all GPS or WAAS equipped aircraft can revert to the LNAV MDA and land safely using GPS only, which is available nearly 100 percent of the time.

LPV identifies WAAS approach with vertical guidance (APV) approach minimums with electronic lateral and vertical guidance capability. LPV is used for approaches constructed with WAAS criteria where the value for the vertical alarm limit is more than 12 meters and less than 50 meters. WAAS avionics equipment approved for LPV approaches is required for this type of approach. The lateral guidance is equivalent to localizer accuracy, and the protected area is considerably smaller than the protected area for the present LNAV and LNAV/VNAV lateral protection. Aircraft can fly this minima line with a statement in the AFM that the installed equipment supports LPV approaches. In Figure 4-12, notice the WAAS information shown in the top left corner of the pilot briefing information on the chart depicted. Below the term WAAS is the WAAS channel number (CH 56202), and the WAAS approach identifier (W35A), indicating Runway 35L in this case, and then a letter to designate the first in a series of procedures to that runway [Fig 4-12].

LNAV/VNAV identifies APV minimums developed to accommodate an RNAV IAP with vertical guidance, usually provided by approach certified Baro-VNAV, but with vertical and lateral integrity limits larger than a precision approach or LPV. Many RNAV systems that have RNP 0.3 or less approach capability are specifically approved in the AFM. Airplanes that are commonly approved in these types of

operations include Boeing 737NG, 767, and 777, as well as the Airbus A300 series. Landing minimums are shown as DAs because the approaches are flown using an electronic glide path. Other RNAV systems require special approval. In some cases, the visibility minimums for LNAV/VNAV might be greater than those for LNAV only. This situation occurs because DA on the LNAV/VNAV vertical descent path is farther away from the runway threshold than the LNAV MDA missed approach point.

Also shown in Figure 4-12, is the LNAV minimums line. This minimum is for lateral navigation only, and the approach minimum altitude is published as a MDA. LNAV provides the same level of service as the present GPS stand alone approaches. LNAV supports the following systems: WAAS, when the navigation solution will not support vertical navigation; and GPS navigation systems which are presently authorized to conduct GPS approaches.

Circling minimums that may be used with any type of approach approved RNAV equipment when publication of straight-in approach minimums is not possible.

Ground-Based Augmentation System (GBAS)

The United States version of the Ground-Based Augmentation System (GBAS) has traditionally been referred to as the Local Area Augmentation System (LAAS). The worldwide community has adopted GBAS as the official term for this type of navigation system. To coincide with international terminology, the FAA is also adopting the term GBAS to be consistent with the international community. GBAS is a ground-based augmentation to GPS that focuses its service on the airport area (approximately a 20–30 mile radius) for precision approach, DPs, and terminal area operations. It broadcasts its correction message via a very high frequency (VHF) radio data link from a ground-based transmitter. GBAS yields the extremely high accuracy, availability, and integrity necessary for Category I, II, and III precision approachesand provides the ability for flexible, curved approach paths. GBAS demonstrated accuracy is less than one meter in both the horizontal and vertical axis. [Figure 4-13]

The GBAS augments the GPS to improve aircraft safety during airport approaches and landings. It is expected that the end state configuration will pinpoint the aircraft's position to within one meter or less with a significant improvement in service flexibility and user operating costs.

GBAS is comprised of ground equipment and avionics. The ground equipment includes four reference receivers, a GBAS ground facility, and a VHF data broadcast transmitter. This ground equipment is complemented by GBAS avionics

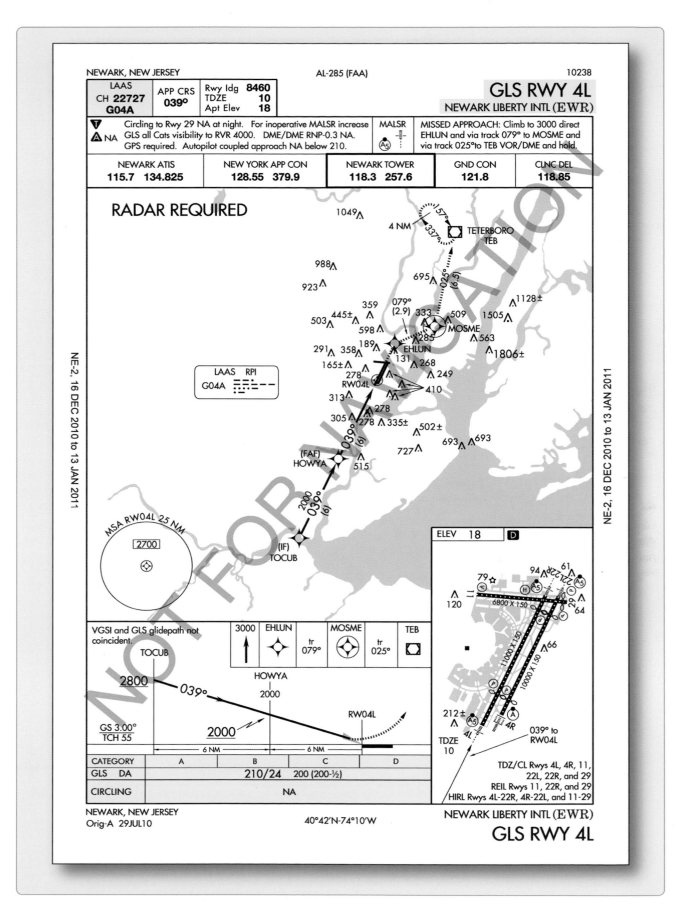

Figure 4-14. GLS approach at Newark, New Jersey.

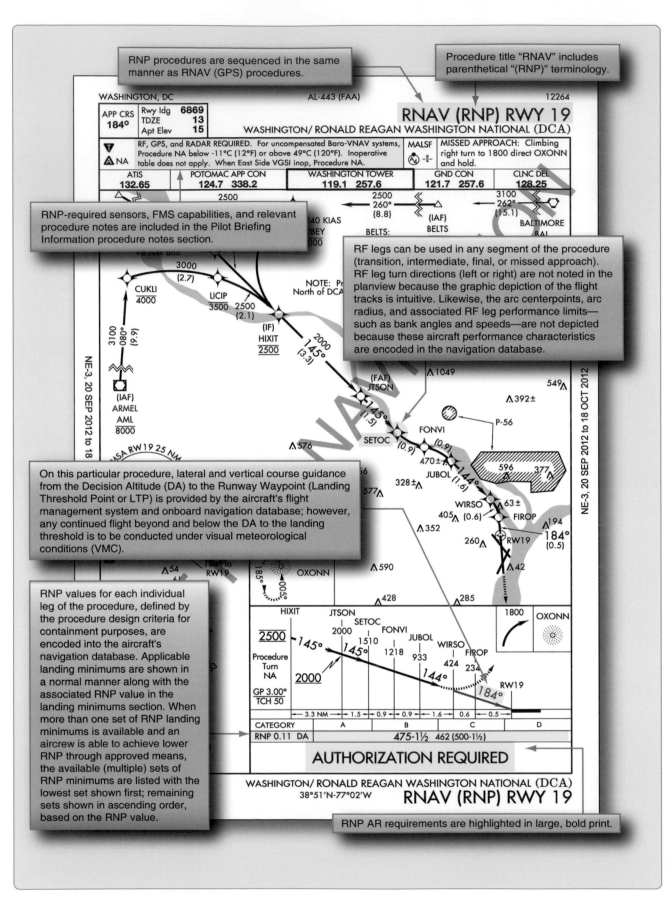

RNP procedures are sequenced in the same manner as RNAV (GPS) procedures.

Procedure title "RNAV" includes parenthetical "(RNP)" terminology.

RNP-required sensors, FMS capabilities, and relevant procedure notes are included in the Pilot Briefing Information procedure notes section.

RF legs can be used in any segment of the procedure (transition, intermediate, final, or missed approach). RF leg turn directions (left or right) are not noted in the planview because the graphic depiction of the flight tracks is intuitive. Likewise, the arc centerpoints, arc radius, and associated RF leg performance limits—such as bank angles and speeds—are not depicted because these aircraft performance characteristics are encoded in the navigation database.

On this particular procedure, lateral and vertical course guidance from the Decision Altitude (DA) to the Runway Waypoint (Landing Threshold Point or LTP) is provided by the aircraft's flight management system and onboard navigation database; however, any continued flight beyond and below the DA to the landing threshold is to be conducted under visual meteorological conditions (VMC).

RNP values for each individual leg of the procedure, defined by the procedure design criteria for containment purposes, are encoded into the aircraft's navigation database. Applicable landing minimums are shown in a normal manner along with the associated RNP value in the landing minimums section. When more than one set of RNP landing minimums is available and an aircrew is able to achieve lower RNP through approved means, the available (multiple) sets of RNP minimums are listed with the lowest set shown first; remaining sets shown in ascending order, based on the RNP value.

RNP AR requirements are highlighted in large, bold print.

Figure 4-15. RNAV RNP approach procedure with curved flight tracks.

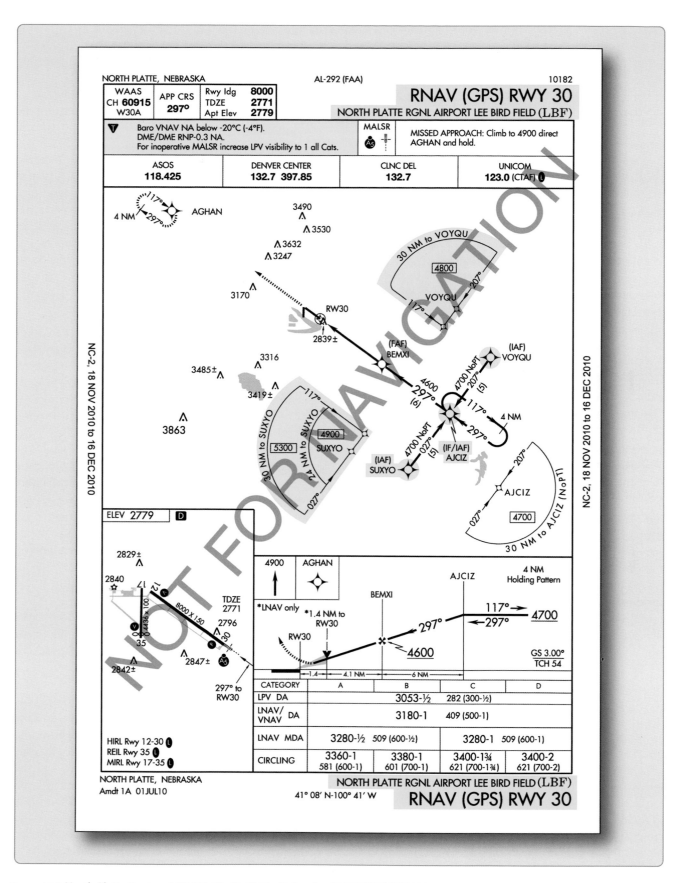

Figure 4-16. North Platte Regional (KLBF), North Platte, Nebraska, RNAV (GPS) RWY 30.

installed on the aircraft. Signals from GPS satellites are received by the GBAS GPS reference receivers (four receivers for each GBAS) at the GBAS equipped airport. The reference receivers calculate their position using GPS. The GPS reference receivers and GBAS ground facility work together to measure errors in GPS provided position.

The GBAS ground facility produces a GBAS correction message based on the difference between actual and GPS calculated position. Included in this message is suitable integrity parameters and approach path information. This GBAS correction message is then sent to a VHF data broadcast (VDB) transmitter. The VDB broadcasts the GBAS signal throughout the GBAS coverage area to avionics in GBAS equipped aircraft. GBAS provides its service to a local area (approximately a 20–30 mile radius). The signal coverage is designed support the aircraft's transition from en route airspace into and throughout the terminal area airspace.

The GBAS equipment in the aircraft uses the corrections provided on position, velocity, and time to guide the aircraft safely to the runway. This signal provides ILS look alike guidance as low as 200 feet above touchdown. GBAS will eventually support landings all the way to the runway surface. Figure 4-14 is an example of a GBAS (LAAS) approach into Newark, New Jersey.

Required Navigation Performance (RNP)

The operational advantages of RNP include accuracy, onboard performance monitoring and alerting which provide increased navigation precision and lower minimums than conventional RNAV. RNP DAs can be as low as 250 feet with visibilities as low as 3/4 SM. Besides lower minimums, the benefits of RNP include improved obstacle clearance limits, as well as reduced pilot workload. When RNP capable aircraft fly an accurate, repeatable path, ATC can be confident that these aircraft are at a specific position, thus maximizing safety and increasing capacity.

To attain the benefits of RNP approach procedures, a key component is curved flight tracks. Constant radius turns around a fix are called "radius-to-fix legs (RF legs)." These turns, which are encoded into the navigation database, allow the aircraft to avoid critical areas of terrain or conflicting airspace while preserving positional accuracy by maintaining precise, positive course guidance along the curved track. The introduction of RF legs into the design of terminal RNAV procedures results in improved use of airspace and allows procedures to be developed to and from runways that are otherwise limited to traditional linear flight paths or, in some cases, not served by an IFR

procedure at all. Navigation systems with RF capability are a prerequisite to flying a procedure that includes an RF leg. Refer to the notes box of the pilot briefing portion of the approach chart in Figure 4-15.

In the United States, operators who seek to take advantage of RNP approach procedures must meet the special RNP requirements outlined in FAA AC 90-101, Approval Guidance for RNP Procedures with Authorization Required (AR). Currently, most new transport category airplanes receive an airworthiness approval for RNP operations. However, differences can exist in the level of precision that each system is qualified to meet. Each individual operator is responsible for obtaining the necessary approval and authorization to use these instrument flight procedures with navigation databases.

RNAV Approach Authorization

Like any other authorization given to air carriers and Part 91 operators, the authorization to use VNAV on a conventional non-precision approach, RNAV approaches, or LNAV/VNAV approaches is found in that operator's OpSpecs, AFM, or other FAA-approved documents. There are many different levels of authorizations when it comes to the use of RNAV approach systems. The type of equipment installed in the aircraft, the redundancy of that equipment, its operational status, the level of flight crew training, and the level of the operator's FAA authorization are all factors that can affect a pilot's ability to use VNAV information on an approach.

Because most Part 121, 125, 135, and 91 flight departments include RNAV approach information in their pilot training programs, a flight crew considering an approach to North Platte, Nebraska, using the RNAV (GPS) RWY 30 approach shown in Figure 4-16, would already know which minimums they were authorized to use. The company's OpSpecs, FOM, and the AFM for the pilot's aircraft would dictate the specific operational conditions and procedures by which this type of approach could be flown.

There are several items of note that are specific to this type of approach that should be considered and briefed. One is the terminal arrival area (TAA) that is displayed in the approach planview. TAAs, discussed later in this chapter, depict the boundaries of specific arrival areas, and the MIA for those areas. The TAAs should be included in an IAP briefing in the same manner as any other IFR transition altitude. It is also important to note that the altitudes listed in the TAAs should be referenced in place of the MSAs on the approach chart for use in emergency situations.

In addition to the obvious differences contained in the planview of Figure 4-16, RNAV (GPS) approach procedure example, pilots should be aware of the issues related to Baro- VNAV and RNP . The notes section of the procedure in the example contains restrictions relating to these topics.

RNP values for each individual leg of the procedure, defined by the procedure design criteria for containment purposes, are encoded into the aircraft's navigation database. Applicable landing minimums are shown in a normal manner along with the associated RNP value in the landing minimums section.

RNP required sensors, FMS capabilities, and relevant procedure notes are included in the Pilot Briefing Information procedure notes section. [Figure 4-15] RNP AR requirements are highlighted in large, bold print. RNP procedures are sequenced in the same manner as RNAV (GPS) procedures. Procedure title "RNAV" includes parenthetical "(RNP)" terminology. RF legs can be used in any segment of the procedure (transition, intermediate, final, or missed approach). RF leg turn directions (left or right) are not noted in the planview because the graphic depiction of the flight tracks is intuitive. Likewise, the arc center points, arc radius, and associated RF leg performance limits, such as bank angles and speeds are not depicted because these aircraft performance characteristics are encoded in the navigation database. RNP values for each individual leg of the procedure, defined by the procedure design criteria for containment purposes, are encoded into the aircraft's navigation database. Applicable landing minimums are shown in a normal manner along with the associated RNP value in the landing minimums section.

When more than one set of RNP landing minimums is available and an aircrew is able to achieve lower RNP through approved means, the available (multiple) sets of RNP minimums are listed with the lowest set shown first; remaining sets shown in ascending order, based on the RNP value. On this particular procedure, lateral and vertical course guidance from the DA to the Runway Waypoint (LTP) is provided by the aircraft's FMS and onboard navigation database; however, any continued flight below the DA to the landing threshold is to be conducted under VMC. [Figure 4-15]

Baro-VNAV

Baro-VNAV is an RNAV system function that uses barometric altitude information from the aircraft's altimeter to compute and present a vertical guidance path to the pilot. The specified vertical path is computed as a geometric path, typically computed between two waypoints or an angle based computation from a single waypoint. Operational approval must also be obtained for Baro–VNAV systems to operate to the LNAV/VNAV minimums. Baro–VNAV may not be authorized on some approaches

due to other factors, such as no local altimeter source being available. Baro–VNAV is not authorized on LPV procedures.

For the RNAV (GPS) RWY 30 approach, the note "DME/DME RNP-0.3 NA" prohibits aircraft that use only DME/DME sensors for RNAV from conducting the approach. [Figure 4-16]

Because these procedures can be flown with an approach approved RNP system and "RNP" is not sensor specific, it was necessary to add this note to make it clear that those aircraft deriving RNP 0.3 using DME/DME only are not authorized to conduct the procedure
.

The least accurate sensor authorized for RNP navigation is DME/DME. The necessary DME NAVAID ground infrastructure may or may not be available at the airport of intended landing. The procedure designer has a computer program for determining the usability of DME based on geometry and coverage. Where FAA flight inspection successfully determines that the coverage and accuracy of DME facilities support RNP, and that the DME signal meets inspection tolerances, although there are none currently published, the note "DME/DME RNP 0.3 Authorized" would be charted. Where DME facility availability is a factor, the

CATEGORY	A	B	C	D
LPV DA		558/24	250 (300-½)	
LNAV/VNAV DA		1572-5	1264 (1300-5)	
LNAV MDA	1180/24 872 (900-½)	1180/40 872 (900-¾)	1180-2 872 (900-2)	1180-2¼ 872 (900-2¼)
CIRCLING	1180-1 870 (900-1)	1180-1¼ 870 (900-1¼)	1180-2½ 870 (900-2½)	1180-2¾ 870 (900-2¾)

Figure 4-17. Example of LNAV and Circling Minima lower than LNAV/VNAV DA. Harrisburg International RNAV (GPS) Runway 13.

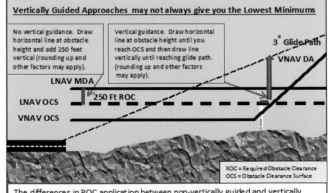

Vertically Guided Approaches may not always give you the Lowest Minimums

No vertical guidance. Draw horizontal line at obstacle height and add 250 feet vertical (rounding up and other factors may apply).

Vertical guidance. Draw horizontal line at obstacle height until you reach OCS and then draw line vertically until reaching glide path. (rounding up and other factors may apply).

3° Glide Path
VNAV DA
LNAV MDA
250 Ft ROC
LNAV OCS
VNAV OCS

ROC = Required Obstacle Clearance
OCS = Obstacle Clearance Surface

The differences in ROC application between non-vertically guided and vertically guided approach procedures may generate differences in minimum altitudes (MDA, DA) that seem illogical. Depending on the location and height of the obstacle, cases can exist where the straight-in (SI) vertically-guided DA must be higher than the SI non-vertically guided MDA.
Note: The vertically-guided DA may also be higher than the circling minima in cases where the circling MDA is not higher than the SI non-vertically guided MDA (typically where the circling maneuver is restricted).

Figure 4-18. Explanation of Minima.

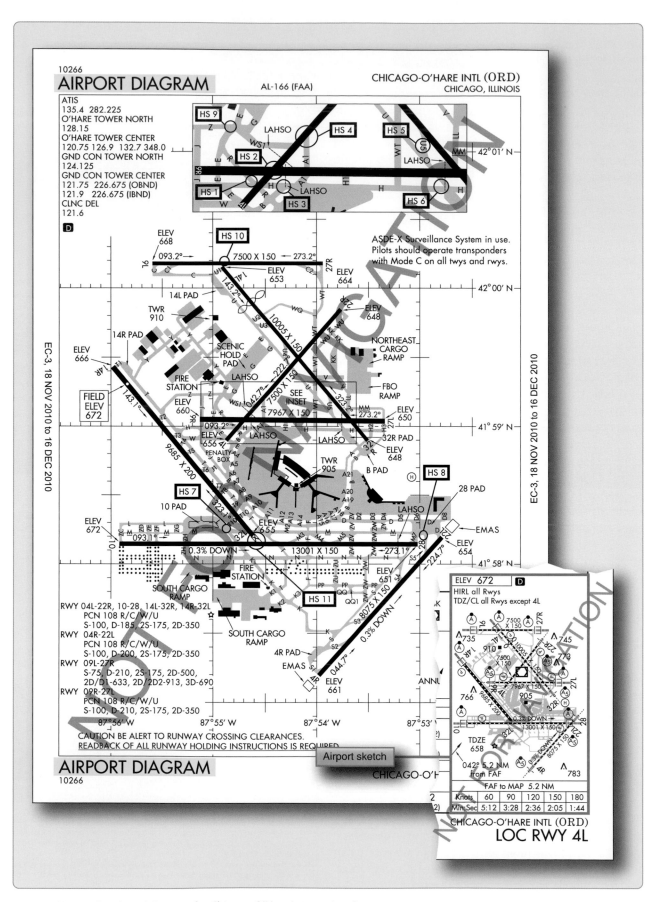

Figure 4-19. Airport sketch and diagram for Chicago O'Hare International.

note would read, "DME/DME RNP 0.3 Authorized; ABC and XYZ required," meaning that ABC and XYZ DME facilities are required to assure RNP 0.3.

Hot and Cold Temperature Limitations

A minimum and maximum temperature limitation is published on procedures that authorize Baro–VNAV operation. These temperatures represent the airport temperature above or below which Baro–VNAV is not authorized to LNAV/VNAV minimums unless temperature compensation can be accomplished. As an example, the limitation will read, uncompensated Baro–VNAV NA below –11 °C (12 °F) or above 49 °C (120 °F). See [Figure 4-15] This information will be found in the upper left hand box of the pilot briefing. When the temperature is above the high temperature or below the low temperature limit, Baro–VNAV may be used to provide a stabilized descent to the LNAV MDA; however, extra caution should be used in the visual segment to ensure a vertical correction is not required. If the VGSI is aligned with the published glide path, and the aircraft instruments indicate on glide path, an above or below glide path indication on the VGSI may indicate that temperature error is causing deviations to the glide path. These deviations should be considered if the approach is continued below the MDA.

Many systems which apply Baro–VNAV temperature compensation only correct for cold temperature. In this case, the high temperature limitation still applies. Also, temperature compensation may require activation by maintenance personnel during installation in order to be functional, even though the system has the feature. Some systems may have a temperature correction capability, but correct the Baro–altimeter all the time, rather than just on the final, which would create conflicts with other aircraft if the feature were activated. Pilots should be aware of compensation capabilities of the system prior to disregarding the temperature limitations. The information can be seen in the notes section in Figure 4-16.

In response to aviation industry concerns over cold weather altimetry errors, the FAA conducted a risk analysis to determine if current 14 CFR Part 97 instrument approach procedures, in the NAS place aircraft at risk during cold temperature operations. This study applied the coldest recorded temperature at the given airports in the last five years and specifically determined if there was a probability that during these non-standard day operations, anticipated altitude errors in a barometric altimetry system could exceed the Required Obstacle Clearance (ROC) used on procedure segment altitudes. If a probability of the ROC being exceeded went above one percent on a segment of the approach, a temperature restriction was applied to that segment. In addition to the low probability that these procedures will be required, the probability of the ROC being exceeded precisely at an obstacle position is extremely low, providing an even greater safety margin.

Pilots need to make an altitude correction to the published, "at", "at or above" and "at or below" altitudes on designated segment(s) of IAPs listed at specific airports, on all published procedures and runways, when the reported airport temperature is at or below the published airport cold temperature restriction.

This list may also be found at the bottom of the, "Terminal Procedures Basic Search" page found at: http://www.faa.gov/air_traffic/flight_info/aeronav/digital_products/dtpp/search/

Pilots without temperature compensating aircraft are responsible to calculate and make a manual cold-temperature altitude correction to the designated segment(s) of the approach using the AIM 7-2-3, ICAO Cold Temperature Error Table.

No extrapolation above the 5000 ft column required. Pilots should use the 5000 feet "height above airport in feet" column for calculating corrections of greater than 5000 feet above reporting station. Pilots will add correction(s) from the table to the segment altitude(s) and fly at the new corrected altitude. PILOTS SHOULD NOT MAKE AN ALTIMETER CHANGE to accomplish an altitude correction.

Pilots with temperature compensating aircraft must ensure the system is on and operating for each segment requiring an altitude correction. Pilots must ensure they are flying at corrected altitude. If the system is not operating, the pilot is responsible to calculate and apply a manual cold weather altitude correction using the AIM 7-2-3 ICAO Cold Temperature Error Table.

Pilots must report cold temperature corrected altitudes to Air Traffic Control (ATC) whenever applying a cold temperature correction on an intermediate segment and/or a published missed approach final altitude. This should be done on initial radio contact with the ATC issuing approach clearance. ATC requires this information in order to ensure appropriate vertical separation between known traffic. ATC will not be providing a cold temperature correction to Minimum Vectoring Altitudes (MVA). Pilots must not apply cold temperature compensation to ATC assigned altitudes or when flying on radar vectors in lieu of a published missed approach procedure unless cleared by ATC.

Pilots should query ATC when vectors to an intermediate segment are lower than the requested intermediate segment altitude corrected for temperature. Pilots are encouraged to self-announce corrected altitude when flying into uncontrolled airfields.

The following are examples of appropriate pilot-to-ATC communication when applying cold-temperature altitude corrections:

On initial check-in with ATC providing approach clearance: Hayden, CO (example below).

Intermediate segment: "Require 10600 ft. for cold temperature operations until BEEAR",

Missed Approach segment: "Require final holding altitude, 10600 ft. on missed approach for cold temperature operations"

Pilots cleared by ATC for an instrument approach procedure; "Cleared the RNAV RWY 28 approach (from any IAF)". Hayden, CO (example below).

Intermediate Segment: "Level 10600 ft. for cold temperature operations inside HIPNA to BEEAR"

Pilots are not required to advise ATC if correcting on the final segment only. Pilots must use the corrected MDA or DA/DH as the minimum for an approach. Pilots must meet the requirements in 14 CFR Part 91.175 in order to operate below the corrected MDA or DA/DH. Pilots must see and avoid obstacles when descending below the MDA.

The temperature restriction at a "Cold Temperature Restricted Airport" is mutually exclusive from the charted temperature restriction published for "uncompensated baro-VNAV systems" on 14 CFR Part 97 RNAV (GPS) and RNAV (RNP) approach plates. The charted temperature restriction for uncompensated baro-VNAV systems is applicable to the final segment LNAV/VNAV minima. The charted temperature restriction must be followed regardless of the cold temperature restricted airport temperature.

Pilots are not required to calculate a cold temperature altitude correction at any airport with a runway length of 2,500 feet or greater that is not included in the airports list found at the URL above. Pilots operating into an airport with a runway length less than 2,500 feet, may make a cold temperature altitude correction in cold temperature conditions.

Cold Temperature Restricted Airports: These airports are listed in the FAA Notices To Airmen Publication (NTAP) found here: https://www.faa.gov/air_traffic/publications/notices/.

Airports are listed by ICAO code, Airport Name, Temperature Restriction in Celsius/Fahrenheit and affected Segment. One temperature may apply to multiple segments. Italicized airports have two affected segments, each with a different temperature restrictions. The warmest temperature will be indicated on Airport IAPs next to a snowflake symbol, ❄-35°C in the United States Terminal Procedure Publication. The ICON will be added to the TPPs incrementally each charting cycle.

LNAV, LNAV/VNAV and Circling Minimums

There are some RNAV procedures with lower non-precision LNAV minimums [Figure 4-17] than vertically-guided LNAV/VNAV minimums. Circling procedures found on the same approach plate may also have lower minimums than the vertically-guided LNAV/VNAV procedure. Each RNAV procedure is evaluated independently and different approach segments have differing required obstacle clearance (ROC) values, obstacle evaluation area (OEA) dimensions and final segment types. Figure 4-18 explains the differences.

Airport/Runway Information

Another important piece of a thorough approach briefing is the discussion of the airport and runway environment. A detailed examination of the runway length (this must include the A/FD for the landing distance available), the intended turnoff taxiway, and the route of taxi to the parking area, are all important briefing items. In addition, runway conditions should be discussed. The effect on the aircraft's performance must be considered if the runway is contaminated.

FAA approach charts include a runway sketch on each approach chart to make important airport information easily accessible to pilots. In addition, at airports that have complex runway/taxiway configurations, a separate full-page airport diagram is published.

The airport diagram also includes the latitude/longitude information required for initial programming of FMS equipment. The included latitude/longitude grid shows the specific location of each parking area on the airport surface for use in initializing FMS. Figure 4-19 shows the airport sketch and diagram for Chicago-O'Hare International Airport (KORD).

Pilots making approaches to airports that have this type of complex runway and taxiway configuration must ensure that they are familiar with the airport diagram prior to initiating an instrument approach. A combination of poor weather, high traffic volume, and high ground controller workload makes the pilot's job on the ground every bit as critical as the one just performed in the air.

Instrument Approach Procedure (IAP) Briefing

A thorough instrument approach briefing greatly increases the likelihood of a successful instrument approach. Most Part 121, 125, and 135 operators designate specific items to be included in an IAP briefing, as well as the order in which those items are briefed.

Before an IAP briefing can begin, flight crews must decide which procedure is most likely to be flown from the information that is available to them. Most often, when the flight is being conducted into an airport that has ATIS information, the ATIS provides the pilots with the approaches that are in use. If more than one approach is in use, the flight crew may have to make an educated guess as to which approach will be issued to them based on the weather, direction of their arrival into the area, any published airport NOTAMs, and previous contact with the approach control facility. Aircrews can query ATC as to which approach is to be expected from the controller. Pilots may request specific approaches to meet the individual needs of their equipment or regulatory restrictions at any time and ATC will, in most cases, be able to accommodate those requests, providing that workload and traffic permit.

If the flight is operating into an airport without a control tower, the flight crew is occasionally given the choice of any available instrument approach at the field. In these cases, the flight crew must choose an appropriate approach based on the expected weather, aircraft performance, direction of arrival, airport NOTAMs, and previous experience at the airport.

Navigation and Communication Radios

Once the anticipated approach and runway have been selected, each crewmember sets up their side of the flight deck. The pilots use information gathered from ATIS, dispatch (if available), ATC, the specific approach chart for the approach selected, and any other sources that are available. Company regulations dictate how certain things are set up and others are left up to pilot technique. In general, the techniques used at most companies are similar. This section addresses two-pilot operations. During single-pilot IFR flights, the same items must be set up and the pilot should still do an approach briefing to verify that everything is set up correctly.

The number of items that can be set up ahead of time depends on the level of automation of the aircraft and the avionics available. In a conventional flight deck, the only things that can be set up, in general, are the airspeed bugs (based on performance calculations), altimeter bug (to DA, DH, or MDA), go around thrust/power setting, the radio altimeter bug (if installed and needed for the approach), and the navigation/communication radios (if a standby frequency selector is available). The standby side of the PF navigation radio should be set to the primary NAVAID for the approach and the PM navigation radio standby selector should be set to any other NAVAIDs that are required or available, and as dictated by company procedures, to add to the overall situational awareness of the crew. The ADF should also be tuned to an appropriate frequency as required by the approach, or as selected by the crew. Aircrews should, as much as possible, set up the instruments for best success in the event of a vacuum or electrical failure. For example, if the aircraft will only display Nav 1 on battery or emergency power, aircrews should ensure that Nav 1 is configured to the primary NAVAID for the final approach to be flown.

Flight Management System (FMS)

In addition to the items that are available on a conventional flight deck aircraft, glass flight deck aircraft, as well as aircraft with an approved RNAV (GPS) system, usually give the crew the ability to set the final approach course for the approach selected and many other options to increase situational awareness. Crews of FMS equipped aircraft have many options available as far as setting up the flight management computer (FMC), depending on the type of approach and company procedures. The PF usually programs the FMC for the approach and the PM verifies the information. A menu of available approaches is usually available to select from based on the destination airport programmed at the beginning of the flight or a new destination selected while en route.

The amount of information provided for the approach varies from aircraft to aircraft, but the crew can make modifications if something is not pre-programmed into the computer, such as adding a MAP or even building an entire approach for situational awareness purposes only. The PF can also program a VNAV profile for the descent and LNAV for segments that were not programmed during preflight, such as a standard terminal arrival route (STAR) or expected route to the planned approach. Any crossing restrictions for the STAR might need to be programmed as well. The most common crossing restrictions, whether mandatory or "to be expected," are usually automatically programmed when the STAR is selected, but can be changed by ATC at any time. Other items that need to be set up are dictated by aircraft-specific procedures, such as autopilot, auto-throttles, auto-brakes, pressurization system, fuel system, seat belt signs, anti-icing/ deicing equipment, and igniters.

Autopilot Modes

In general, an autopilot can be used to fly approaches even if the FMC is inoperative (refer to the specific airplane's minimum equipment list (MEL) to determine authorization for operating with the FMC inoperative). Whether or not the FMC is available, use of the autopilot should be discussed during the approach briefing, especially regarding the use of the altitude pre-selector and auto-throttles, if equipped. The AFM for the specific airplane outlines procedures and limitations required for the use of the autopilot during an instrument approach in that aircraft.

There are just as many different autopilot modes to climb or descend the airplane, as there are terms for these modes. Some examples are level change (LVL CHG), vertical speed (V/S), VNAV, and takeoff/go around (TO/GA). The pilot controls the airplane through the autopilot by selecting pitch modes and/or roll modes, as well as the associated auto-throttle modes. This panel, sometimes called a mode control panel, is normally accessible to both pilots. Most aircraft with sophisticated auto-flight systems and auto-throttles have the capability to select modes that climb the airplane with maximum climb thrust and descend the airplane with the throttles at idle (LVL CHG, flight level change (FL CHG), and manage level). They also have the capability to capture, or level off at pre-selected altitudes, as well as track a LOC and glideslope (G/S) or a VOR course. If the airplane is RNAV-equipped, the autopilot also tracks the RNAV- generated course. Most of these modes are used at some point during an instrument approach using the autopilot. Additionally, these modes can be used to provide flight director (FD) guidance to the pilot while hand-flying the aircraft.

For the purposes of this precision approach example, the auto-throttles are engaged when the autopilot is engaged and specific airspeed and configuration changes are not discussed. The PF controls airspeed with the speed selector on the mode control panel and calls for flaps and landing gear as needed, which the PM selects. The example in Figure 4-20 begins with the airplane 5 NM northwest of KNUCK at 4,500 feet with the autopilot engaged, and the flight has been cleared to track the Rwy 12 LOC inbound. The current roll mode is LOC with the PF's NAV radio tuned to the LOC frequency of 109.3; and the current pitch mode is altitude hold (ALT HOLD). Approach control clears the airplane for the approach. The PF makes no immediate change to the autopilot mode to prevent the aircraft from capturing a false glideslope; but the PM resets the altitude selector to 1,700 feet. The aircraft remains level because the pitch mode remains in ALT HOLD until another pitch mode is selected. Upon reaching KNUCK, the PF selects LVL CHG as the pitch mode. The auto-throttles retard to idle as the airplane begins a descent. Approaching 1,700 feet, the pitch mode automatically changes to altitude acquire (ALT ACQ) then to ALT HOLD as the airplane levels at 1,700 feet. In addition to slowing the airplane and calling for configuration changes, the PF selects approach mode (APP). The roll mode continues to track the LOC and the pitch mode remains in ALT HOLD; however, the G/S mode

arms. Selecting APP once the aircraft has leveled at the FAF altitude is a suggested technique to ensure that the airplane captures the glideslope from below and that a false glideslope is not being tracked.

The PF should have the aircraft fully configured for landing before intercepting the glideslope to ensure a stabilized approach. As the airplane intercepts the glideslope the pitch mode changes to G/S. Once the glideslope is captured by the autopilot, the PM can select the missed approach altitude in the altitude pre-selector, as requested by the PF. The airplane continues to track the glideslope. The minimum altitude at which the PF is authorized to disconnect the autopilot is airplane specific. For example, 50 feet below DA, DH, or MDA but not less than 50 feet AGL. The PF can disconnect the autopilot at any time prior to reaching this altitude during a CAT I approach. The initial missed approach is normally hand flown with FD guidance unless both autopilots are engaged for auto-land during a CAT II or III approach.

The differences when flying the underlying non-precision approach begin when the aircraft has leveled off at 1,700 feet. Once ALT HOLD is annunciated, the MDA is selected by the PM as requested by the PF. It is extremely important for both pilots to be absolutely sure that the correct altitude is selected for the MDA so that the airplane does not inadvertently descend below the MDA. For aircraft that the altitude pre-selector can only select 100 foot increments, the MDA for this approach must be set at 700 feet instead of 660 feet.

Vertical speed mode is used from the FAF inbound to allow for more precise control of the descent. If the pilots had not selected the MDA in the altitude pre-selector window, the PF would not be able to input a V/S and the airplane would remain level. The autopilot mode changes from ALT ACQ to ALT HOLD as the airplane levels at 700 feet. Once ALT HOLD is annunciated, the PF calls for the missed approach altitude of 5,000 feet to be selected in the altitude pre-selector window. This step is very important because accurate FD guidance is not available to the PF during a missed approach if the MDA is left in the window.

NOTE: See "Maximum Acceptable Descent Rates" under the heading "Descent Rates and Glide paths for Non-precision Approaches."

Descents

Stabilized Approach

In IMC, you must continuously evaluate instrument information throughout an approach to properly maneuver the aircraft or monitor autopilot performance and to decide on the proper course of action at the decision point (DA, DH, or MAP). Significant speed and configuration changes during an approach can seriously degrade situational awareness and complicate the decision of the proper action to take at the decision point. The swept wing handling characteristics at low airspeeds and slow engine response of many turbojets further complicate pilot tasks during approach and landing operations. You must begin to form a decision concerning the probable success of an approach before reaching the decision point. Your decision-making process requires you to be able to determine displacements from the course or glideslope/glidepath centerline, to mentally project the aircraft's three-dimensional flight path by referring to flight instruments, and then apply control inputs as necessary to achieve and maintain the desired approach path. This process is simplified by maintaining a constant approach speed, descent rate, vertical flight path, and configuration during the final stages of an approach. This is referred to as the stabilized approach concept.

A stabilized approach is essential for safe turbojet operations and commercial turbojet operators must establish and use procedures that result in stabilized approaches. A stabilized approach is also strongly recommended for propeller-driven airplanes and helicopters. You should limit configuration changes at low altitudes to those changes that can be easily accommodated without adversely affecting your workload. For turbojets, the airplane must be in an approved configuration for landing or circling, if appropriate, with the engines spooled up, and on the correct speed and flight path with a descent rate of less than 1,000 feet per minute (fpm) before descending below the following minimum stabilized approach heights:

- For all straight-in instrument approaches, to include contact approaches in IFR weather conditions, the approach must be stabilized before descending below 1,000 feet above the airport or TDZE.

- For visual approaches and straight-in instrument approaches in VFR weather conditions, the approach must be stabilized before descending below 500 feet above the airport elevation.

- For the final segment of a circling approach maneuver, the approach must be stabilized 500 feet above the airport elevation or at the MDA, whichever is lower. These conditions must be maintained throughout the approach until touchdown for the approach to be considered a stabilized approach. This also helps you to recognize a wind shear situation should abnormal indications exist during the approach.

Descent Rates and Glidepaths for Nonprecision Approaches

Maximum Acceptable Descent Rates

Operational experience and research have shown that a descent rate of greater than approximately 1,000 fpm is unacceptable during the final stages of an approach (below 1,000 feet AGL). This is due to a human perceptual limitation that is independent of the type of airplane or helicopter. Therefore, the operational practices and techniques must ensure that descent rates greater than 1,000 fpm are not permitted in either the instrument or visual portions of an approach and landing operation.

For short runways, arriving at the MDA at the MAP when the MAP is located at the threshold may require a missed approach for some airplanes. For non-precision approaches, a descent rate should be used that ensures the airplane reaches the MDA at a distance from the threshold that allows landing in the TDZ. On many IAPs, this distance is annotated by a VDP. To determine the required rate of descent, subtract the TDZE from the FAF altitude and divide this by the time inbound. For example, if the FAF altitude is 2,000 feet MSL, the TDZE is 400 feet MSL and the time inbound is 2 minutes, an 800 fpm rate of descent should be used.

To verify the airplane is on an approximate three degree glidepath, use a calculation of 300 feet to 1 NM. The glidepath height above TDZE is calculated by multiplying the NM distance from the threshold by 300. For example, at 10 NM the aircraft should be 3,000 feet above the TDZE, at 5 NM the aircraft should be 1,500 feet above the TDZE, at 2 NM the aircraft should be 600 feet above the TDZE, and at 1.5 NM the aircraft should be 450 feet above the TDZE until a safe landing can be made. Using the example in the previous text, the aircraft should arrive at the MDA (800 feet MSL) approximately 1.3 NM from the threshold and in a position to land within the TDZ. Techniques for deriving a 300-to-1 glide path include using DME, distance advisories provided by radar-equipped control towers, RNAV, GPS, dead reckoning, and pilotage when familiar features on the approach course are visible. The runway threshold should be crossed at a nominal height of 50 feet above the TDZE.

Transition to a Visual Approach

The transition from instrument flight to visual flight during an instrument approach can be very challenging, especially during low visibility operations. Aircrews should use caution when transitioning to a visual approach at times of shallow fog. Adequate visibility may not exist to allow

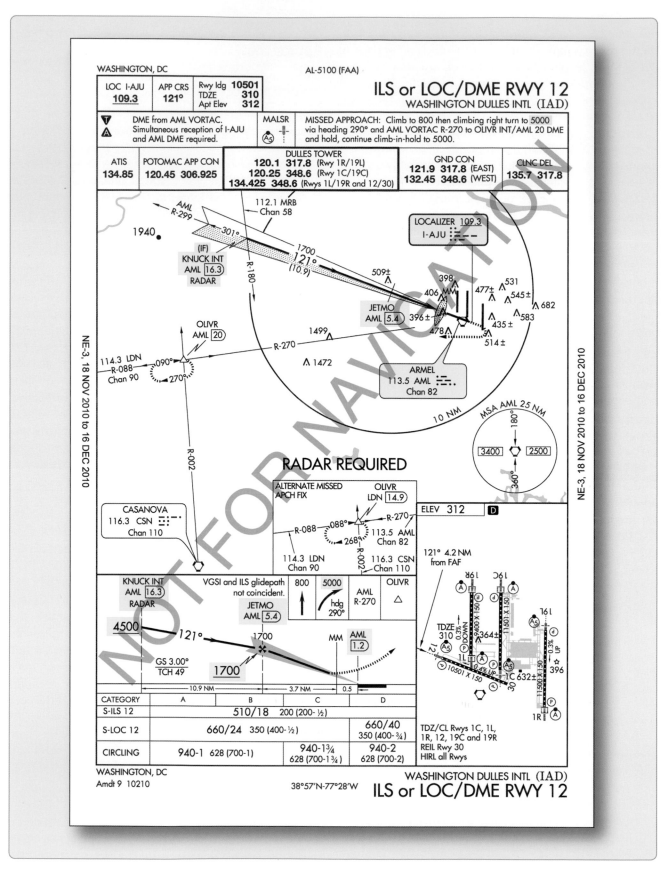

Figure 4-20. Example approaches using autopilot.

flaring of the aircraft. Aircrews must always be prepared to execute a missed approach/go-around. Additionally, single-pilot operations make the transition even more challenging. Approaches with vertical guidance add to the safety of the transition to visual because the approach is already stabilized upon visually acquiring the required references for the runway. 100 to 200 feet prior to reaching the DA, DH, or MDA, most of the PM's attention should be outside of the aircraft in order to visually acquire at least one visual reference for the runway, as required by the regulations. The PF should stay focused on the instruments until the PM calls out any visual aids that can be seen, or states "runway in sight." The PF should then begin the transition to visual flight. It is common practice for the PM to call out the V/S during the transition to confirm to the PF that the instruments are being monitored, thus allowing more of the PF's attention to be focused on the visual portion of the approach and landing. Any deviations from the stabilized approach criteria should also be announced by the PM.

Single-pilot operations can be much more challenging because the pilot must continue to fly by the instruments while attempting to acquire a visual reference for the runway. While it is important for both pilots of a two-pilot aircraft to divide their attention between the instruments and visual references, it is even more critical for the single- pilot operation. The flight visibility must also be at least the visibility minimum stated on the instrument approach chart, or as required by regulations. CAT II and III approaches have specific requirements that may differ from CAT I precision or non-precision approach requirements regarding transition to visual and landing. This information can be found in the operator's OpSpecs or FOM.

The visibility published on an approach chart is dependent on many variables, including the height above touchdown for straight-in approaches or height above airport elevation for circling approaches. Other factors include the approach light system coverage, and type of approach procedure, such as precision, non-precision, circling or straight-in. Another factor determining the minimum visibility is the penetration of the 34:1 and 20:1 surfaces. These surfaces are inclined planes that begin 200 feet out from the runway and extend outward to the DA point (for approaches with vertical guidance), the VDP location (for non-precision approaches) and 10,000 feet for an evaluation to a circling runway. If there is a penetration of the 34:1 surface, the published visibility can be no lower than three-fourths SM. If there is penetration of the 20:1 surface, the published visibility can be no lower than 1 SM with a note prohibiting approaches to the affected runway at night (both straight-in and circling). [Figure 4-21] Circling may be permitted at night if penetrating obstacles are marked and lighted.

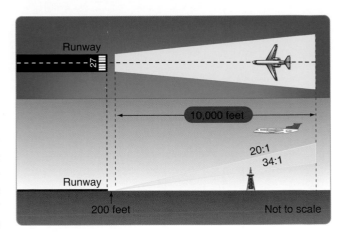

Figure 4-21. Determination of visibility minimums.

If the penetrating obstacles are not marked and lighted, a note is published that night circling is "Not Authorized." Pilots should be aware of these penetrating obstacles when entering the visual and/or circling segments of an approach and take adequate precautions to avoid them. For RNAV approaches only, the presence of a grey shaded line from the MDA to the runway symbol in the profile view is an indication that the visual segment below the MDA is clear of obstructions on the 34:1 slope. Absence of the gray shaded area indicates the 34:1 OCS is not free of obstructions. [Figure 4-22]

Missed Approach

Many reasons exist for executing a missed approach. The primary reasons, of course, are that the required flight visibility prescribed in the IAP being used does not exist when natural vision is used under 14 CFR Part 91, section 91.175(c), the required enhanced flight visibility is less than that prescribed in the IAP when an EFVS is used under 14 CFR Part 91, section 91.175 (l), or the required visual references for the runway cannot be seen upon arrival at the DA, DH, or MAP. In addition, according to 14 CFR Part 91, the aircraft must continuously be in a position from which a descent to a landing on the intended runway can be made at a normal rate of descent using normal maneuvers, and for operations conducted under Part 121 or 135, unless that descent rate allows touchdown to occur within the TDZ of the runway of intended landing. [Figure 4-23] CAT II and III approaches call for different visibility requirements as prescribed by the FAA Administrator.

Prior to initiating an instrument approach procedure, the pilot should assess the actions to be taken in the event of a balked (rejected) landing beyond the missed approach point or below the MDA or DA (H) considering the anticipated weather conditions and available aircraft performance. 14 CFR 91.175(e) authorizes the pilot to fly an appropriate missed approach procedure that ensures obstruction clearance, but it does not necessarily

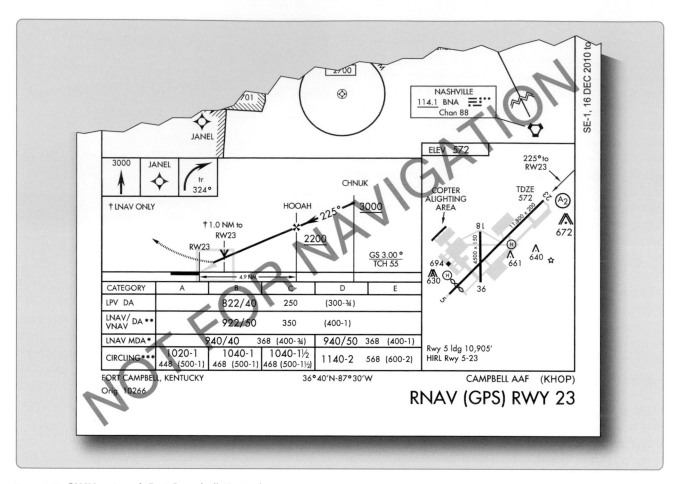

Figure 4-22. RNAV approach Fort Campbell, Kentucky.

consider separation from other air traffic. The pilot must consider other factors such as the aircraft's geographical location with respect to the prescribed missed approach point, direction of flight, and/ or the minimum turning altitudes in the prescribed missed approach procedure. The pilot must also consider aircraft performance, visual climb restrictions, charted obstacles, published obstacle departure procedure, takeoff visual climb requirements as expressed by nonstandard takeoff minima, other traffic expected to be in the vicinity, or other factors not specifically expressed by the approach procedures.

A clearance for an instrument approach procedure includes a clearance to fly the published missed approach procedure, unless otherwise instructed by ATC. Once descent below the DA, DH, or MDA is begun, a missed approach must be executed if the required visibility is lost or the runway environment is no longer visible, unless the loss of sight of the runway is a result of normal banking of the aircraft during a circling approach. A MAP is also required upon the execution of a rejected landing for any reason, such as men and equipment or animals on the runway, or if the approach becomes unstabilized and a normal landing cannot be performed. After the MAP in the visual segment

of a non-precision approach, there may be hazards when executing a missed approach below the MDA. The published missed approach procedure provides obstacle clearance only when the missed approach is conducted on the missed approach segment from or above the missed approach point, and assumes a climb rate of 200 FPNM or higher, as published. If the aircraft initiates a missed approach at a point other than the missed approach point, from below MDA or DA (H), or on a circling approach, obstacle clearance is not provided by following the published missed approach procedure, nor is separation assured from other air traffic in the vicinity.

The missed approach climb is normally executed at the MAP. If such a climb is initiated at a higher altitude prior to the MAP, pilots must be aware of any published climb-altitude limitations, which must be accounted for when commencing an early climb. Figure 4-24 gives an example of an altitude restriction that would prevent a climb between the FAF and MAP. In this situation, the Orlando Executive ILS or LOC RWY 7 approach altitude is restricted at the BUVAY 3 DME fix to prevent aircraft from penetrating the overlying protected airspace for approach routes into Orlando International Airport. If a missed approach is initiated before reaching BUVAY, a pilot may be required to continue descent to 1,200 feet before proceeding to the MAP and executing the missed approach climb instructions. In addition to the missed approach notes on the chart, the Pilot Briefing Information icons in the profile view indicate the initial vertical and lateral missed approach guidance.

The missed approach course begins at the MAP and continues until the aircraft has reached the designated fix and a holding pattern has been entered. [Figure 4-25] In these circumstances, ATC normally issues further instructions before the aircraft reaches the final fix of the missed approach course. It is also common for the designated fix to be an IAF so that another approach attempt can be made without having to fly from the holding fix to an IAF.

In the event a balked (rejected) landing occurs at a position other than the published missed approach point, the pilot should contact ATC as soon as possible to obtain an amended clearance. If unable to contact ATC for any reason, the pilot should attempt to re–intercept a published segment of the missed approach and comply with route and altitude instructions. If unable to contact ATC, and in the pilot's judgment it is no longer appropriate to fly the published missed approach procedure, then consider either maintaining visual conditions (if possible) and reattempt a landing, or a circle–climb over the airport. Should a missed approach become necessary when operating to an airport that is not served by an operating control tower, continuous contact with an air traffic facility may not be possible. In this case, the pilot should execute the appropriate go–around/missed approach procedure without delay and contact ATC when able to do so.

As shown in Figure 4-26 , there are many different ways that the MAP can be depicted, depending on the type of approach. On all approach charts, it is depicted in the profile and plan views by the end of the solid course line and the beginning of the dotted missed approach course line for the top-line/ lowest published minima. For a precision approach, the MAP is the point at which the aircraft reaches the DA or DH while on the glideslope/ glidepath. MAPs on non-precision approaches can be determined in many different ways. If the primary NAVAID is on the airport, and either a VOR or NDB approach is being executed, the MAP is normally the point at which the aircraft passes the NAVAID.

On some non-precision approaches, the MAP is given as a fixed distance with an associated time from the FAF to the MAP based on the groundspeed of the aircraft. A table on the lower right or left hand side of the approach chart shows the distance in NM from the FAF to the MAP and the time it takes at specific groundspeeds, given in 30 knot increments. Pilots must determine the approximate groundspeed and time based on the approach speed and true airspeed of their aircraft and the current winds along the final approach course. A clock or stopwatch should be started at the FAF of an approach requiring this method. Many non-precision approaches designate a specific fix as the MAP. These can be identified by a course (LOC or VOR) and DME, a cross radial from a VOR, or an RNAV (GPS) waypoint.

Obstacles or terrain in the missed approach segment may require a steeper climb gradient than the standard 200 FPNM. If a steeper climb gradient is required, a note is published on the approach chart plan view with the penetration description and examples of the required FPM rate of climb for a given groundspeed (future charting uses climb gradient). An alternative is normally charted that allows using the standard climb gradient. [Figure 4-26] In this example, if the missed approach climb requirements cannot be met for the Burbank ILS RWY 8 chart, the alternative is to use the LOC RWY 8 that is charted separately. The LOC RWY 8, S-8 procedure has a MDA that is 400 feet higher than the ILS RWY 8, S-LOC 8 MDA and meets the standard climb gradient requirement over the terrain. For some approaches a new charting standard is requiring two sets of minimums to be published when a climb gradient greater than 200 FPNM is required. The first set of minimums is the lower of the two, requiring a climb gradient greater than 200 FPNM. The second set of minimums is higher, but doesn't require a climb gradient. Shown in Figure 4-27, Barstow-Daggett (KDAG) RNAV (GPS) RWY 26 is an example where there are two LPV lines of minimums.

Example Approach Briefing

During an instrument approach briefing, the name of the airport and the specific approach procedure should be identified to allow other crewmembers the opportunity to cross-reference the chart being used for the brief. This ensures that pilots intending to conduct an instrument approach have collectively reviewed and verified the information pertinent to the approach. Figure 4-28 gives an example of the items to be briefed and their sequence. Although the following example is based on multi-crew aircraft, the process is also applicable to single-pilot operations. A complete instrument approach and operational briefing example follows.

The approach briefing begins with a general discussion of the ATIS information, weather, terrain, NOTAMs, approaches

91.175 TAKEOFF AND LANDING UNDER IFR
- (c) Operation below DA/ DH or MDA. Except as provided in paragraph (l) of this section, where a DA/DH or MDA is applicable, no pilot may operate an aircraft, except a military aircraft of the United States, below the authorized MDA or continue an approach below the authorized DA/DH unless—
  - (1) The aircraft is continuously in a position from which a descent to a landing on the intended runway can be made at a normal rate of descent using normal maneuvers, and for operations conducted under part 121 or part 135 unless that descent rate will allow touchdown to occur within the touchdown zone of the runway of intended landing;
  - (2) The flight visibility is not less than the visibility prescribed in the standard instrument approach being used; and
  - (3) Except for a Category II or Category III approach where any necessary visual reference requirements are specified by the Administrator, at least one of the following visual references for the intended runway is distinctly visible and identifiable to the pilot:
    - (i) The approach light system, except that the pilot may not descend below 100 feet above the touchdown zone elevation using the approach lights as a reference unless the red terminating bars or the red side row bars are also distinctly visible and identifiable.
    - (ii) The threshold.
    - (iii) The threshold markings.
    - (iv) The threshold lights.
    - (v) The runway end identifier lights.
    - (vi) The visual approach slope indicator.
    - (vii) The touchdown zone or touchdown zone markings.
    - (viii) The touchdown zone lights.
    - (ix) The runway or runway markings.
    - (x) The runway lights.
- (l) Approach to straight-in landing operations below DH, or MDA using an enhanced flight vision system (EFVS).
  For straight-in instrument approach procedures other than Category II or Category III, no pilot operating under this section or §§121.651, 125.381, and 135.225 of this chapter may operate an aircraft at any airport below the authorized MDA or continue an approach below the authorized DH and land unless—
  - (1) The aircraft is continuously in a position from which a descent to a landing on the intended runway can be made at a normal rate of descent using normal maneuvers, and, for operations conducted under part 121 or part 135 of this chapter, the descent rate will allow touchdown to occur within the touchdown zone of the runway of intended landing;
  - (2) The pilot determines that the enhanced flight visibility observed by use of a certified enhanced flight vision system is not less than the visibility prescribed in the standard instrument approach procedure being used;
  - (3) The following visual references for the intended runway are distinctly visible and identifiable to the pilot using the enhanced flight vision system:
    - (i) The approach light system (if installed); or
    - (ii) The following visual references in both paragraphs (l)(3)(ii)(A) and (B) of this section:
      - (A) The runway threshold, identified by at least one of the following:
        - (1) The beginning of the runway landing surface;
        - (2) The threshold lights; or
        - (3) The runway end identifier lights.
      - (B) The touchdown zone, identified by at least one of the following:
        - (1) The runway touchdown zone landing surface;
        - (2) The touchdown zone lights;
        - (3) The touchdown zone markings; or
        - (4) The runway lights.
  - (4) At 100 feet above the touchdown zone elevation of the runway of intended landing and below that altitude, the flight visibility must be sufficient for the following to be distinctly visible and identifiable to the pilot without reliance on the enhanced flight vision system to continue to a landing:
    - (i) The lights or markings of the threshold; or
    - (ii) The lights or markings of the touchdown zone;
  - (5) The pilot(s) is qualified to use an EFVS as follows—
    - (i) For parts 119 and 125 certificate holders, the applicable training, testing and qualification provisions of parts 121, 125, and 135 of this chapter;
    - (ii) For foreign persons, in accordance with the requirements of the civil aviation authority of the State of the operator; or
    - (iii) For persons conducting any other operation, in accordance with the applicable currency and proficiency requirements of part 61 of this chapter;

Figure 4-23. Takeoff and landing under IFR.

in use, runway conditions, performance considerations, expected route to the final approach course, and the traffic situation. As the discussion progresses, the items and format of the briefing become more specific. The briefing can also be used as a checklist to ensure that all items have been set up correctly. Most pilots verbally brief the specific MAP so that it is fresh in their minds and there is no confusion as to who is doing what during a missed approach. Also, it is a very good idea to brief the published missed approach even if the tower is most likely to give you alternate instructions in the event of a missed approach. A typical approach briefing might sound like the following example for a flight inbound to the Monroe Regional Airport (KMLU):

ATIS: "Monroe Regional Airport Information Bravo, time 2253 Zulu, wind 360 at 10, visibility 1 mile, mist, ceiling 300 overcast, temperature 4, dew point 3, altimeter 29.73, ILS Runway 4 approach in use, landing and departing Runway 4, advise on initial contact that you have information Bravo."

PF: "We're planning an ILS approach to Runway 4 at Monroe Regional Airport, page 270, effective date 22 Sep 11 to 20 Oct 11. Localizer frequency is 109.5, SABAR Locator Outer Marker is 392, Monroe VOR is 117.2, final approach course is 042°. We'll cross SABAR at 1,483 feet barometric, decision altitude is 278 feet barometric, touchdown zone elevation is 78 feet with an airport elevation of 79 feet. MAP is climb to 2,000 feet, then climbing right turn to 3,000 feet direct Monroe VOR and hold. The MSA is 2,200 feet to the north and along our missed approach course, and 3,100 feet to the south along the final approach course. ADF or DME is required for the approach and the airport has pilot controlled lighting when the tower is closed, which does not apply to this approach. The runway has a medium intensity approach lighting system with runway alignment indicator lights and a precision approach path indicator (PAPI). We need a half- mile visibility so with one mile we should be fine. Runway length is 7,507 feet. I'm planning a flaps 30 approach, auto- brakes 2, left turn on Alpha or Charlie 1 then Alpha, Golf to the ramp. With a left crosswind, the runway should be slightly to the right. I'll use the autopilot until we break out and, after landing, I'll slow the aircraft straight ahead until you say you have control and I'll contact ground once we are clear of the runway. In the case of a missed approach, I'll press TOGA (Take-off/Go-Around button used on some turbojets), call 'go-around thrust, flaps 15, positive climb, gear up, set me up,' climb straight ahead to 2,000 feet then climbing right turn to 3,000 feet toward Monroe or we'll follow the tower's instructions. Any questions?"

PM: "I'll back up the auto-speedbrakes. Other than that, I don't have any questions."

Instrument Approach Procedure Segments

An instrument approach may be divided into as many as four approach segments: initial, intermediate, final, and missed approach. Additionally, feeder routes provide a transition from the en route structure to the IAF. FAA Order 8260.3 (TERPS) criteria provides obstacle clearance for each segment of an approach procedure as shown in Figure 4-29.

Feeder Routes

By definition, a feeder route is a route depicted on IAP charts to designate routes for aircraft to proceed from the en route structure to the IAF. [Figure 4-30] Feeder routes, also referred to as approach transitions, technically are not considered approach segments but are an integral part of many IAPs. Although an approach procedure may have several feeder routes, pilots normally choose the one closest to the en route arrival point. When the IAF is part of the en route structure, there may be no need to designate additional routes for aircraft to proceed to the IAF.

When a feeder route is designated, the chart provides the course or bearing to be flown, the distance, and the minimum altitude. En route airway obstacle clearance criteria apply to feeder routes, providing 1,000 feet of obstacle clearance (2,000 feet in mountainous areas).

Terminal Routes

In cases where the IAF is part of the en route structure and feeder routes are not required, a transition or terminal route is still needed for aircraft to proceed from the IAF to the intermediate fix (IF). These routes are initial approach segments because they begin at the IAF. Like feeder routes, they are depicted with course, minimum altitude, and distance to the IF. Essentially, these routes accomplish the same thing as feeder routes but they originate at an IAF, whereas feeder routes terminate at an IAF. [Figure 4-31]

DME Arcs

DME arcs also provide transitions to the approach course, but DME arcs are actually approach segments while feeder routes, by definition, are not. When established on a DME arc, the aircraft has departed the en route phase and has begun the approach and is maneuvering to enter an intermediate or final segment of the approach. DME arcs may also be used as an intermediate or a final segment, although they are extremely rare as final approach segments.

An arc may join a course at or before the IF. When joining a course at or before the IF, the angle of intersection of the arc and the course is designed so it does not exceed 120°. When the angle exceeds 90°, a radial that provides at least 2

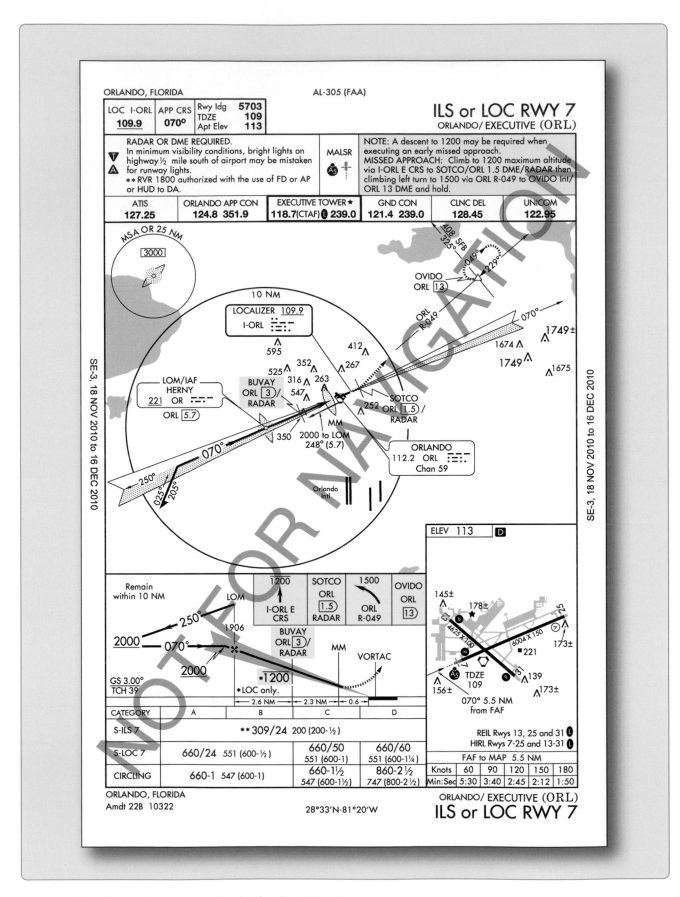

Figure 4-24. Orlando Executive Airport, Orlando, Florida, ILS RWY 7.

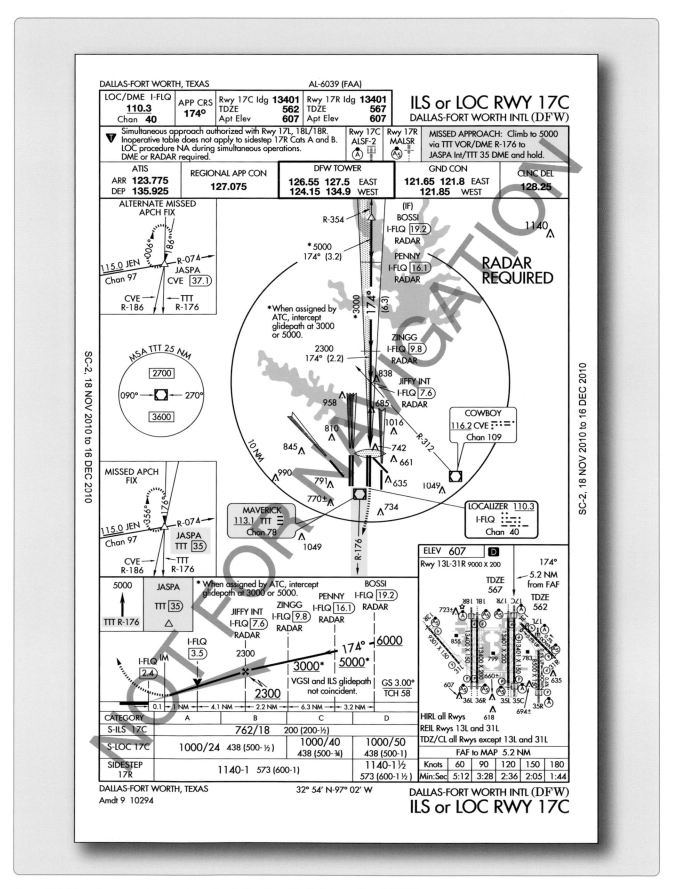

Figure 4-25. Missed approach procedures for Dallas-Fort Worth International (DFW).

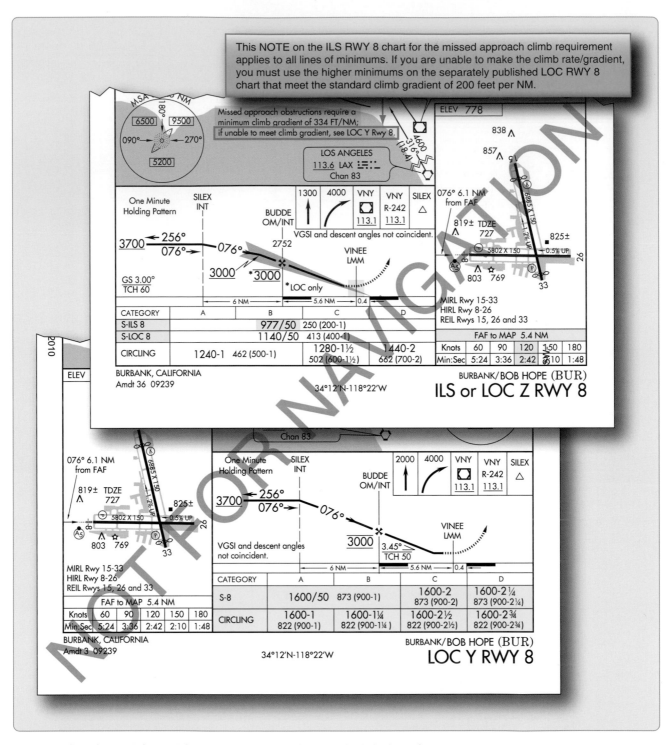

Figure 4-26. Missed approach point depiction and steeper than standard climb gradient requirements.

NM of lead will be identified to assist in leading the turn on to the intermediate course. DME arcs are predicated on DME collocated with a facility providing omnidirectional course information, such as a VOR. A DME arc cannot be based on an ILS or LOC DME source because omnidirectional course information is not provided.

The ROC along the arc depends on the approach segment.

For an initial approach segment, a ROC of 1,000 feet is required in the primary area, which extends to 4 NM on either side of the arc. For an intermediate segment primary area, the ROC is 500 feet. The initial and intermediate segment secondary areas extend 2 NM from the primary boundary area edge. The ROC starts at the primary area boundary edge at 500 feet and tapers to zero feet at the secondary area outer edge. [Figure 4-32]

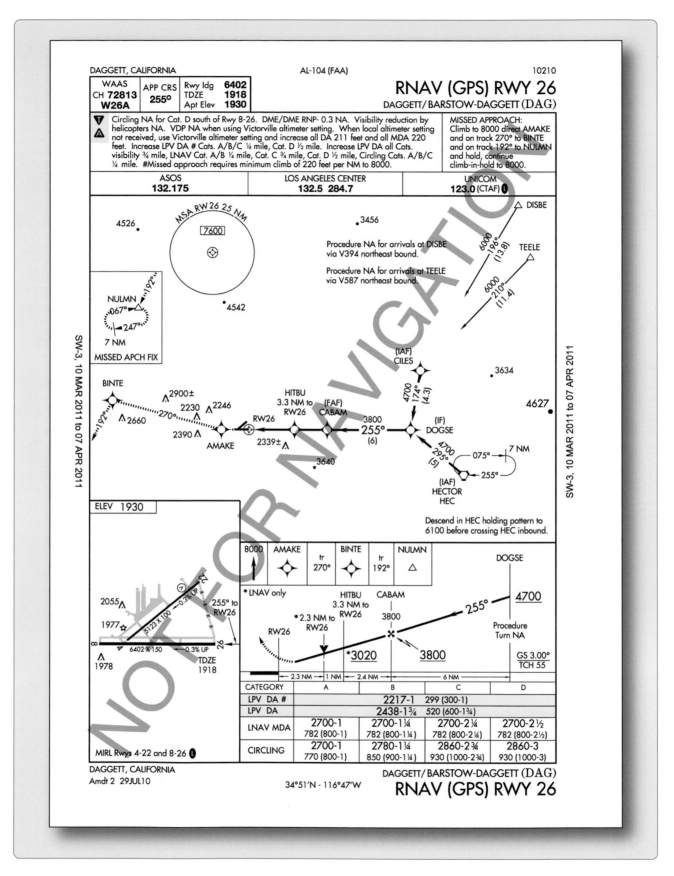

Figure 4-27. Two sets of minimums required when a climb gradient greater than 200 FPNM is required.

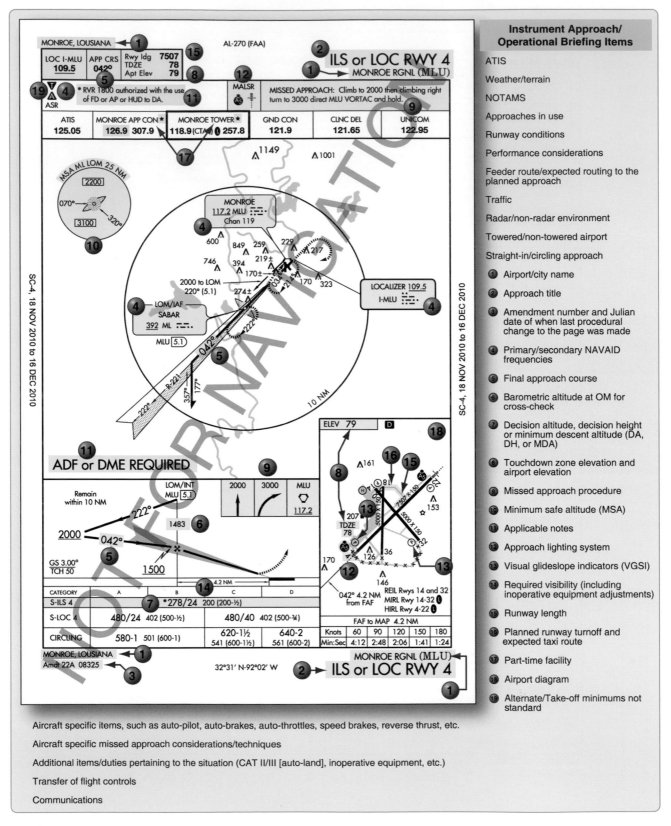

Figure 4-28. Example of approach chart briefing sequence.

Instrument Approach/Operational Briefing Items

ATIS

Weather/terrain

NOTAMS

Approaches in use

Runway conditions

Performance considerations

Feeder route/expected routing to the planned approach

Traffic

Radar/non-radar environment

Towered/non-towered airport

Straight-in/circling approach

1. Airport/city name
2. Approach title
3. Amendment number and Julian date of when last procedural change to the page was made
4. Primary/secondary NAVAID frequencies
5. Final approach course
6. Barometric altitude at OM for cross-check
7. Decision altitude, decision height or minimum descent altitude (DA, DH, or MDA)
8. Touchdown zone elevation and airport elevation
9. Missed approach procedure
10. Minimum safe altitude (MSA)
11. Applicable notes
12. Approach lighting system
13. Visual glideslope indicators (VGSI)
14. Required visibility (including inoperative equipment adjustments)
15. Runway length
16. Planned runway turnoff and expected taxi route
17. Part-time facility
18. Airport diagram
19. Alternate/Take-off minimums not standard

Aircraft specific items, such as auto-pilot, auto-brakes, auto-throttles, speed brakes, reverse thrust, etc.

Aircraft specific missed approach considerations/techniques

Additional items/duties pertaining to the situation (CAT II/III [auto-land], inoperative equipment, etc.)

Transfer of flight controls

Communications

Course Reversal

Some approach procedures do not permit straight-in approaches unless pilots are being radar vectored. In these situations, pilots are required to complete a procedure turn (PT) or other course reversal, generally within 10 NM of the PT fix, to establish the aircraft inbound on the intermediate or final approach segment.

If Category E airplanes are using the PT or there is a descent gradient problem, the PT distance available can be as much as 15 NM. During a procedure turn, a maximum speed of 200 knots indicated airspeed (KIAS) should be observed from first crossing the course reversal IAF through the procedure turn maneuver to ensure containment within the obstruction clearance area. Unless a holding pattern or teardrop procedure is published, the point where pilots begin the turn and the type and rate of turn are optional. If above the procedure turn minimum altitude, pilots may begin descent as soon as they cross the IAF outbound.

A procedure turn is the maneuver prescribed to perform a course reversal to establish the aircraft inbound on an intermediate or final approach course. The procedure turn or hold-in-lieu-of procedure turn is a required maneuver when it is depicted on the approach chart. However, the procedure turn or the hold-in-lieu-of PT is not permitted when the symbol "No PT" is depicted on the initial segment being flown, when a RADAR VECTOR to the final approach course is provided, or when conducting a timed approach from a holding fix.

The altitude prescribed for the procedure turn is a minimum altitude until the aircraft is established on the inbound course. The maneuver must be completed within the distance specified in the profile view. This distance is usually 10 miles. This may be reduced to 5 miles where only Category A or helicopter aircraft are operated. This distance may be increased to as much as 15 miles to accommodate high performance aircraft.

The pilot may elect to use the procedure turn or hold-in-lieu-of PT when it is not required by the procedure, but must first receive an amended clearance from ATC. When ATC is radar vectoring to the final approach course, or to the intermediate fix as may occur with RNAV standard instrument approach procedures, ATC may specify in the approach clearance "CLEARED STRAIGHT-IN (type) APPROACH" to ensure that the pilot understands that the procedure turn or hold-in- lieu-of PT is not to be flown. If the pilot is uncertain whether ATC intends for a procedure turn or a straight-in approach to be flown, the pilot will immediately request clarification from ATC.

On U.S. Government charts, a barbed arrow indicates the maneuvering side of the outbound course on which the procedure turn is made. Headings are provided for course reversal using the 45° type procedure turn. However, the point at which the turn may be commenced and the type and rate of turn is left to the discretion of the pilot (limited by the charted remain within XX NM distance). Some of the options are the 45° procedure turn, the racetrack pattern, the teardrop procedure turn, or the 80° procedure turn, or the 80° ↔ 260° course reversal. Racetrack entries should be conducted on the maneuvering side where the majority of protected airspace resides. If an entry places the pilot on the non-maneuvering side of the PT, correction to intercept the outbound course ensures remaining within protected airspace.

Some procedure turns are specified by procedural track. These turns must be flown exactly as depicted. These requirements are necessary to stay within the protected airspace and maintain adequate obstacle clearance. [Figure 4-33] A minimum of 1,000 feet of obstacle clearance is provided in the procedure turn primary area. [Figure 4-34] In the secondary area, 500 feet of obstacle clearance is provided at the inner edge, tapering uniformly to 0 feet at the outer edge.

The primary and secondary areas determine obstacle clearance in both the entry and maneuvering zones. The use of entry and maneuvering zones provides further relief from obstacles. The entry zone is established to control the obstacle clearance prior to proceeding outbound from the procedure turn fix. The maneuvering zone is established to control obstacle clearance after proceeding outbound from the procedure turn fix.

Descent to the PT completion altitude from the PT fix altitude (when one has been published or assigned by ATC) must not begin until crossing over the PT fix or abeam and proceeding outbound. Some procedures contain a note in the chart profile view that says "Maintain (altitude) or above until established outbound for procedure turn." Newer procedures simply depict an "at or above" altitude at the PT fix without a chart note. Both are there to ensure required obstacle clearance is provided in the procedure turn entry zone. Absence of a chart note or specified minimum altitude adjacent to the PT fix is an indication that descent to the procedure turn altitude can commence immediately upon crossing over the PT fix, regardless of the direction of flight. This is because the minimum altitudes in the PT entry zone and the PT maneuvering zone are the same.

A holding pattern-in-lieu-of procedure turn may be specified for course reversal in some procedures. In

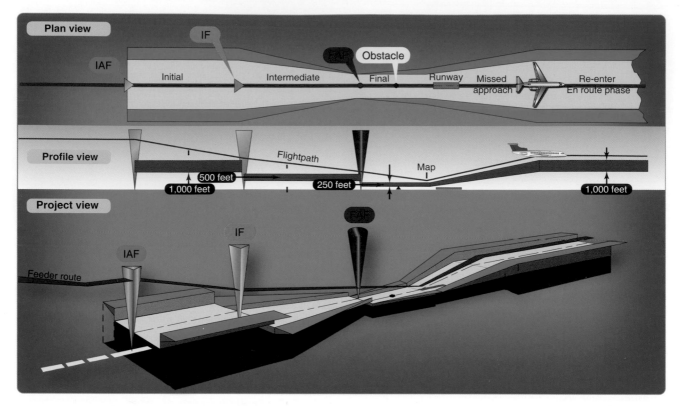

Figure 4-29. Approach segments and obstacle clearance.

such cases, the holding pattern is established over an intermediate fix or a FAF. The holding pattern distance or time specified in the profile view must be observed. For a hold-in-lieu-of PT, the holding pattern direction must be flown as depicted and the specified leg length/timing must not be exceeded. Maximum holding airspeed limitations as set forth for all holding patterns apply. The holding pattern maneuver is completed when the aircraft is established on the inbound course after executing the appropriate entry. If cleared for the approach prior to returning to the holding fix and the aircraft is at the prescribed altitude, additional circuits of the holding pattern are not necessary nor expected by ATC. If pilots elect to make additional circuits to lose excessive altitude or to become better established on course, it is their responsibility to so advise ATC upon receipt of their approach clearance. Refer to the AIM section 5-4-9 for additional information on holding procedures.

Initial Approach Segment

The purposes of the initial approach segment are to provide a method for aligning the aircraft with the intermediate or final approach segment and to permit descent during the alignment. This is accomplished by using a DME arc, a course reversal, such as a procedure turn or holding pattern, or by following a terminal route that intersects the final approach course. The initial approach segment begins at an IAF and usually ends where it joins the intermediate approach segment or at an IF. The letters IAF

on an approach chart indicate the location of an IAF and more than one may be available. Course, distance, and minimum altitudes are also provided for initial approach segments. A given procedure may have several initial approach segments. When more than one exists, each joins a common intermediate segment, although not necessarily at the same location.

Many RNAV approaches make use of a dual-purpose IF/IAF associated with a hold-in-lieu-of PT (HILO) anchored at the Intermediate Fix. The HILO forms the Initial Approach Segment when course reversal is required.

When the PT is required, it is only necessary to enter the holding pattern to reverse course. The dual purpose fix functions as an IAF in that case. Once the aircraft has entered the hold and is returning to the fix on the inbound course, the dual-purpose fix becomes an IF, marking the beginning of the intermediate segment.

ATC may provide a vector to an IF at an angle of 90 degrees or less and specify "Cleared Straight-in (type) Approach". In those cases, the radar vector is providing the initial approach segment and the pilot should not fly the PT without a clearance from ATC.

Occasionally, a chart may depict an IAF, although there is no initial approach segment for the procedure. This usually

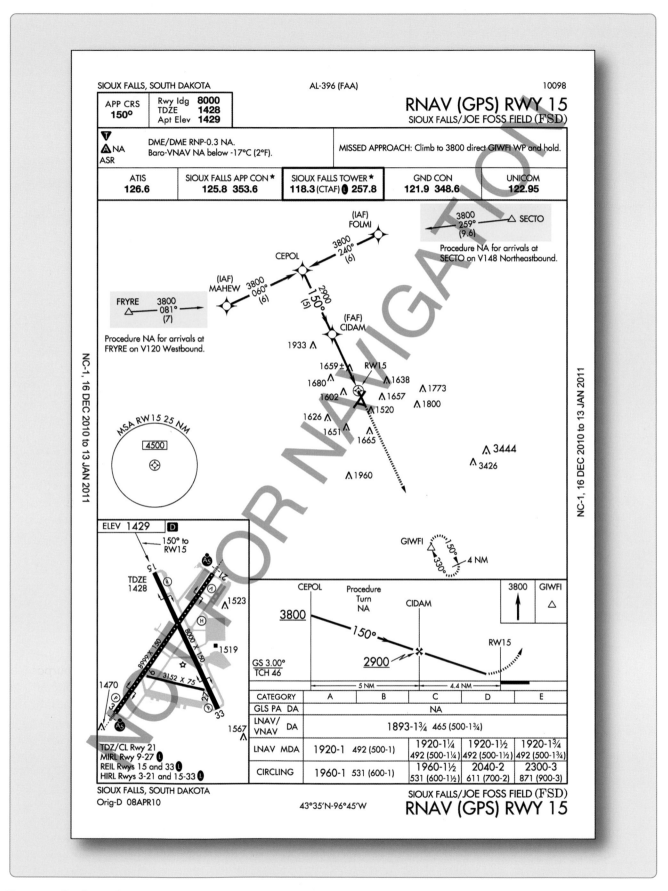

Figure 4-30. Feeder routes.

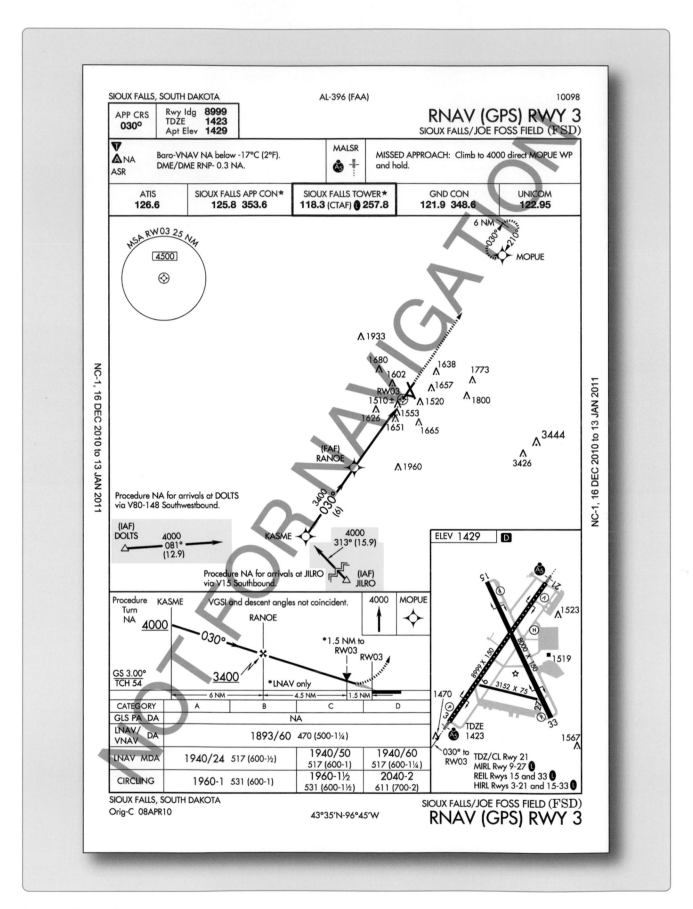

Figure 4-31. Terminal routes.

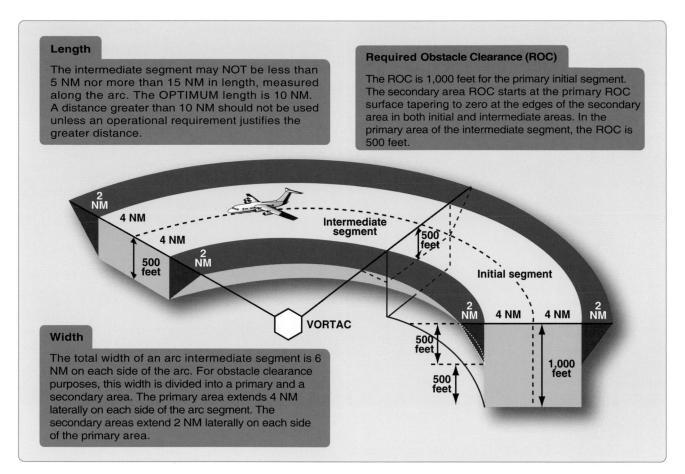

Length

The intermediate segment may NOT be less than 5 NM nor more than 15 NM in length, measured along the arc. The OPTIMUM length is 10 NM. A distance greater than 10 NM should not be used unless an operational requirement justifies the greater distance.

Required Obstacle Clearance (ROC)

The ROC is 1,000 feet for the primary initial segment. The secondary area ROC starts at the primary ROC surface tapering to zero at the edges of the secondary area in both initial and intermediate areas. In the primary area of the intermediate segment, the ROC is 500 feet.

Width

The total width of an arc intermediate segment is 6 NM on each side of the arc. For obstacle clearance purposes, this width is divided into a primary and a secondary area. The primary area extends 4 NM laterally on each side of the arc segment. The secondary areas extend 2 NM laterally on each side of the primary area.

Figure 4-32. DME arc obstruction clearance.

occurs at a point located within the en route structure where the intermediate segment begins. In this situation, the IAF signals the beginning of the intermediate segment.

Intermediate Approach Segment

The intermediate segment is designed primarily to position the aircraft for the final descent to the airport. Like the feeder route and initial approach segment, the chart depiction of the intermediate segment provides course, distance, and minimum altitude information.

The intermediate segment, normally aligned within 30° of the final approach course, begins at the IF, or intermediate point, and ends at the beginning of the final approach segment. In some cases, an IF is not shown on an approach chart. In this situation, the intermediate segment begins at a point where you are proceeding inbound to the FAF, are properly aligned with the final approach course, and are located within the prescribed distance prior to the FAF. An instrument approach that incorporates a procedure turn is the most common example of an approach that may not have a charted IF. The intermediate segment in this example begins when you intercept the inbound course after completing the procedure turn. [Figure 4-35]

Final Approach Segment

The final approach segment for an approach with vertical guidance or a precision approach begins where the glideslope/glidepath intercepts the minimum glideslope/glidepath intercept altitude shown on the approach chart. If ATC authorizes a lower intercept altitude, the final approach segment begins upon glideslope/glidepath interception at that altitude. For a non-precision approach, the final approach segment begins either at a designated FAF, which is depicted as a cross on the profile view, or at the point where the aircraft is established inbound on the final approach course. When a FAF is not designated, such as on an approach that incorporates an on-airport VOR or NDB, this point is typically where the procedure turn intersects the final approach course inbound. This point is referred to as the final approach point (FAP). The final approach segment ends at either the designated MAP or upon landing.

There are three types of procedures based on the final approach course guidance:

- Precision approach (PA)—an instrument approach based on a navigation system that provides course

4-53

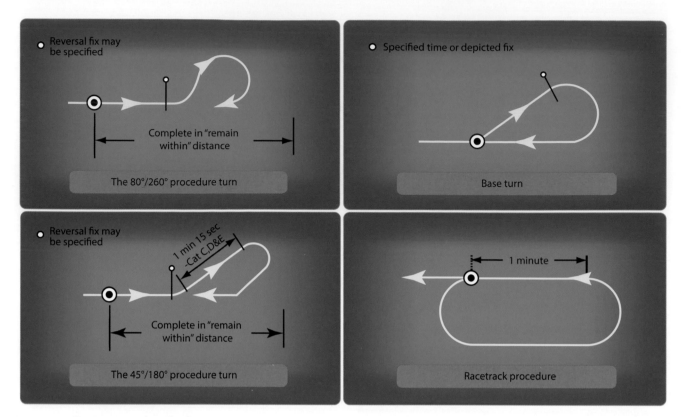

Figure 4-33. Course reversal methods.

and glidepath deviation information meeting precision standards of ICAO Annex 10. For example, PAR, ILS, and GLS are precision approaches.

- Approach with vertical guidance (APV) —an instrument approach based on a navigation system that is not required to meet the precision approach standards of ICAO Annex 10, but provides course and glidepath deviation information. For example, Baro-VNAV, LDA with glidepath, LNAV/VNAV and LPV are APV approaches.

- Non-precision approach (NPA)—an instrument approach based on a navigation system that provides course deviation information but no glidepath deviation information. For example, VOR, TACAN, LNAV, NDB, LOC, and ASR approaches are examples of NPA procedures.

Missed Approach Segment

The missed approach segment begins at the MAP and ends at a point or fix where an initial or en route segment begins. The actual location of the MAP depends upon the type of approach you are flying. For example, during a precision or an APV approach, the MAP occurs at the DA or DH on the glideslope/glidepath. For non-precision approaches, the MAP is either a fix, NAVAID, or after a specified period of time has elapsed after crossing the FAF.

Approach Clearance

According to FAA Order 7110.65, ATC clearances authorizing instrument approaches are issued on the basis that if visual contact with the ground is made before the approach is completed, the entire approach procedure is followed unless the pilot receives approval for a contact approach, is cleared for a visual approach, or cancels the IFR flight plan.

Approach clearances are issued based on known traffic. The receipt of an approach clearance does not relieve the pilot of his or her responsibility to comply with applicable parts of the CFRs and notations on instrument approach charts, which impose on the pilot the responsibility to comply with or act on an instruction, such as "procedure not authorized at night." The name of the approach, as published, is used to identify the approach. Approach name items within parentheses are not included in approach clearance phraseology.

Vectors To Final Approach Course

The approach gate is an imaginary point used within ATC as a basis for vectoring aircraft to the final approach course. The gate is established along the final approach course one mile from the FAF on the side away from the airport and is no closer than 5 NM from the landing threshold. Controllers are also required to ensure the assigned altitude conforms to the following:

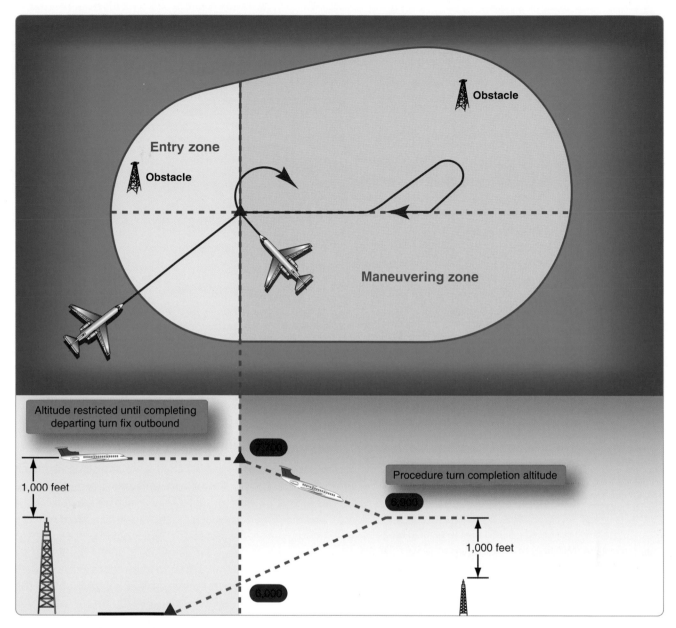

Figure 4-34. Procedure turn obstacle clearance.

- For a precision approach, at an altitude not above the glideslope/glidepath or below the minimum glideslope/glidepath intercept altitude specified on the approach procedure chart.

- For a non-precision approach, at an altitude that allows descent in accordance with the published procedure.

  Further, controllers must assign headings that intercept the final approach course no closer than the following table:

A typical vector to the final approach course and associated approach clearance is as follows:

"…four miles from LIMAA, turn right heading three four zero, maintain two thousand until established on the localizer, cleared ILS runway three six approach."

Other clearance formats may be used to fit individual circumstances, but the controller should always assign an altitude to maintain until the aircraft is established on a segment of a published route or IAP. The altitude assigned must guarantee IFR obstruction clearance from the point at which the approach clearance is issued until the aircraft is established on a published route. 14 CFR Part 91, section 91.175 (j) prohibits a pilot from making a procedure turn when vectored to a FAF or course, when conducting a timed approach, or when the procedure specifies "NO PT."

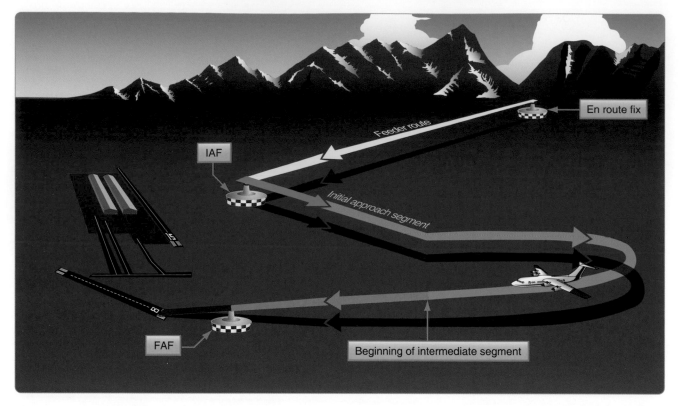

Figure 4-35. Approach without a designated IF.

When vectoring aircraft to the final approach course, controllers are required to ensure the intercept is at least 2 NM outside the approach gate. Exceptions include the following situations, but do not apply to RNAV aircraft being vectored for a GPS or RNAV approach:

- When the reported ceiling is at least 500 feet above the MVA/MIA and the visibility is at least 3 SM (may be a pilot report (PIREP) if no weather is reported for the airport), aircraft may be vectored to intercept the final approach course closer than 2 NM outside the approach gate but no closer than the approach gate.

- If specifically requested by the pilot, aircraft may be vectored to intercept the final approach course inside the approach gate but no closer than the FAF.

Distance from interception point to approach gate	Maximum interception angle
Less than 2 miles or triple simultaneous ILS approaches in use	20°
2 miles or more	30 ° (45 °for helicopters)

Nonradar Environment

In the absence of radar vectors, an instrument approach begins at an IAF. An aircraft that has been cleared to a holding fix that, prior to reaching that fix, is issued a clearance for an approach, but not issued a revised routing, such as, "proceed direct to…" is expected to proceed via the last assigned route, a feeder route if one is published on the approach chart, and then to commence the approach as published. If, by following the route of flight to the holding fix, the aircraft would overfly an IAF or the fix associated with the beginning of a feeder route to be used, the aircraft is expected to commence the approach using the published feeder route to the IAF or from the IAF as appropriate. The aircraft would not be expected to overfly and return to the IAF or feeder route.

For aircraft operating on unpublished routes, an altitude is assigned to maintain until the aircraft is established on a segment of a published route or IAP. (Example: "Maintain 2,000 until established on the final approach course outbound, cleared VOR/DME runway 12.") The FAA definition of established on course requires the aircraft to be established on the route centerline. Generally, the controller assigns an altitude compatible with glideslope/glidepath intercept prior to being cleared for the approach.

Types of Approaches

In the NAS, there are approximately 1,105 VOR stations, 916 NDB stations, and 1,194 ILS installations, including 25 LOC-type directional aids (LDAs), 11 simplified directional facilities (SDFs), and 235 LOC only facilities. As time progresses, it is the intent of the FAA to reduce navigational dependence on VOR, NDB, and other ground-based NAVAIDs and, instead, to increase the use of satellite-based navigation.

To expedite the use of RNAV procedures for all instrument pilots, the FAA has begun an aggressive schedule to develop RNAV procedures. As of 2010, the number of RNAV/ GPS approaches published in the NAS numbered 10,212 - with additional procedures published every revision cycle. While it had originally been the plan of the FAA to begin decommissioning VORs, NDBs, and other ground-based NAVAIDs, the overall strategy has been changed to incorporate a majority dependence on augmented satellite navigation while maintaining a satisfactory backup system. This backup system includes retaining all CAT II and III ILS facilities and close to one-half of the existing VOR network.

Each approach is provided obstacle clearance based on the Order 8260.3 TERPS design criteria as appropriate for the surrounding terrain, obstacles, and NAVAID availability. Final approach obstacle clearance is different for every type of approach but is guaranteed from the start of the final approach segment to the runway (not below the MDA for non-precision approaches) or MAP, whichever occurs last within the final approach area. It is dependent upon the pilot to maintain an appropriate flight path within the boundaries of the final approach area and maintain obstacle clearance.

There are numerous types of instrument approaches available for use in the NAS including RNAV (GPS), ILS, MLS, LOC, VOR, NDB, SDF, and radar approaches. Each approach has separate and individual design criteria, equipment requirements, and system capabilities.

Visual and Contact Approaches

To expedite traffic, ATC may clear pilots for a visual approach in lieu of the published approach procedure if flight conditions permit. Requesting a contact approach may be advantageous since it requires less time than the published IAP and provides separation from IFR and special visual flight rules (SVFR) traffic. A contact or visual approach may be used in lieu of conducting a SIAP, and both allow the flight to continue as an IFR flight to landing while increasing the efficiency of the arrival.

Visual Approaches

When it is operationally beneficial, ATC may authorize pilots to conduct a visual approach to the airport in lieu of the published IAP. A pilot, or the controller, can initiate a visual approach. Before issuing a visual approach clearance, the controller must verify that pilots have the airport, or a preceding aircraft that they are to follow, in sight. In the event pilots have the airport in sight but do not see the aircraft they are to follow, ATC may issue the visual approach clearance but maintain responsibility for aircraft and wake turbulence separation. Once pilots report the aircraft in sight, they assume the responsibilities for their own separation and wake turbulence avoidance.

A visual approach is an ATC authorization for an aircraft on an IFR flight plan to proceed visually to the airport of intended landing; it is not an IAP. Also, there is no missed approach segment. An aircraft unable to complete a visual approach must be handled as any other go-around and appropriate separation must be provided. A vector for a visual approach may be initiated by ATC if the reported ceiling at the airport of intended landing is at least 500 feet above the MVA/MIA and the visibility is 3 SM or greater. At airports without weather reporting service, there must be reasonable assurance through area weather reports and PIREPs that descent and approach to the airport can be made visually, and the pilot must be informed that weather information is not available.

The visual approach clearance is issued to expedite the flow of traffic to an airport. It is authorized when the ceiling is reported or expected to be at least 1,000 feet AGL and the visibility is at least 3 SM. Pilots must remain clear of the clouds at all times while conducting a visual approach. At an airport with a control tower, pilots may be cleared to fly a visual approach to one runway while others are conducting VFR or IFR approaches to another parallel, intersecting, or converging runway. Also, when radar service is provided, it is automatically terminated when the controller advises pilots to change to the tower or advisory frequency. While conducting a visual approach, the pilot is responsible for providing safe obstacle clearance.

Contact Approaches

If conditions permit, pilots can request a contact approach, which is then authorized by the controller. A contact approach cannot be initiated by ATC. This procedure may be used instead of the published procedure to expedite arrival, as long as the airport has a SIAP the reported ground visibility is at least 1 SM, and pilots are able to remain clear of clouds with at least one statute mile flight visibility throughout the approach. Some advantages of a contact approach are that it usually requires less time than the published instrument procedure, it allows pilots to retain the IFR clearance, and provides separation from IFR and SVFR traffic. On the other hand, obstruction clearances and VFR traffic avoidance becomes the pilot's responsibility. Unless otherwise restricted, the pilot may find it necessary to descend, climb, or fly a circuitous route to the airport to maintain cloud clearance or terrain/ obstruction clearance.

The main differences between a visual approach and a contact approach are: a pilot must request a contact approach, while a visual approach may be assigned by ATC or requested by the pilot; and a contact approach may be approved with 1 mile visibility if the flight can remain clear of clouds, while a visual approach requires the pilot to have the airport in sight, or a preceding aircraft to be followed, and the ceiling must be at least 1,000 feet AGL with at least 3 SM visibility.

Charted Visual Flight Procedures

A charted visual flight procedure (CVFP) may be established at some airports with control towers for environmental or noise considerations, as well as when necessary for the safety and efficiency of air traffic operations. Designed primarily for turbojet aircraft, CVFPs depict prominent landmarks, courses, and recommended altitudes to specific runways. When pilots are flying the Roaring Fork Visual RWY 15, shown in Figure 4-36 , mountains, rivers, and towns provide guidance to Aspen, Colorado's Sardy Field instead of VORs, NDBs, and DME fixes.

Pilots must have a charted visual landmark or a preceding aircraft in sight, and weather must be at or above the published minimums before ATC will issue a CVFP clearance. ATC will clear pilots for a CVFP if the reported ceiling at the airport of intended landing is at least 500 feet above the MVA/MIA, and the visibility is 3 SM or more, unless higher minimums are published for the particular CVFP. When accepting a clearance to follow a preceding aircraft, pilots are responsible for maintaining a safe approach interval and wake turbulence separation. Pilots must advise ATC if unable at any point to continue a charted visual approach or if the pilot loses sight of the preceding aircraft.

RNAV Approaches

Because of the complications with database coding, naming conventions were changed in January 2001 to accommodate all approaches using RNAV equipment into one classification which is RNAV. This classification includes both ground- based and satellite dependent systems. Eventually all approaches that use some type of RNAV will reflect RNAV in the approach title.

This changeover is being made to reflect two shifts in instrument approach technology. The first shift is the use of the RNP concept outlined in Chapter 1, Departure Procedures, in which a single performance standard concept is being implemented for departure/approach procedure design. Through the use of RNP, the underlying system of navigation may not be required, provided the aircraft can maintain the appropriate RNP standard. The second shift is advanced avionics systems, such as FMS, used by most airlines, needed a new navigation standard by which RNAV could be fully integrated into the instrument approach system.

An FMS uses multi-sensor navigation inputs to produce a composite position. Essentially, the FMS navigation function automatically blends or selects position sensors to compute aircraft position. Instrument approach charts and RNAV databases needed to change to reflect these issues. A complete discussion of airborne navigation databases is included in Chapter 6, Airborne Navigation Databases. Due to the multi- faceted nature of RNAV, new approach criteria have been developed to accommodate the design of RNAV instrument approaches. This includes criteria for terminal arrival areas (TAAs), RNAV basic approach criteria, and specific final approach criteria for different types of RNAV approaches.

Terminal Arrival Areas

The Terminal Arrival Area (TAA) provides a transition from the en route structure to the terminal environment with little required pilot/air traffic control interface for aircraft equipped with Area Navigation (RNAV) systems. TAAs provide minimum altitudes with standard obstacle clearance when operating within the TAA boundaries. TAAs are primarily used on RNAV approaches but may be used on an ILS approach when RNAV is the sole means for navigation to the IF; however, they are not normally used in areas of heavy concentration of air traffic . [Figure 4-37]

The basic design of the RNAV procedure underlying the TAA is normally the "T" design (also called the "Basic T"). The "T" design incorporates two IAFs plus a dual purpose IF/IAF that functions as both an intermediate fix and an

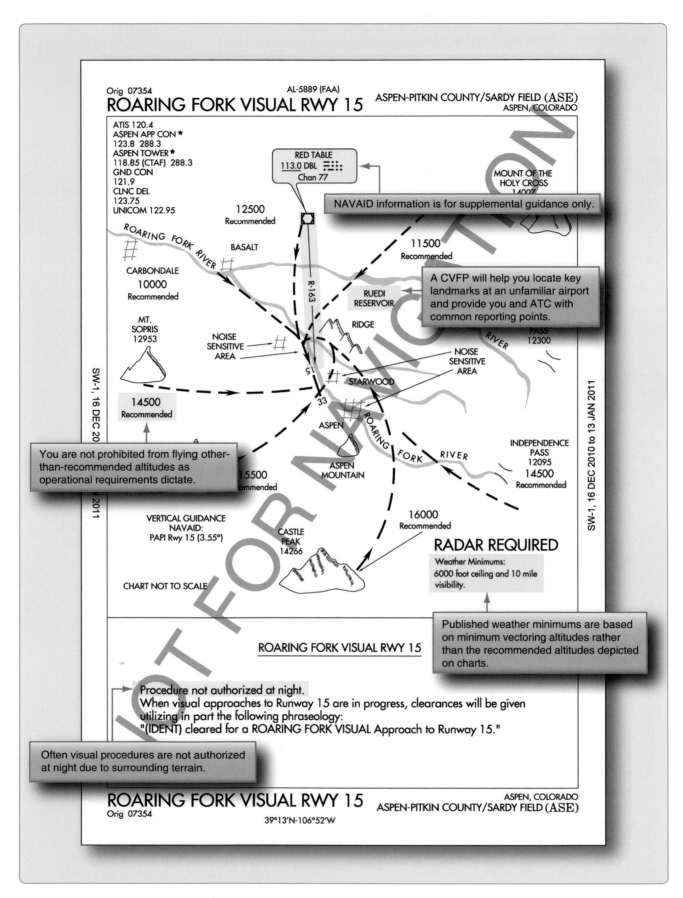

Figure 4-36. Charted visual flight procedures (CVFP).

initial approach fix. The T configuration continues from the IF/IAF to the FAF and then to the MAP. The two base leg IAFs are typically aligned in a straight-line perpendicular to the intermediate course connecting at the IF/IAF. A Hold-in-Lieu-of Procedure Turn (HILO) is anchored at the IF/IAF and depicted on U.S. Government publications using the "hold–in–lieu–of–PT" holding pattern symbol. When the HILO is necessary for course alignment and/or descent, the dual purpose IF/IAF serves as an IAF during the entry into the pattern. Following entry into the HILO pattern and when flying a route or sector labeled "NoPT," the dual-purpose fix serves as an IF, marking the beginning of the Intermediate Segment.

The standard TAA based on the "T" design consists of three areas defined by the IAF legs and the intermediate segment course beginning at the IF/IAF. These areas are called the straight–in, left–base, and right–base areas. [FIG 4-37] TAA area lateral boundaries are identified by magnetic courses TO the IF/IAF. The straight–in area can be further divided into pie–shaped sectors with the boundaries identified by magnetic courses TO the IF/ IAF, and may contain step-down sections defined by arcs based on RNAV distances from the IF/IAF.

Entry from the terminal area onto the procedure is normally accomplished via a no procedure turn (NoPT) routing or via a course reversal maneuver. The published procedure will be annotated "NoPT" to indicate when the course reversal is not authorized when flying within a particular TAA sector (See Figures 4-37 and 4-38). Otherwise, the pilot is expected to execute the course reversal under the provisions of 14 CFR Section 91.175. The pilot may elect to use the course reversal pattern when it is not required by the procedure, but must receive clearance from air traffic control before beginning the procedure.

ATC should not clear an aircraft to the left base leg or right base leg IAF within a TAA at an intercept angle exceeding 90 degrees. Pilots must not execute the HILO course reversal when the sector or procedure segment is labeled "NoPT."

ATC may clear aircraft direct to the fix labeled IF/IAF if the course to the IF/IAF is within the straight-in sector labeled "NoPT" and the intercept angle does not exceed 90 degrees. Pilots are expected to proceed direct to the IF/IAF and accomplish a straight-in approach. Do not execute HILO course reversal. Pilots are also expected to fly the straight–in approach when ATC provides radar vectors and monitoring to the IF/IAF and issues a "straight-in" approach clearance; otherwise, the pilot is expected to execute the HILO course reversal. (See AIM Paragraph 5–4–6, Approach Clearance)

On rare occasions, ATC may clear the aircraft for an approach at the airport without specifying the approach procedure by name or by a specific approach (e.g., "cleared RNAV Runway 34 approach") without specifying a particular IAF. In either case, the pilot should proceed direct to the IAF or to the IF/IAF associated with the sector that the aircraft will enter the TAA and join the approach course from that point and if required by that sector (i.e., sector is not labeled "NoPT"), complete the HILO course reversal.

NOTE–If approaching with a TO bearing that is on a sector boundary, the pilot is expected to proceed in accordance with a "NoPT" routing unless otherwise instructed by ATC.

Altitudes published within the TAA replace the MSA altitude. However, unlike MSA altitudes the TAA altitudes are operationally usable altitudes. These altitudes provide at least 1,000 feet of obstacle clearance, and more in mountainous areas. It is important that the pilot knows which area of the TAA that the aircraft will enter in order to comply with the minimum altitude requirements. The pilot can determine which area of the TAA the aircraft will enter by determining the magnetic bearing of the aircraft TO the fix labeled IF/IAF. The bearing should then be compared to the published lateral boundary bearings that define the TAA areas. Do not use magnetic bearing to the right-base or left-base IAFs to determine position.

An ATC clearance direct to an IAF or to the IF/IAF without an approach clearance does not authorize a pilot to descend to a lower TAA altitude. If a pilot desires a lower altitude without an approach clearance, request the lower TAA altitude from ATC. Pilots not sure of the clearance should confirm their clearance with ATC or request a specific clearance. Pilots entering the TAA with two–way radio communications failure (14 CFR Section 91.185, IFR Operations: Two–way Radio Communications Failure), must maintain the highest altitude prescribed by Section 91.185(c)(2) until arriving at the appropriate IAF.

Once cleared for the approach, pilots may descend in the TAA sector to the minimum altitude depicted within the defined area/subdivision, unless instructed otherwise by air traffic control. Pilots should plan their descent within the TAA to permit a normal descent from the IF/IAF to the FAF.

U.S. Government charts depict TAAs using icons located in the plan view outside the depiction of the actual approach procedure. Use of icons is necessary to avoid obscuring any portion of the "T" procedure (altitudes, courses, minimum altitudes, etc.). The icon for each TAA area will be located and oriented on the plan view with respect

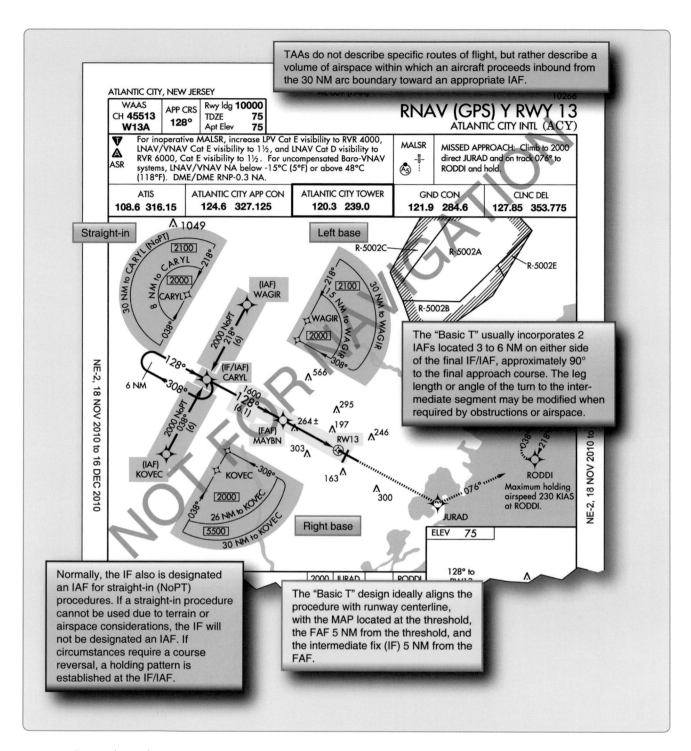

Figure 4-37. Terminal arrival area (TAA) design "basic T."

to the direction of arrival to the approach procedure, and will show all TAA minimum altitudes and sector/radius subdivisions. The IAF for each area of the TAA is included on the icon where it appears on the approach to help the pilot orient the icon to the approach procedure. The IAF name and the distance of the TAA area boundary from the IAF are included on the outside arc of the TAA area icon.

TAAs may be modified from the standard size and shape to accommodate operational or ATC requirements. Some areas may be eliminated, while the other areas are expanded. The "T" design may be modified by the procedure designers where required by terrain or ATC considerations. For instance, the "T" design may appear more like a regularly or irregularly shaped "Y," an upside down "L," or an "I."

4-61

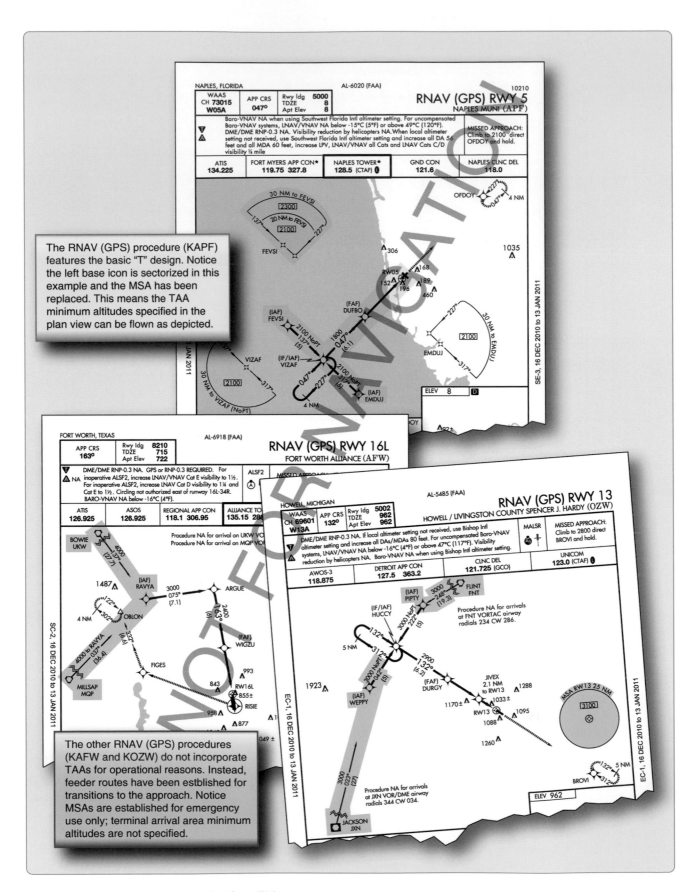

The RNAV (GPS) procedure (KAPF) features the basic "T" design. Notice the left base icon is sectorized in this example and the MSA has been replaced. This means the TAA minimum altitudes specified in the plan view can be flown as depicted.

The other RNAV (GPS) procedures (KAFW and KOZW) do not incorporate TAAs for operational reasons. Instead, feeder routes have been estblished for transitions to the approach. Notice MSAs are established for emergency use only; terminal arrival area minimum altitudes are not specified.

Figure 4-38. RNAV approaches with and without TAAs.

When an airway does not cross the lateral TAA boundaries, a feeder route will be established from an airway fix or NAVAID to the TAA boundary to provide a transition from the en route structure to the appropriate IAF. Each feeder route will terminate at the TAA boundary and will be aligned along a path pointing to the associated IAF. Pilots should descend to the TAA altitude after crossing the TAA boundary and cleared for the approach by ATC.

Each waypoint on the "T" is assigned a pronounceable 5–letter name, except the missed approach waypoint. These names are used for ATC communications, RNAV databases, and aeronautical navigation products. The missed approach waypoint is assigned a pronounceable name when it is not located at the runway threshold.

RNAV Approach Types

RNAV encompasses a variety of underlying navigation systems and, therefore, approach criteria. This results in different sets of criteria for the final approach segment of various RNAV approaches. RNAV instrument approach criteria address the following procedures:

- GPS overlay of pre-existing nonprecision approaches.
- VOR/DME based RNAV approaches.
- Stand-alone RNAV (GPS) approaches.
- RNAV (GPS) approaches with vertical guidance (APV).
- RNAV (GPS) precision approaches (WAAS and LAAS).

GPS Overlay of Nonprecision Approach

The original GPS approach procedures provided authorization to fly non-precision approaches based on conventional, ground-based NAVAIDs. Many of these approaches have been converted to stand-alone approaches, and the few that remain are identified by the name of the procedure and "or GPS." These GPS non-precision approaches are predicated upon the design criteria of the ground-based NAVAID used as the basis of the approach. As such, they do not adhere to the RNAV design criteria for stand-alone GPS approaches, and are not considered part of the RNAV (GPS) approach classification for determining design criteria. [Figure 4-39]

GPS Stand-Alone/RNAV (GPS) Approach

The number of GPS stand-alone approaches continues to decrease as they are replaced by RNAV approaches. RNAV (GPS) approaches are named so that airborne navigation databases can use either GPS or RNAV as the title of the approach. This is required for non-GPS approach systems, such as VOR/DME based RNAV systems. In the past, these approaches were often referred to as "stand-alone GPS" approaches. They are considered non-precision approaches, offering only LNAV and circling minimums. Precision minimums are not authorized, although LNAV/VNAV minimums may be published and used as long as the on-board system is capable of providing approach approved VNAV. The RNAV (GPS) Runway 14 approach for Lincoln, Nebraska, incorporates only LNAV and circling minimums [Figure 4-40].

For a non-vertically guided straight-in RNAV (GPS) approach, the final approach course must be aligned within 15° of the extended runway centerline. The final approach segment should not exceed 10 NM, and when it exceeds 6 NM, a stepdown fix is typically incorporated. A minimum of 250 feet obstacle clearance is also incorporated into the final approach segment for straight-in approaches, and a maximum 400-FPNM descent gradient is permitted.

The approach design criteria are different for approaches that use vertical guidance provided by a Baro-VNAV system. Because the Baro-VNAV guidance is advisory and not primary, Baro-VNAV approaches are not authorized in areas of hazardous terrain, nor are they authorized when a remote altimeter setting is required. Due to the inherent problems associated with barometric readings and cold temperatures, these procedures are also temperature limited. Additional approach design criteria for RNAV Approach Construction Criteria can be found in the appropriate FAA Order 8260-series orders.

RNAV (GPS) Approach Using WAAS

WAAS was commissioned in July 2003, with IOC. Although precision approach capability is still in the future, WAAS currently provides a type of APV known as LPV. WAAS can support the following minima types: LPV, LNAV/VNAV, LP, and LNAV. Approach minima as low as 200 feet HAT and 1/2 SM visibility is possible, even though LPV is not considered a precision approach. WAAS covers 95 percent of the country 95 percent of the time.

NOTE: WAAS avionics receive an airworthiness approval in accordance with Technical Standard Order (TSO) C145, Airborne Navigation Sensors Using the Global Positioning System (GPS) Augmented by the Satellite Based Augmentation System (SBAS), or TSO-146, Stand-Alone Airborne Navigation Equipment Using the Global Positioning System (GPS) Augmented by the Satellite Based Augmentation System (SBAS), and installed in accordance with AC 20-138C, Airworthiness Approval of Positioning and Navigation Systems.

Precision approach capability will become available as more GBAS (LAAS) approach types become operational. GBAS

(LAAS) further increases the accuracy of GPS and improves signal integrity warnings. Precision approach capability requires obstruction planes and approach lighting systems to meet Part 77 standards for ILS approaches. This delays the implementation of RNAV (GPS) precision approach capability due to the cost of certifying each runway.

ILS Approaches

Notwithstanding emerging RNAV technology, the ILS is the most precise and accurate approach NAVAID currently in use throughout the NAS. An ILS CAT I precision approach allows approaches to be made to 200 feet above the TDZE and with visibilities as low as 1,800 RVR; with CAT II and CAT III approaches allowing descents and visibility minimums that are even lower. Non-precision approach alternatives cannot begin to offer the precision or flexibility offered by an ILS. In order to further increase the approach capacity of busy airports and exploit the maximum potential of ILS technology, many different applications are in use.

A single ILS system can accommodate 29 arrivals per hour on a single runway. Two or three parallel runways operating consecutively can double or triple the capacity of the airport. For air commerce, this means greater flexibility in scheduling passenger and cargo service. Capacity is increased through the use of parallel (dependent) ILS, simultaneous parallel (independent) ILS, simultaneous close parallel (independent) ILS, precision runway monitor (PRM), and converging ILS approaches. A parallel (dependent) approach differs from a simultaneous (independent) approach in that the minimum distance between parallel runway centerlines is reduced; there is no requirement for radar monitoring or advisories; and a staggered separation of aircraft on the adjacent localizer/azimuth course is required.

In order to successfully accomplish parallel, simultaneous parallel, and converging ILS approaches, flight crews and ATC have additional responsibilities. When multiple instrument approaches are in use, ATC advises flight crews either directly or through ATIS. It is the pilot's responsibility to inform ATC if unable or unwilling to execute a simultaneous approach. Pilots must comply with all ATC requests in a timely manner and maintain strict radio discipline, including using complete aircraft call signs. It is also incumbent upon the flight crew to notify ATC immediately of any problems relating to aircraft communications or navigation systems. At the very least, the approach procedure briefing should cover the entire approach procedure including the approach name, runway number, frequencies, final approach course, glideslope intercept altitude, DA or DH, and the missed approach instructions. The review of autopilot procedures is also appropriate when making coupled ILS or MLS approaches.

As with all approaches, the primary navigation responsibility falls upon the pilot in command. ATC instructions will be limited to ensuring aircraft separation. Additionally, MAPs are normally designed to diverge in order to protect all involved aircraft. ILS approaches of all types are afforded the same obstacle clearance protection and design criteria, no matter how capacity is affected by multiple ILS approaches. [Figure 4-41]

ILS Approach Categories

There are three general classifications of ILS approaches: CAT I, CAT II, and CAT III (autoland). The basic ILS approach is a CAT I approach and requires only that pilots be instrument rated and current, and that the aircraft be equipped appropriately. CAT II and CAT III ILS approaches typically have lower minimums and require special certification for operators, pilots, aircraft, and airborne/ground equipment. Because of the complexity and high cost of the equipment, CAT III ILS approaches are used primarily in air carrier and military operations. [Figure 4-42]

CAT II and III Approaches

The primary authorization and minimum RVRs allowed for an air carrier to conduct CAT II and III approaches can be found in OpSpecs Part C. CAT II and III operations allow authorized pilots to make instrument approaches in weather that would otherwise be prohibitive.

While CAT I ILS operations permit substitution of midfield RVR for TDZ RVR (when TDZ RVR is not available), CAT II ILS operations do not permit any substitutions for TDZ RVR. The TDZ RVR system is required and must be used. The TDZ RVR is controlling for all CAT II ILS operations.

The weather conditions encountered in CAT III operations range from an area where visual references are adequate for manual rollout in CAT IIIa, to an area where visual references are inadequate even for taxi operations in CAT IIIc. To date, no U.S. operator has received approval for CAT IIIc in OpSpecs. Depending on the auto-flight systems, some airplanes require a DH to ensure that the airplane is going to land in the TDZ and some require an Alert Height as a final cross-check of the performance of the auto-flight systems. These heights are based on radio altitude (RA) and can be found in the specific aircraft's AFM. [Figure 4-43]

Both CAT II and III approaches require special ground and airborne equipment to be installed and operational, as well as special aircrew training and authorization. The OpSpecs of individual air carriers detail the requirements of these types of approaches, as well as their performance criteria. Lists of locations where each operator is approved to conduct CAT II and III approaches can also be found in the OpSpecs.

Special Authorization CAT I and Special Authorization CAT III are approaches designed to take advantage of advances in flight deck avionics and technologies like Head-Up Displays (HUD) and automatic landings. There are extensive ground infrastructures and lighting requirements for standard CAT II/III, and the Special Authorization approaches mitigate the lack of some lighting with the modern avionics found in many aircraft today. Similar to standard CAT II/III, an air carrier must be specifically authorized to conduct Special Authorization CAT I/II in OpSpecs Part C.

Approaches To Parallel Runways

Airports that have two or three parallel runways may be authorized to use parallel approaches to maximize the capacity of the airport. There are three classifications of parallel approaches, depending on the runway centerline separation and ATC procedures.

NOTE:

1. Simultaneous approaches involving an RNAV approach may only be conducted when (GPS) appears in the approach title or a chart note states that GPS is required.

2. Simultaneous dependent approaches may only be conducted where instrument approach sharts specifically authorize simultaneous approaches to adjacent runways.

Parallel (Dependent) Approaches

Parallel (dependent) approaches are allowed at airports with parallel runways that have centerlines separated by at least 2,500 feet. Aircraft are allowed to fly and other approaches to parallel runways; however, the aircraft must be staggered by a minimum of 1 1/2 NM diagonally. Aircraft are staggered by 2 NM diagonally for runway centerlines that are separated by more than 4,300 feet and up to but not including 9,000 feet, and that do not have final monitor air traffic controllers. Radar separation is provided between aircraft participating in parallel (dependent) approach operations. [Figure 4-44] At some airports, dependent instrument approaches can be conducted with runways spaced less than 2,500 feet with specific centerline separations and threshold staggers. The lead aircraft of the dependent pair is restricted to being small or large aircraft weight type and is cleared to the lower approach. The geometry of the approach, aircraft weight type, and lateral separation between the two approaches provide necessary wake turbulence avoidance for this type of operation.

Where this type of approach is approved, each approach plate indicates the other runway with which simultaneous approaches can be conducted. For example, "Simultaneous approaches authorized with runway 12L". Until the approach plates for all such runway pairs can be modified to include this note, P-NOTAMS will be issued identifying such operations. ATC normally communicates an advisory over ATIS that parallel approach procedures are in effect. For example, pilots flying into Sacramento, California, may encounter parallel approach procedures. [Figure 4-45]

Simultaneous Parallel Approaches

Simultaneous parallel approaches are used at authorized airports that have between 4,300 feet and 9,000 feet separation between runway centerlines. A dedicated final monitor controller is required to monitor separation for this type of approach, which eliminates the need for staggered approaches. Final monitor controllers track aircraft positions and issue instructions to pilots of aircraft observed deviating from the final approach course. [Figure 4-46] As of March 2010, RNAV approach procedures with vertical guidance were permitted to conduct simultaneous parallel approach operations.

Triple simultaneous approaches are authorized provided the runway centerlines are separated by at least 5,000 feet, or 4,300 feet with PRM, and are below 1,000 feet MSL airport elevation. Additionally, for triple parallel approaches above airport elevations of 1,000 feet MSL, ASR with high-resolution final monitor aids or high update RADAR with associated final monitor aids is required.

As a part of the simultaneous parallel approach approval, normal operating zones (NOZ) and no-transgression zones (NTZ) must be established to ensure proper flight track boundaries for all aircraft. The NOZ is the operating zone within which aircraft remain during normal approach operations. The NOZ is typically no less than 1,400 feet wide, with 700 feet of space on either side of the runway centerline. A NTZ is a 2,000-foot wide area located between the parallel runway final approach courses. It is equidistant between the runways and indicates an area within which flight is not authorized. [Figure 4-47] Any time an aircraft breaches the NTZ, ATC issues instructions for all aircraft to break off the approach to avoid potential conflict.

Simultaneous Close Parallel Precision Runway Monitor Approaches

Simultaneous close parallel (independent) PRM approaches are authorized for use at airports that have parallel runways separated by at least 3,400 feet and no more than 4,300 feet. [Figure 4-48] They are also approved for airports with parallel runways separated by at least 3,000 feet with an offset LOC where the offset angle is at least 2.5° but no more than 3°. Other offset approaches to lesser runway spacing are referred to as Simultaneous Offset Instrument Approaches (SOIA) and are discussed in depth later in this chapter.

The PRM system provides the ability to accomplish simultaneous close parallel (independent) approaches and enables reduced delays and fuel savings during reduced visibility operations. It is also the safest method of increasing approach capacity through the use of parallel approaches. The PRM system incorporates high-update radar with one second or better update time and a high resolution ATC radar display that contains automated tracking software that can track aircraft in real time. Position and velocity is updated each second and a ten second projected position is displayed. The system also incorporates visual and aural alerts for the controllers.

Approval for PRM approaches requires the airport to have a precision runway monitoring system and a final monitor controller who can only communicate with aircraft on the final approach course. Additionally, two tower frequencies are required to be used and the controller broadcasts over both frequencies to reduce the chance of instructions being missed. Pilot training is also required for pilots using the PRM system. Part 121 and 135 operators are required to complete training that includes the viewing of videos. The FAA PRM website (http://www.faa.gov/training_testing/training/prm/) contains training information for PRM approaches and is a location from which the videos can be seen and/or downloaded.

When pilots or flight crews wish to decline a PRM approach, ATC must be notified immediately and the flight will be transitioned into the area at the convenience of ATC. Pilots who are unable to accept a PRM approach may be subject to delays.

The approach chart for the PRM approach typically requires two pages and outlines pilot, aircraft, and procedure requirements necessary to participate in PRM operations. [Figure 4-49] Pilots need to be aware of the differences associated with this type of approach which are listed below:

- Immediately follow break out instructions as soon as safety permits.

- Listen concurrently to the tower and the PRM monitor to avoid missed instructions from stuck mikes or blocked trans missions. The final ATC controller can override the radio frequency if necessary.

- Broadcast only over the main tower frequency.

- Disengage the autopilot for breakouts because hand- flown breakouts are quicker.

- Set the Traffic Alert and Collision Avoidance System (TCAS) to the appropriate TA (traffic advisory) or RA (resolution advisory) mode in compliance with current operational guidance on the attention all users page (AAUP), or other authorized guidance (i.e., approved flight manual, flight operations manual). It is important to note that descending breakouts may be issued. Additionally, flight crews are never issued breakout instructions that clear them below the MVA, and they are not required to descend at more than 1,000 fpm.

Simultaneous Offset Instrument Approaches (SOIAs)

SOIAs allow simultaneous approaches to two parallel runways spaced at least 750 feet apart, but less than 3,000 feet. The SOIA procedure utilizes an ILS/PRM approach to one runway and an offset localizer-type directional aid (LDA)/PRM approach with glideslope to the adjacent runway. The use of PRM technology is also required with these operations; therefore, the approach charts will include procedural notes, such as "Simultaneous approach authorized with LDA PRM RWY XXX." San Francisco had the first published SOIA approach. [Figure 4-50]

The training, procedures, and system requirements for SOIA ILS/PRM and LDA/PRM approaches are identical with those used for simultaneous close parallel ILS/PRM approaches until near the LDA/PRM approach MAP, where visual acquisition of the ILS aircraft by the LDA aircraft must be accomplished. If visual acquisition is not accomplished prior to reaching the LDA MAP , a missed approach must be executed. A visual segment for the LDA/PRM approach is established between the LDA MAP and the runway threshold. Aircraft transition in visual conditions from the LDA course, beginning at the LDA MAP, to align with the runway and can be stabilized by 500 feet AGL on the extended runway centerline. Pilots are reminded that they are responsible for collision avoidance and wake turbulence mitigation between the LDA MAP and the runway.

The FAA website has additional information about PRM and SOIA, including instructional videos at http://www.faa.gov/training_testing/training/prm/.

Converging ILS Approaches

Another method by which ILS approach capacity can be increased is through the use of converging approaches. Converging approaches may be established at airports that have runways with an angle between 15° and 100°

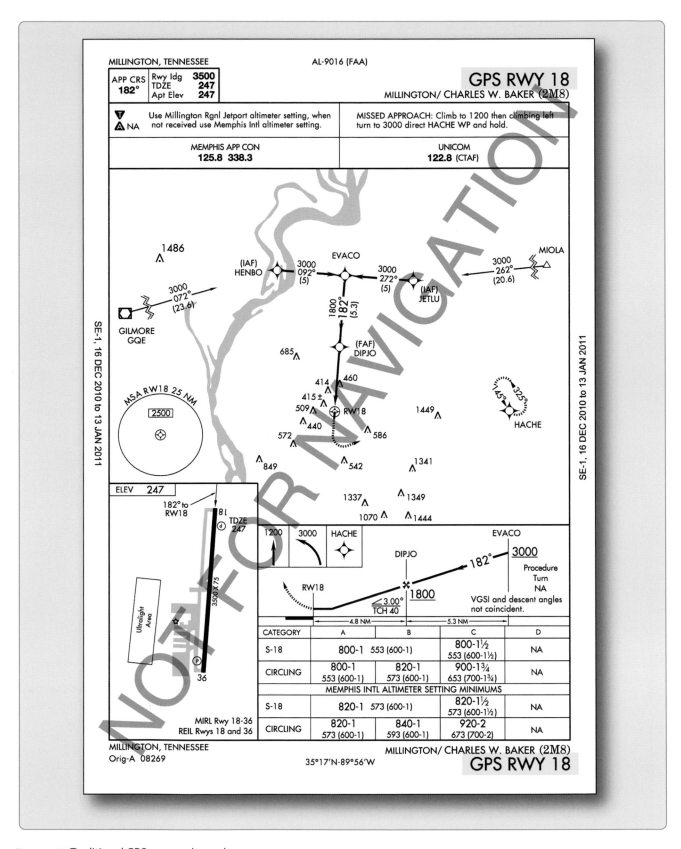

Figure 4-39. Traditional GPS approach overlay.

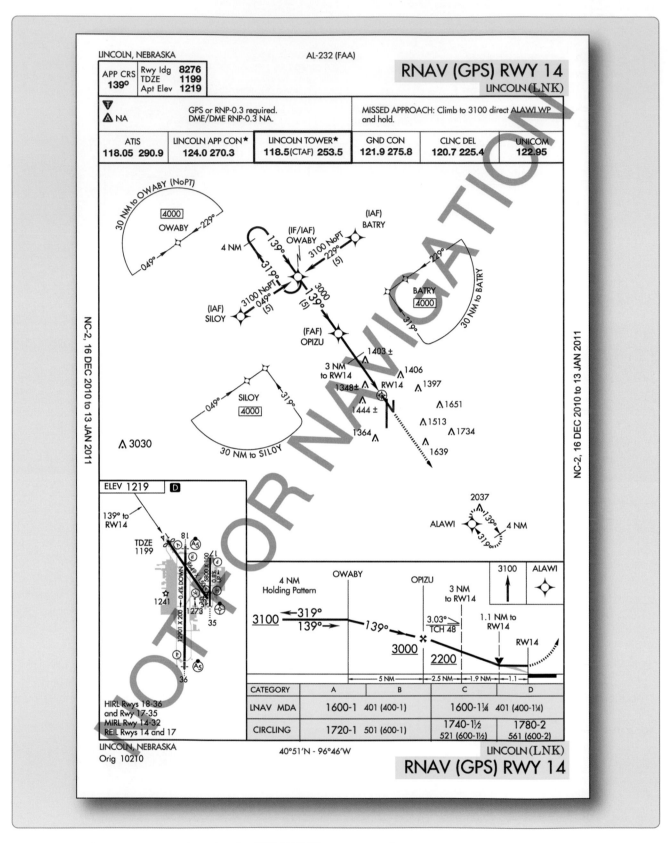

Figure 4-40. Lincoln Muni KLNK Lincoln, Nebraska, RNAV GPS RWY 14 approach.

and each runway must have an ILS. Additionally, separate procedures must be established for each approach, and each approach must have a MAP at least 3 NM apart with no overlapping of the protected missed approach airspace. Only straight-in approaches are approved for converging ILS procedures. If the runways intersect, the controller must be able to visually separate intersecting runway traffic.

Approaches to intersecting runways generally have higher minimums, commonly with 600-foot ceiling and 1 1/4 to 2 mile visibility requirements. Pilots are informed of the use of converging ILS approaches by the controller upon initial contact or through ATIS. [Figure 4-51]

Dallas/Fort Worth International airport is one of the few airports that makes use of converging ILS approaches because its runway configuration has multiple parallel runways and two offset runways. [Figure 4-52] The approach chart title indicates the use of converging approaches and the notes section highlights other runways that are authorized for converging approach procedures. Note the lsight different in charting titles on the IAPs. Soon all Converging ILS procedures will be charted in the newer format shown in Figure 4-51, with the use of "V" in the title, and "CONVERGING" in parenthesis

VOR Approach

The VOR is one of the most widely used non-precision approach types in the NAS. VOR approaches use VOR facilities both on and off the airport to establish approaches and include the use of a wide variety of equipment, such as DME and TACAN. Due to the wide variety of options included in a VOR approach, TERPS outlines design criteria for both on and off airport VOR facilities, as well as VOR approaches with and without a FAF. Despite the various configurations, all VOR approaches are non-precision approaches, require the presence of properly operating VOR equipment, and can provide MDAs as low as 250 feet above the runway. VOR also offers a flexible advantage in that an approach can be made toward or away from the navigational facility.

The VOR approach into Fort Rucker, Alabama, is an example of a VOR approach where the VOR facility is on the airport and there is no specified FAF. [Figure 4-53] For a straight-in approach, the final approach course is typically aligned to intersect the extended runway centerline 3,000 feet from the runway threshold, and the angle of convergence between the two does not exceed 30°. This type of VOR approach also includes a minimum of 300 feet of obstacle clearance in the final approach area. The final approach area criteria include a 2 NM wide primary area at the facility that expands to 6 NM wide at a distance of 10 NM from

the facility. Additional approach criteria are established for courses that require a high altitude teardrop approach penetration.

When DME is included in the title of the VOR approach, operable DME must be installed in the aircraft in order to fly the approach from the FAF. The use of DME allows for an accurate determination of position without timing, which greatly increases situational awareness throughout the approach. Alexandria, Louisiana, is an excellent example of a VOR/DME approach in which the VOR is off the airport and a FAF is depicted. [Figure 4-54] In this case, the final approach course is a radial or straight-in final approach and is designed to intersect the runway centerline at the runway threshold with the angle of convergence not exceeding 30°.

The criteria for an arc final approach segment associated with a VOR/DME approach is based on the arc being beyond 7 NM and no farther than 30 NM from the VOR and depends on the angle of convergence between the runway centerline and the tangent of the arc. Obstacle clearance in the primary area, which is considered the area 4 NM on either side of the arc centerline, is guaranteed by at least 500 feet.

NDB Approach

Like the VOR approach, an NDB approach can be designed using facilities both on and off the airport, with or without a FAF, and with or without DME availability. At one time, it was commonplace for an instrument student to learn how to fly an NDB approach, but with the growing use of GPS, many pilots no longer use the NDB for instrument approaches. New RNAV approaches are also rapidly being constructed into airports that are served only by NDB. The long-term plan includes the gradual phase out of NDB facilities, and eventually, the NDB approach becomes nonexistent. Until that time, the NDB provides additional availability for instrument pilots into many smaller, remotely located airports.

The NDB Runway 35 approach at Carthage/Panola County Sharpe Field is an example of an NDB approach established with an on-airport NDB that does not incorporate a FAF. [Figure 4-55] In this case, a procedure turn or penetration turn is required to be a part of the approach design. For the NDB to be considered an on-airport facility, the facility must be located within one mile of any portion of the landing runway for straight-in approaches and within one mile of any portion of usable landing surface for circling approaches. The final approach segment of the approach is designed with a final approach area that is 2.5 NM wide at the facility and increases to 8 NM wide at 10 NM from the facility. Additionally, the final approach course and

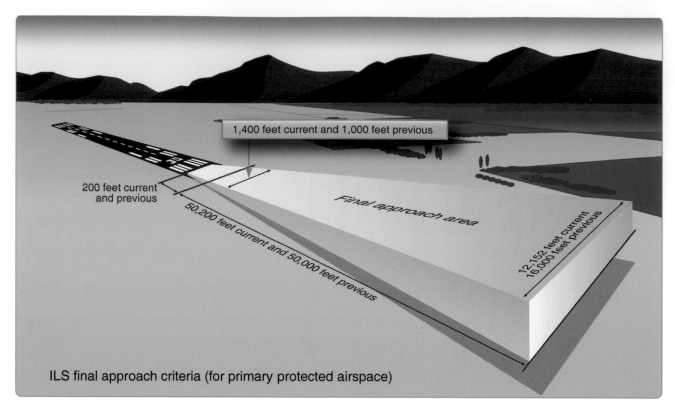

Figure 4-41. ILS final approach segment design criteria.

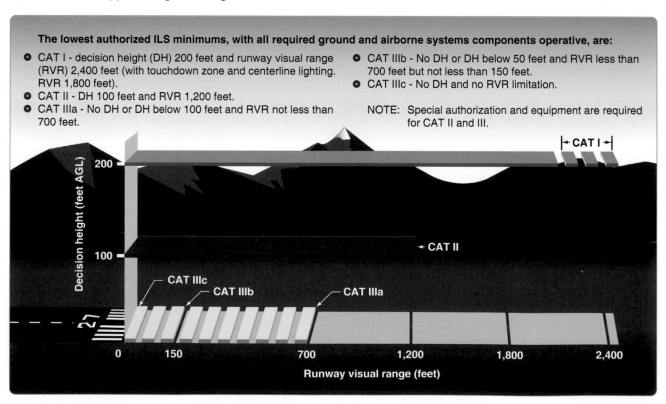

The lowest authorized ILS minimums, with all required ground and airborne systems components operative, are:

- CAT I - decision height (DH) 200 feet and runway visual range (RVR) 2,400 feet (with touchdown zone and centerline lighting. RVR 1,800 feet).
- CAT II - DH 100 feet and RVR 1,200 feet.
- CAT IIIa - No DH or DH below 100 feet and RVR not less than 700 feet.
- CAT IIIb - No DH or DH below 50 feet and RVR less than 700 feet but not less than 150 feet.
- CAT IIIc - No DH and no RVR limitation.

NOTE: Special authorization and equipment are required for CAT II and III.

Figure 4-42. ILS approach categories.

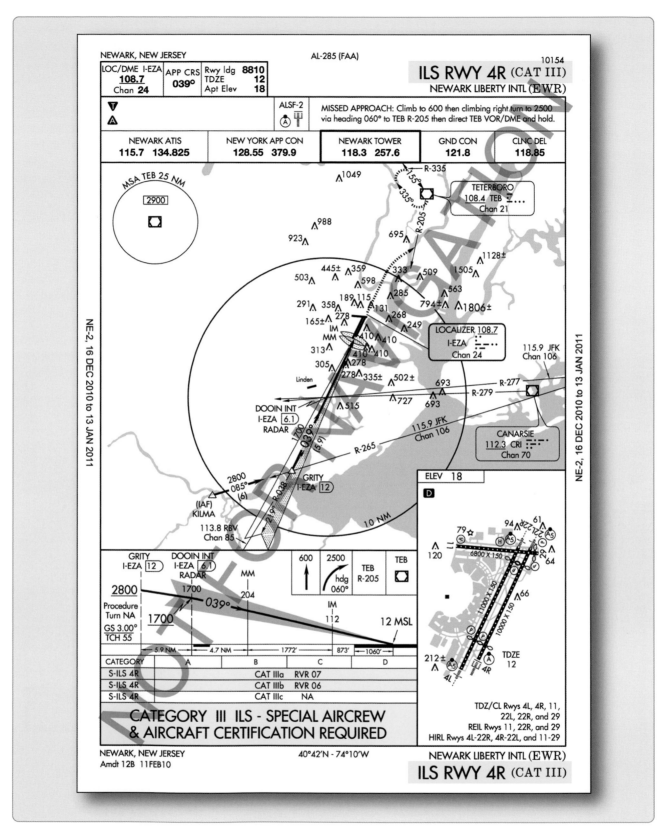

Figure 4-43. Category III approach procedure.

the extended runway centerline angle of convergence cannot exceed 30° for straight-in approaches. This type of NDB approach is afforded a minimum of 350 feet obstacle clearance.

When a FAF is established for an NDB approach, the approach design criteria changes. It also takes into account whether or not the NDB is located on or off the airport. Additionally, this type of approach can be made both moving toward or away from the NDB facility. The Tuscon Ryan Field, NDB/DME RWY 6 is an approach with a FAF using an on-airport NDB facility that also incorporates the use of DME. [Figure 4-56] In this case, the NDB has DME capabilities from the LOC approach system installed on the airport. While the alignment criteria and obstacle clearance remain the same as an NDB approach without a FAF, the final approach segment area criteria changes to an area that is 2.5 NM wide at the facility and increases to 5 NM wide, 15 NM from the NDB.

Radar Approaches

The two types of radar approaches available to pilots when operating in the NAS are precision approach radar (PAR) and airport surveillance radar (ASR). Radar approaches may be given to any aircraft at the pilot's request. ATC may also offer radar approach options to aircraft in distress regardless of the weather conditions or as necessary to expedite traffic. Despite the control exercised by ATC in a radar approach environment, it remains the pilot's responsibility to ensure the approach and landing minimums listed for the approach are appropriate for the existing weather conditions considering personal approach criteria certification and company OpSpecs.

Perhaps the greatest benefit of either type of radar approach is the ability to use radar to execute a no gyro approach. Assuming standard rate turns, ATC can indicate when to begin and end turns. If available, pilots should make use of

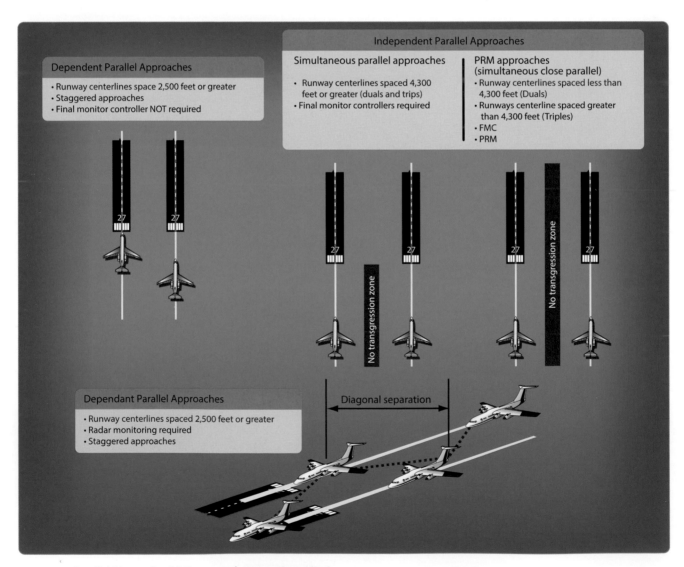

Figure 4-44. Parallel (dependent) ILS approach separation criteria.

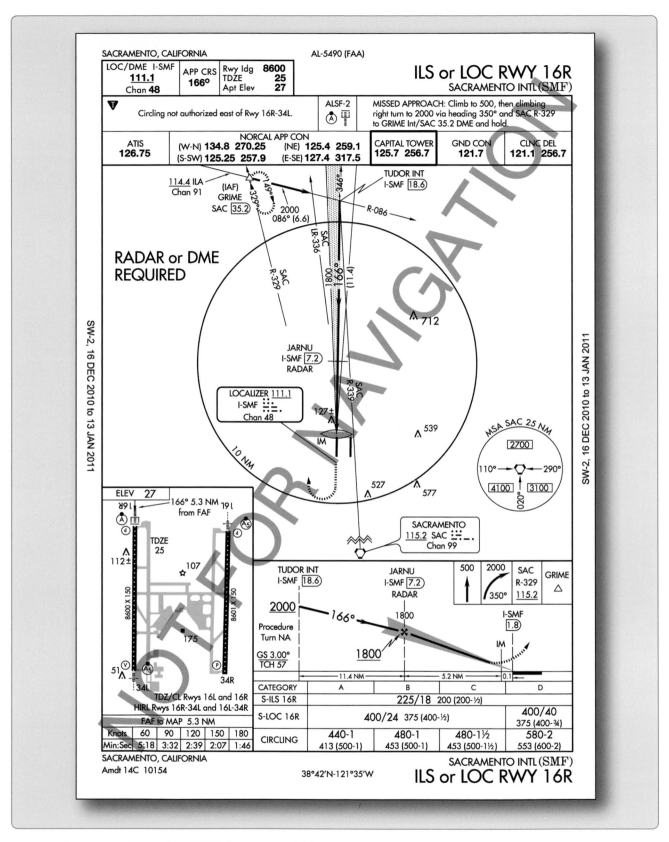

Figure 4-45. Sacramento International KSMF, Sacramento, California, ILS RWY 16L.

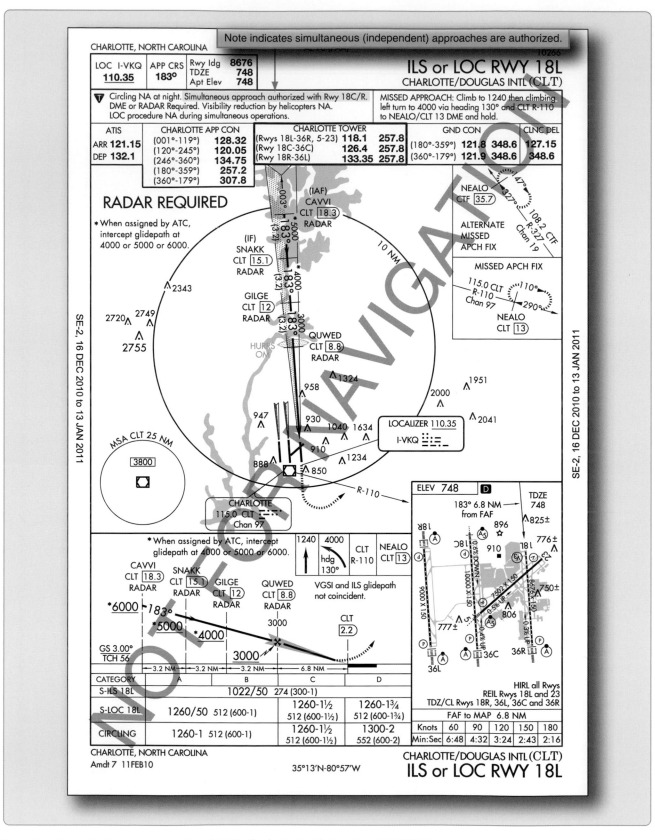

Figure 4-46. Charlotte Douglas International KCLT, Charlotte, North Carolina, ILS RWY 18.

this approach when the heading indicator has failed and partial panel instrument flying is required.

Information about radar approaches is published in tabular form in the front of the TPP booklet. PAR, ASR, and circling approach information including runway, DA, DH, or MDA, height above airport (HAA), HAT, ceiling, and visibility criteria are outlined and listed by specific airport.

Regardless of the type of radar approach in use, ATC monitors aircraft position and issues specific heading and altitude information throughout the entire approach. Particularly, lost communications procedures should be briefed prior to execution to ensure pilots have a comprehensive understanding of ATC expectations if radio communication were lost. ATC also provides additional information concerning weather and missed approach instructions when beginning a radar approach. [Figure 4-57]

Precision Approach Radar (PAR)

PAR provides both vertical and lateral guidance, as well as range, much like an ILS, making it the most precise radar approach available. The radar approach, however, is not able to provide visual approach indications in the flight deck. This requires the flight crew to listen and comply with controller instructions. PAR approaches are rare, with most of the approaches used in a military setting; any opportunity to practice this type of approach is beneficial to any flight crew.

The final approach course of a PAR approach is normally aligned with the runway centerline, and the associated glideslope is typically no less than 2.5° and no more than 3°. Obstacle clearance for the final approach area is based on the particular established glideslope angle and the exact formula is outlined in Order 8260.3, Volume 3, Chapter 3. [Figure 4-58]

Airport Surveillance Radar (ASR)

ASR approaches are typically only approved when necessitated for an ATC operational requirement or in an unusual or emergency situation. This type of radar only provides heading and range information, although the

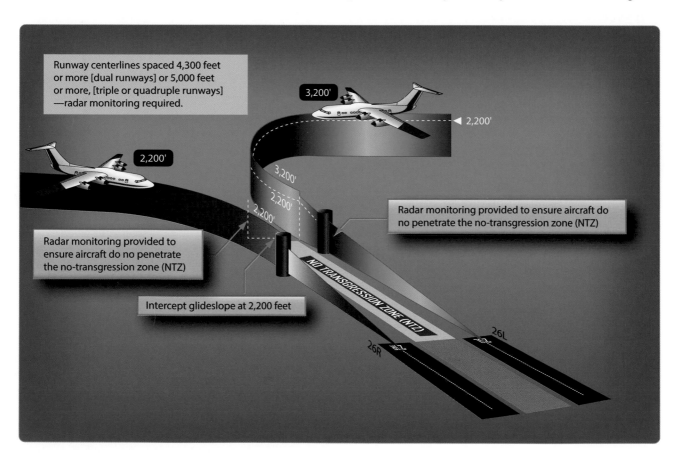

Figure 4-47. Simultaneous parallel ILS approach criteria.

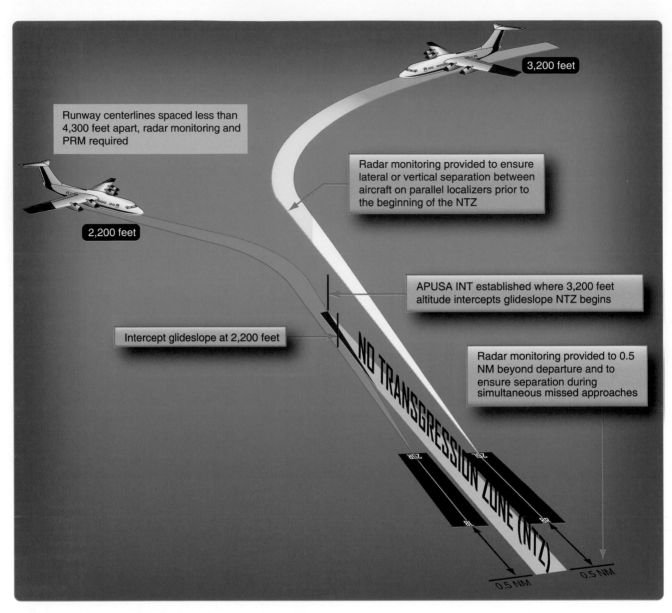

Figure 4-48. Simultaneous close parallel ILS approach ILS PRM criteria.

controller can advise the pilot of the altitude where the aircraft should be based on the distance from the runway. An ASR approach procedure can be established at any radar facility that has an antenna within 20 NM of the airport and meets the equipment requirements outlined in FAA Order 8200.1, U.S. Standard Flight Inspection Manual. ASR approaches are not authorized for use when Center Radar ARTS processing (CENRAP) procedures are in use due to diminished radar capability.

The final approach course for an ASR approach is aligned with the runway centerline for straight-in approaches and aligned with the center of the airport for circling approaches. Within the final approach area, the pilot is also guaranteed a minimum of 250 feet obstacle clearance. ASR

descent gradients are designed to be relatively flat, with an optimal gradient of 150 feet per mile and never exceeding 300 feet per mile.

Localizer Approaches

As an approach system, the localizer is an extremely flexible approach aid that, due to its inherent design, provides many applications for a variety of needs in instrument flying. An ILS glideslope installation may be impossible due to surrounding terrain. For whatever reason, the localizer is able to provide four separate applications from one approach system:

- Localizer approach
- Localizer/DME approach

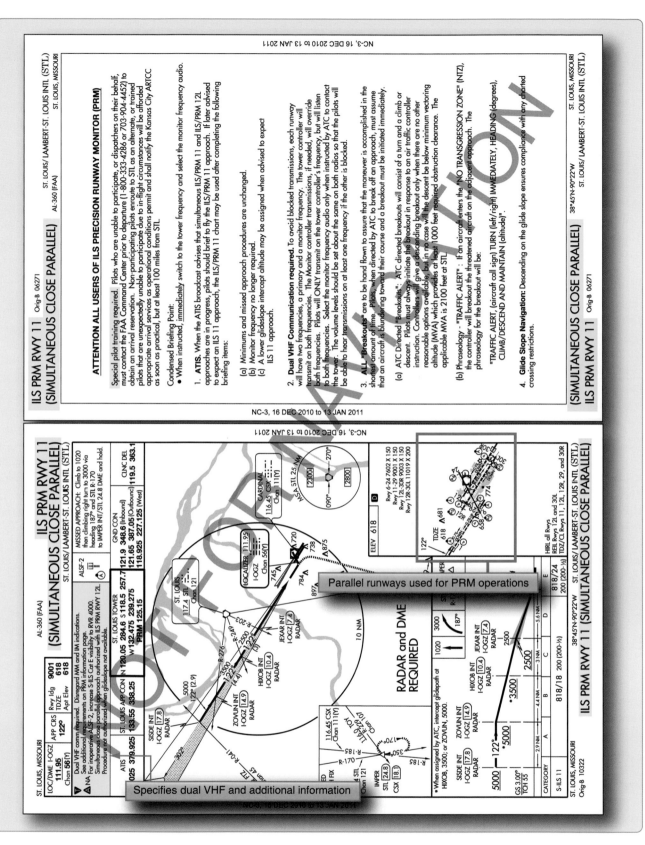

Figure 4-49. St Louis, Missouri, ILS PRM RWY 11.

- Localizer back course approach
- Localizer-type directional aid (LDA)

Localizer and Localizer DME

The localizer approach system can provide both precision and non-precision approach capabilities to a pilot. As a part of the ILS system, the localizer provides horizontal guidance for a precision approach. Typically, when the localizer is discussed, it is thought of as a non-precision approach due to the fact that either it is the only approach system installed, or the glideslope is out of service on the ILS. In either case, the localizer provides a non-precision approach using a localizer transmitter installed at a specific airport. [Figure 4-59]

TERPS provides the same alignment criteria for a localizer approach as it does for the ILS, since it is essentially the same approach without vertical guidance stemming from the glideslope. A localizer is always aligned within 3° of the runway, and it is afforded a minimum of 250 feet obstacle clearance in the final approach area. In the case of a localizer DME (LOC DME) approach, the localizer installation has a collocated DME installation that provides distance information required for the approach. [Figure 4-60]

Localizer Back Course

In cases where an ILS is installed, a back course may be available in conjunction with the localizer. Like the localizer, the back course does not offer a glideslope, but remember that the back course can project a false glideslope signal and the glideslope should be ignored. Reverse sensing occurs on the back course using standard VOR equipment. With a horizontal situation indicator (HSI) system, reverse sensing is eliminated if it is set appropriately to the front course. [Figure 4-61]

Localizer-Type Directional Aid (LDA)

The LDA is of comparable use and accuracy to a localizer but is not part of a complete ILS. The LDA course usually provides a more precise approach course than the similar simplified directional facility (SDF) installation, which may have a course width of 6° or 12°.

The LDA is not aligned with the runway. Straight-in minimums may be published where alignment does not exceed 30° between the course and runway. Circling minimums only are published where this alignment exceeds 30°.

A very limited number of LDA approaches also incorporate a glideslope. These are annotated in the plan view of the instrument approach chart with a note, "LDA/Glideslope."

These procedures fall under a newly defined category of approaches called Approach (Procedure) with Vertical Guidance (aviation) APVs. LDA minima for with and without glideslope is provided and annotated on the minima lines of the approach chart as S–LDA/GS and S–LDA. Because the final approach course is not aligned with the runway centerline, additional maneuvering is required compared to an ILS approach. [Figure 4-62]

Simplified Directional Facility (SDF)

The SDF provides a final approach course similar to that of the ILS localizer. It does not provide glideslope information. A clear understanding of the ILS localizer and the additional factors listed below completely describe the operational characteristics and use of the SDF. [Figure 4-63]

The approach techniques and procedures used in an SDF instrument approach are essentially the same as those employed in executing a standard localizer approach except the SDF course may not be aligned with the runway and the course may be wider, resulting in less precision. Like the LOC type approaches, the SDF is an alternative approach that may be installed at an airport for a variety of reasons, including terrain. The final approach is provided a minimum of 250 feet obstacle clearance for straight-in approaches while in the final approach area, which is an area defined for a 6° course: 1,000 feet at or abeam the runway threshold expanding to 19,228 feet (10 NM) from the threshold. The same final approach area for a 12° course is larger. This type of approach is also designed with a maximum descent gradient of 400 feet per NM, unless circling only minimums are authorized.

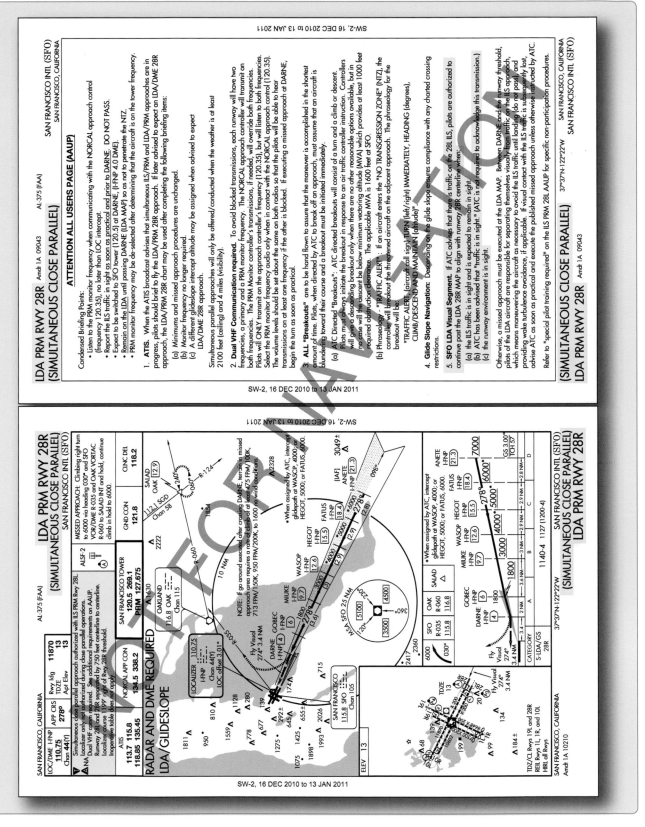

Figure 4-50. Simultaneous offset instrument approach procedure.

4-79

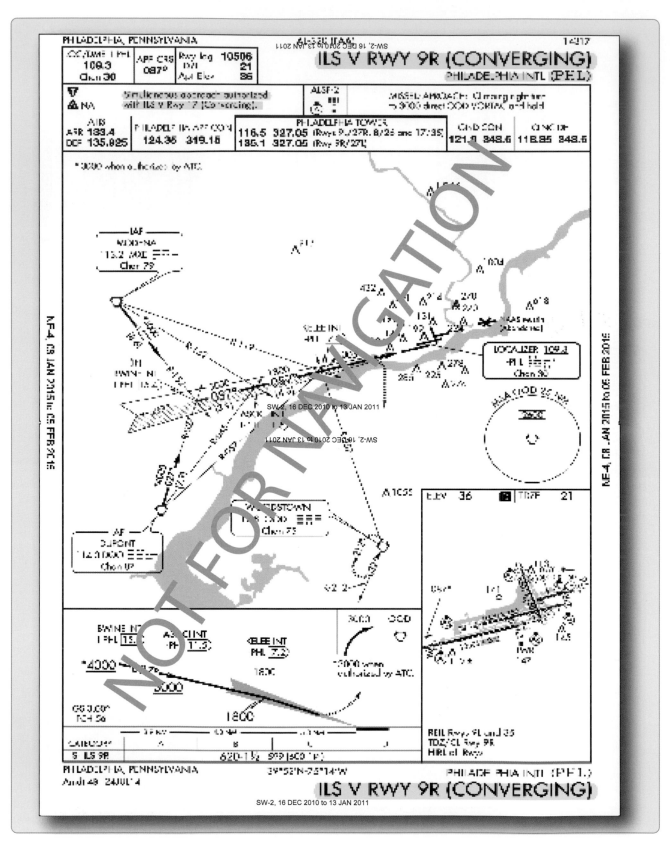

Figure 4-51. Converging approach criteria.

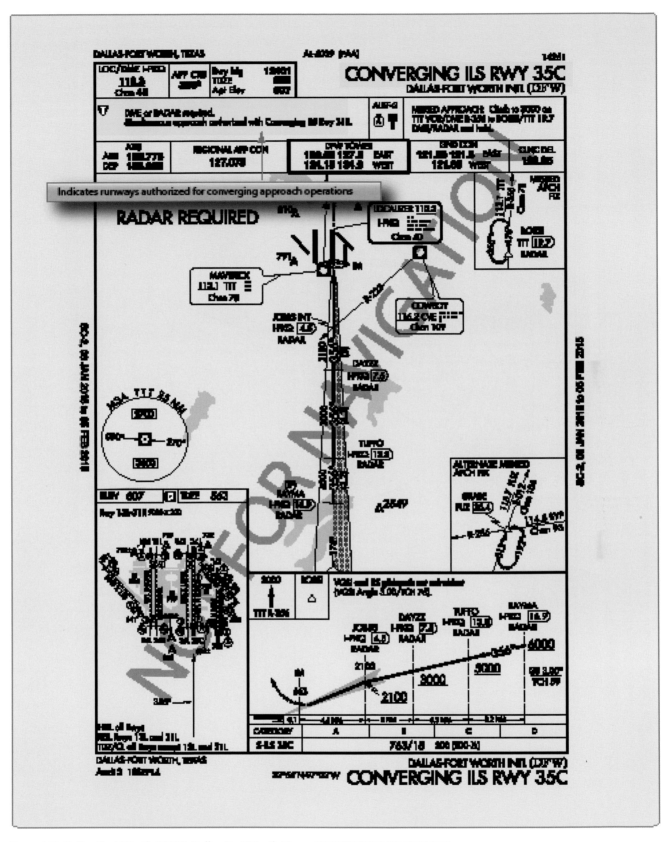

Figure 4-52. Dallas-Fort Worth KDFW, Dallas-Fort Worth, Texas, CONVERGING ILS RWY 35C.

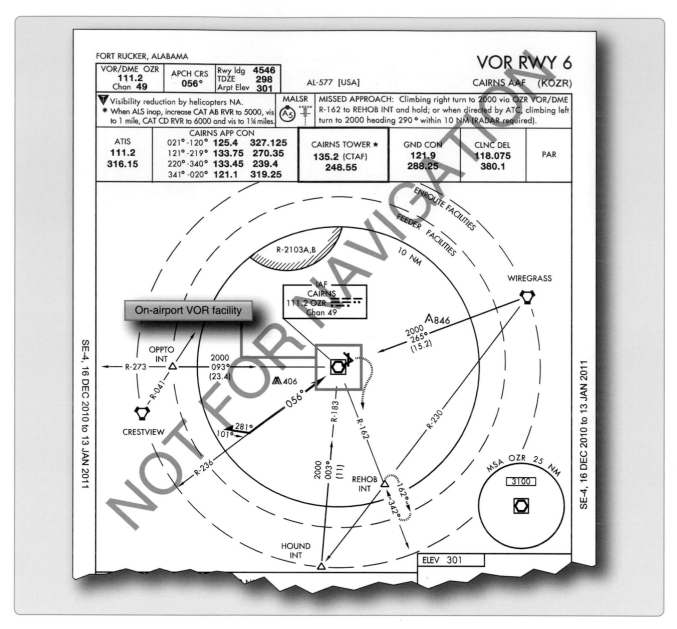

Figure 4-53. Fort Rucker, Alabama, KOZR VOR RWY 6.

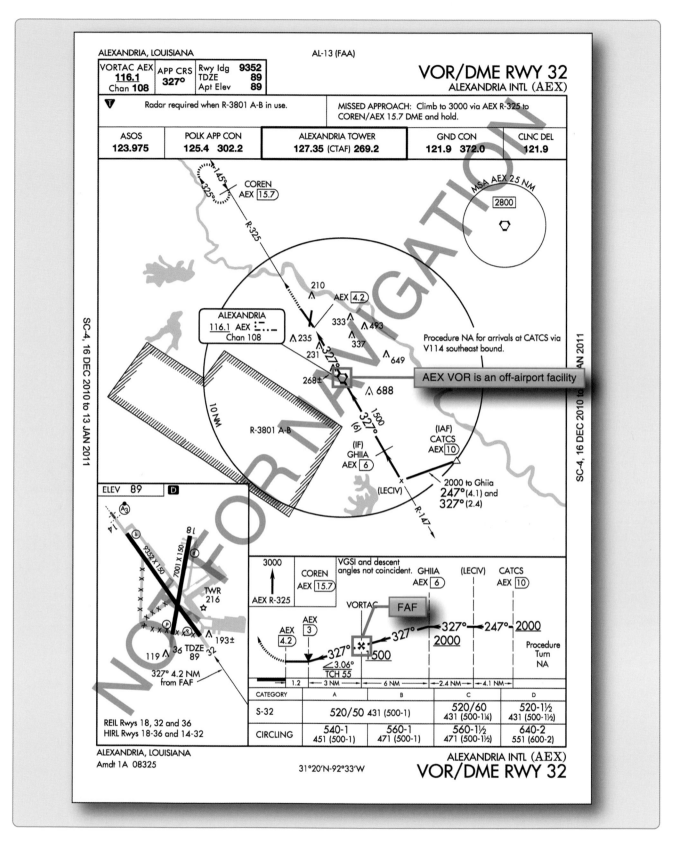

Figure 4-54. Alexandria International (AEX), Alexandria, Louisiana, KAEX VOR DME RWY 32.

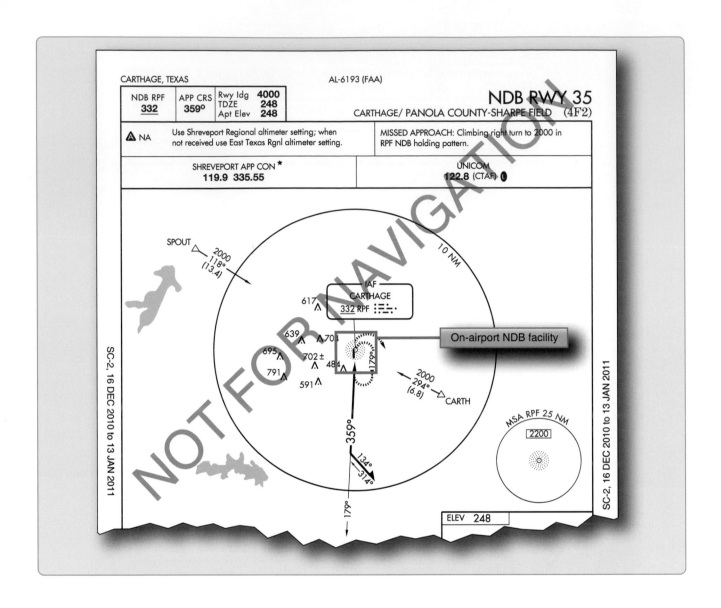

Figure 4-55. Carthage/Panola County-Sharpe Field, Carthage, Texas, (K4F2), NDB RWY 35.

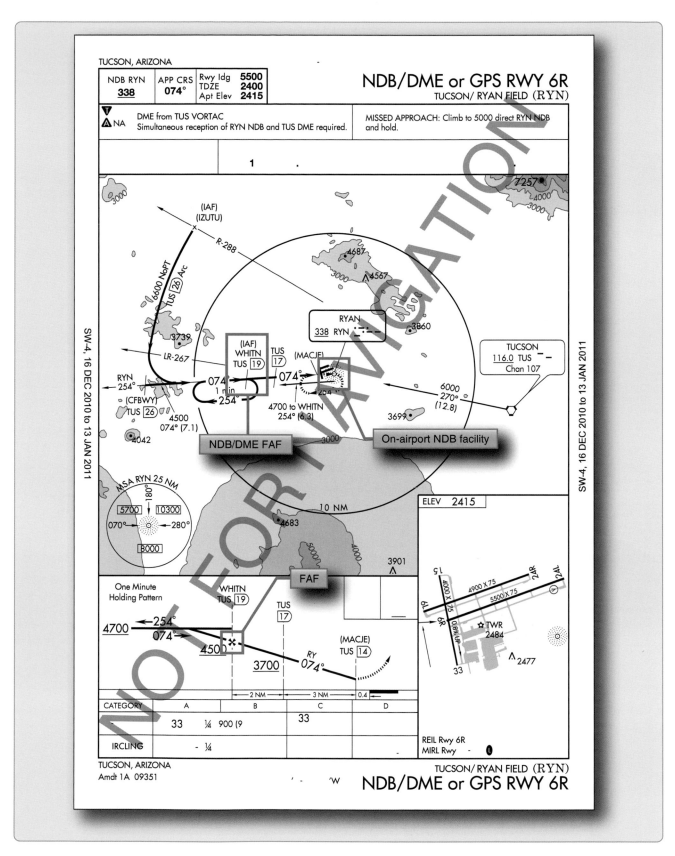

Figure 4-56. Tucson/Ryan Field, Tuscson, Arizona, (KRYN), NDB/DME or GPS RWY 6R.

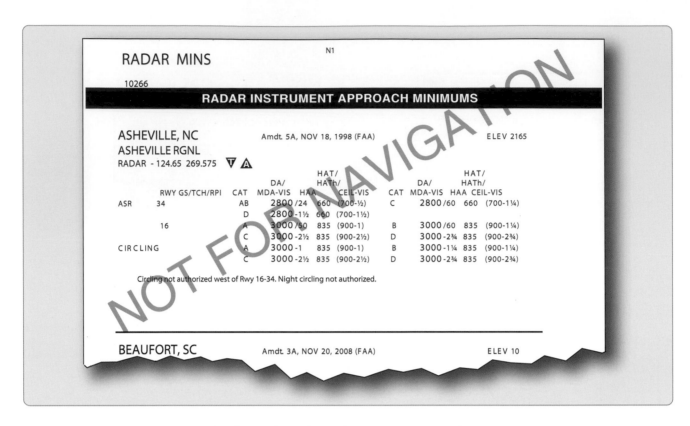

Figure 4-57. Asheville Regional KAVL, Asheville, North Carolina, radar instrument approach minimums.

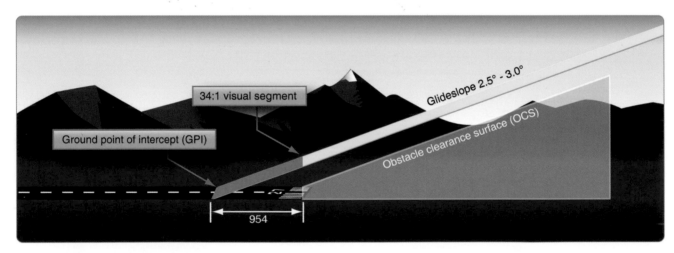

Figure 4-58. PAR final approach area criteria.

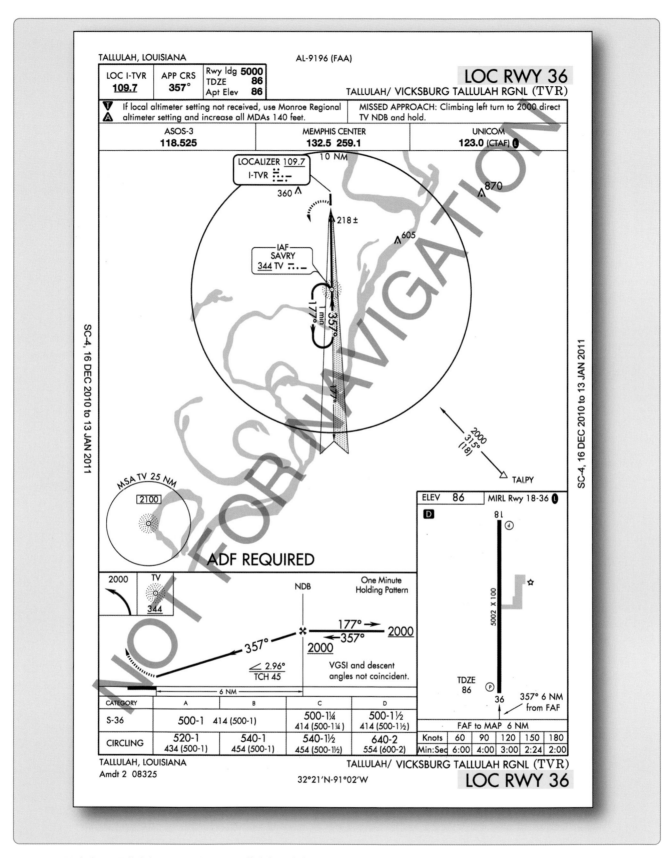

Figure 4-59. Vicksburg Tallulah Regional KTVR, Tallulah Vicksburg, Louisiana, LOC RWY 36.

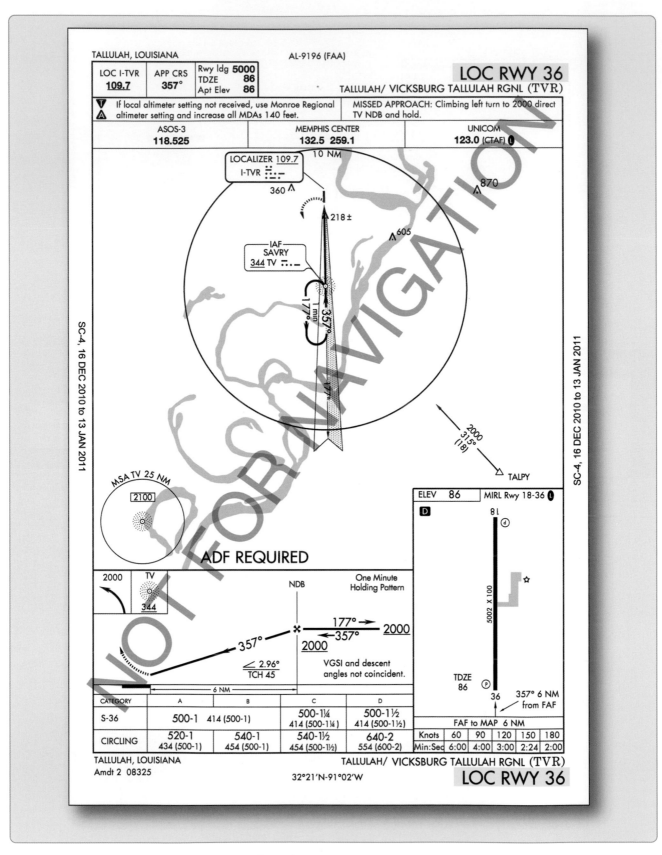

Figure 4-60. Davidson County KEXX, Lexington, North Carolina, LOC DME RWY 6.

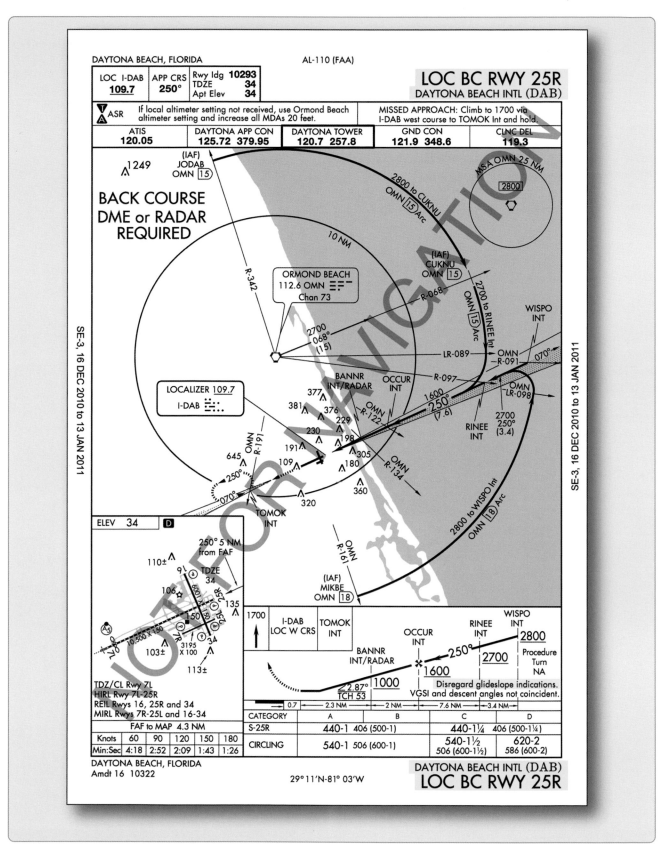

Figure 4-61. Dayton Beach International DAB, Dayton Beach, Florida, LOC BC RWY 25R.

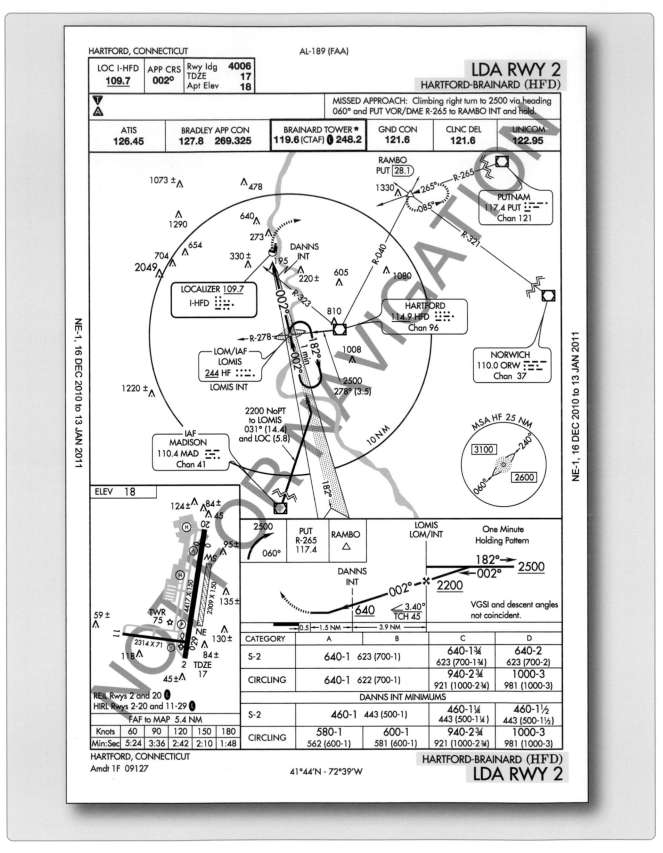

Figure 4-62. Hartford Brainard KHFD, Hartford, Connecticut, LDA RWY 2.

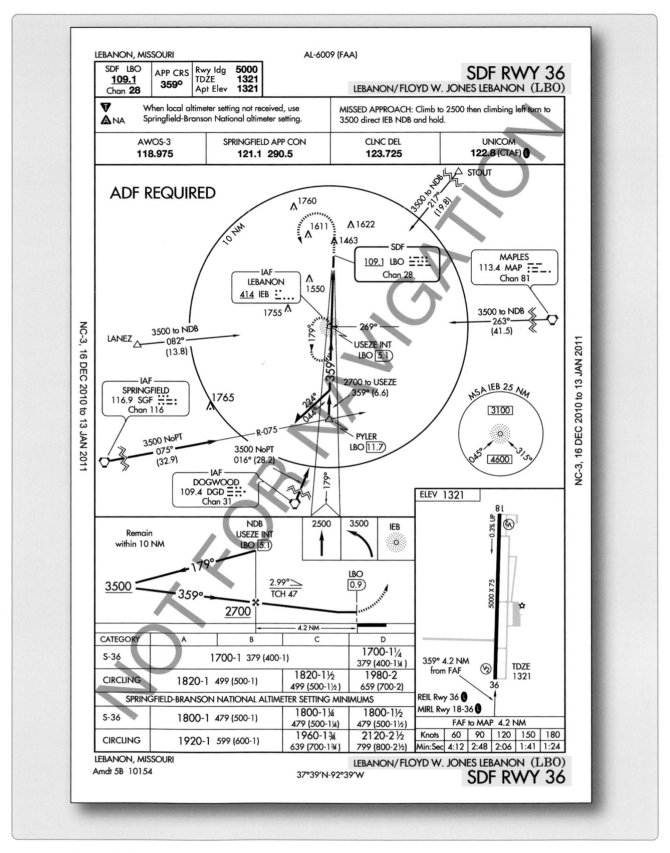

Figure 4-63. Lebanon Floyd W Jones, Lebonon, Missouri, SDF RWY 36.

Chapter 5
Improvement Plans

Introduction

In the upcoming years, exciting new technologies will be developed and implemented to help ease air traffic congestion, add to system capacity, and enhance safety. Some of these seamless changes will be invisible to pilots. Others will entail learning new procedures, aircraft equipment, and systems that will introduce powerful new capabilities and dramatically increase the safety of all flight operations.

Next Generation Air Transportation (NextGen) System

Next Generation Air Transportation System (NextGen) is a comprehensive overhaul of the National Airspace System (NAS) designed to make air travel more convenient and dependable, while ensuring flights are as safe and secure as possible. It moves away from ground-based surveillance and navigation to new and more dynamic satellite-based systems and procedures, and introduces new technological innovations in areas such as weather forecast, digital communications, and networking. [Figure 5-1] When fully implemented, NextGen will safely allow aircraft to fly more closely together on more direct routes, reducing delays, and providing unprecedented benefits for the environment and the economy through reductions in carbon emissions, fuel consumption, and noise. [Figure 5-2]

Implementation in stages across the United States is due between 2012 and 2025. In order to implement NextGen, the Federal Aviation Administration (FAA) will undertake

a wide-range transformation of the entire United States air transportation system. NextGen consists of the following five systems:

1. Automatic dependent surveillance-broadcast (ADS-B)—will use the global positioning system (GPS) satellite signals to provide air traffic control (ATC) and pilots with more accurate information that keeps aircraft safely separated in the sky and on runways. [Figure 5-3] Aircraft transponders receive GPS signals and use them to determine the aircraft's precise position in the sky. This and other data is then broadcast to other aircraft and ATC. Once fully established, both pilots and ATC will, for the first time, see the same real-time display of air traffic, substantially improving safety. The FAA will mandate the avionics necessary for implementing ADS-B.

2. System wide information management (SWIM)— will provide a single infrastructure and information management system to deliver high quality, timely data to many users and applications. By reducing

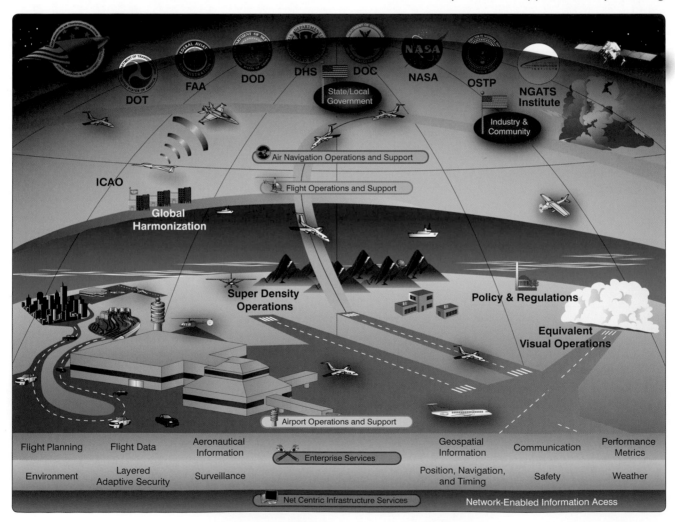

Figure 5-1. Next Generation Air Transportation System (NEXGEN) introduces new technological innovations for weather forecasting, digital communications, and networking.

Figure 5-2. Satellite-based navigation and tracking allows more aircraft to fly closely together on more direct routes.

Figure 5-3. Automatic Dependent Surveillance-Broadcast (ADS-B) systems.

the number and types of interfaces and systems, SWIM will reduce data redundancy and better facilitate multi- user information sharing. SWIM will also enable new modes of decision-making as information is more easily accessed. [Figure 5-4]

3. Next generation data communications—current communications between aircrew and ATC, and between air traffic controllers, are largely realized through voice communications. Initially, the introduction of data communications will provide an additional means of two-way communication for ATC clearances, instructions, advisories, flight crew requests, and reports. With the majority of aircraft data link equipped, the exchange of routine controller-pilot messages and clearances via data link will enable controllers to handle more traffic. This will improve ATC productivity, enhancing capacity and safety. [Figure 5-5]

4. Next generation network enabled weather (NNEW)— seventy percent of NAS delays are attributed to weather every year. The goal of NNEW is to cut weather-related delays at least in half. Tens of thousands of global weather observations and sensor reports from ground, airborne, and space-based sources will fuse into a single national weather information system updated in real time. NNEW will provide a common weather picture across the NAS and enable better air transportation decision-making. [Figure 5-6]

5. NAS voice switch (NVS)—there are currently seventeen different voice switching systems in the NAS; some in use for more than twenty years. NVS will replace these systems with a single air/ground and ground/ground voice communications system. [Figure 5-7]

NextGen Existing Improvements

The goal of NextGen is to provide new capabilities that make air transportation safer and more reliable while improving the capacity of the NAS and reducing aviation's

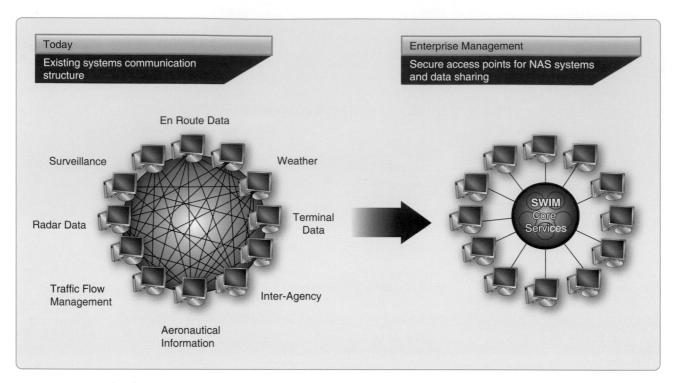

Figure 5-4. System wide information management (SWIM)—an information management system that helps deliver high quality, timely data to improve the efficiency of the national airspace.

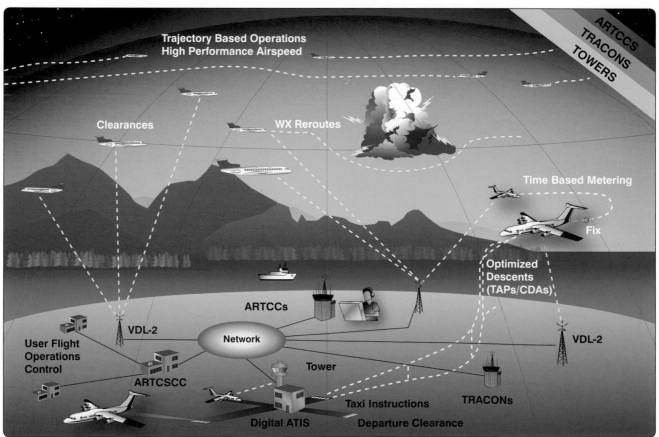

Figure 5-5. Next generation data communications provides an additional means of two-way communication for ATC clearances, instructions, advisories, flight crew requests, and reports.

Figure 5-6. Next generation network enabled weather (NNEW) provides a common weather picture across the NAS.

Figure 5-7. National airspace voice switch (NVS) will replace existing voice switching systems with single air/ground and ground/ground voice communication systems.

impact on the environment. Below is a list of some of the capabilities for operational use that have already been implemented through NextGen.

1. Starting in December 2009, the FAA began controlling air traffic over the Gulf of Mexico, an area of active airspace where surveillance was never before possible, using the satellite-based technology of ADS-B. [Figure 5-8] Having a real-time visual representation of aircraft flying over the Gulf of Mexico, where no radar coverage was available, means that ATC can safely and more efficiently separate air traffic. It also provides pilots with more safety benefits such as improved situational awareness (SA), new weather information, and additional voice communications.

2. Initial operating capability was achieved for ADS-B at Louisville, Kentucky, where ADS-B was integrated into the Common Automated Radar Terminal System.

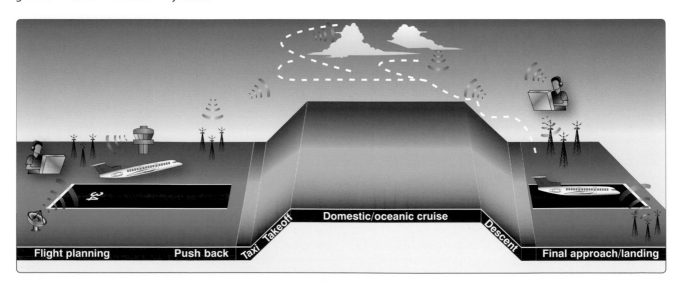

Figure 5-8. ADS-B, is being used to provide ATC surveillance over the Gulf of Mexico.

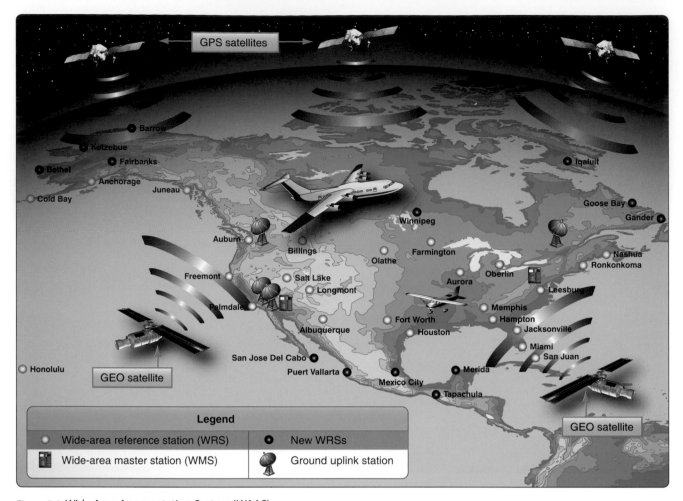

GPS satellites

GEO satellite

GEO satellite

Legend		
⊙ Wide-area reference station (WRS)	⊙	New WRSs
🖳 Wide-area master station (WMS)	📡	Ground uplink station

Figure 5-9. Wide Area Augmentation System (WAAS).

3. Satellite-based technologies, including the Wide Area Augmentation System (WAAS), are improving access to runways at both large and small airports. [Figure 5-9] Directions and maps have been published for more than 500 precision-like approaches enabled by WAAS. Localizer performance with vertical guidance (LPV) procedures improves access to airports in lower visibility conditions and where obstacles are present. These procedures are particularly valuable for smaller airports used by general aviation. There are now oer 2,300 LPV procedures available at runways where no instrument landing system (ILS) is present.

4. The Ground-Based Augmentation System (GBAS) has been approved for Category I operations and the first satellite-based system has been approved for this category of precision approach which enables instrument-based operations down to 200 feet above the surface even during reduced visibility. [Figure 5-10] GBAS was installed at Houston, Texas and Newark, New Jersey airport in 2009.

5. Multilateration, a ground-based surveillance technology, is being implemented to help improve runway access. The FAA installed and is now using wide area multilateration (WAM) systems to control air traffic in Juneau, Alaska, and at four airports in Colorado. This allows air traffic to be safely separated by five miles whereas before each aircraft had to clear the airspace around the airport before the next could enter.

6. New runways at Chicago O'Hare, Washington Dulles, and Seattle-Tacoma Airports opened in November of 2008, which are now beginning to have a reduction in delays.

Benefits of NextGen

The implementation of NextGen will allow pilots and dispatchers to select their own direct flightpaths, rather than follow the existing Victor, Jet, and LF/MF airways. Each airplane will transmit and receive precise information about the time at which it and others will cross key points along their paths. Pilots and air traffic managers on the ground

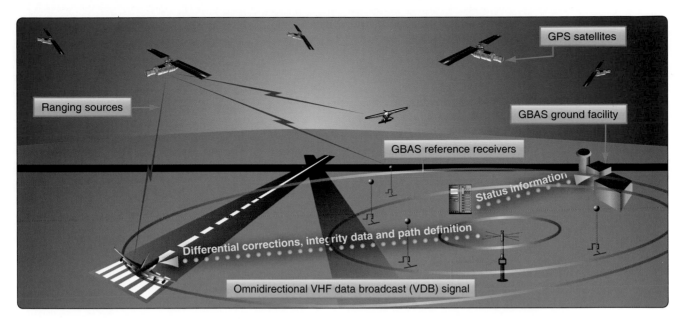

Figure 5-10. Ground-Based Augmentation System (GBAS).

will have the same precise information transmitted via data communications.

Major demand and capacity imbalances will be worked collaboratively between FAA air traffic managers and flight operations. The increased scope, volume, and widespread distribution of information by SWIM will improve decision- making and let more civil aviation authorities participate. The impact of weather on flight operations will be reduced through the use of improved information sharing, new technology to sense and mitigate the impacts of the weather, and to improve weather forecasts and decision-making. Better forecasts, coupled with greater automation, will minimize airspace limitations and traffic restrictions.

The new procedures of NextGen will improve airport surface movements, reduce spacing and separation

Figure 5-11. NextGen improves airport surface movements, reduces spacing and separation requirements, and better manages the overall flows into and out of busy airports.

requirements, and better manage the overall flows into and out of busy airspace, as well as provide maximum use of busy airports. [Figure 5-11] Targeting NextGen at the whole of the NAS, rather than just the busiest airports, will uncover untapped capacity across the whole system. During busy traffic periods, NextGen will rely on aircraft to fly precise routes into and out of many airports to increase throughput. For more information on NextGen, visit www. faa.gov/nextgen.

Head-Up Displays (HUD)

As aircraft became more sophisticated and electronic instrument landing systems (ILS) were developed in the 1930s and 1940s, it was necessary while landing in poor weather for one pilot to monitor the instruments to keep the aircraft aligned with radio beams while the second pilot divided time between monitoring the instruments and the outside environment. The pilot monitoring reported the runway environment in sight and the flying pilot completed the approach visually. This is still the standard practice used for passenger carrying aircraft in commercial service while making ILS landings. As single-piloted aircraft became more complex, it became very difficult for pilots to focus on flying the aircraft while also monitoring a large number of navigation, flight, and systems instruments. To overcome this problem, the head-up display (HUD) was developed. By showing airspeed, altitude, heading, and aircraft attitude on the HUD glass, pilots were able to keep their eyes outside of the flight deck rather than have to continuously scan from outside to inside to view the flight instruments. [Figure 5-12] Collimators make the image on the glass appear to be far out in front of the aircraft so that the pilot need not change eye focus to view the relatively

Figure 5-12. Head-up guidance system (HGS).

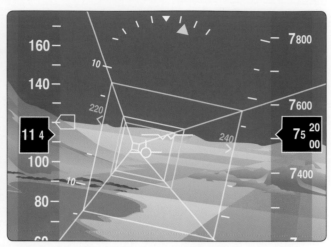

Figure 5-14. A synthetic vision system (SVS) is an electronic means to display a synthetic vision image of the external scene topography to the flight crew to assist during takeoffs, landings, and en route operations.

Figure 5-13. HGS using a holographic display.

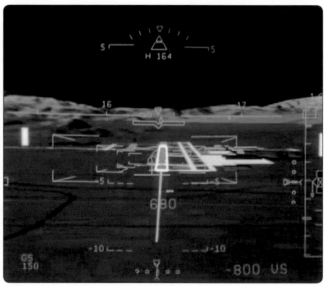

Figure 5-15. An aircraft on an approach equipped with a SVS.

nearby HUD. Today's head-up guidance systems (HGS) use holographic displays. [Figure 5-13] Everything from weapons status to approach information can be shown on current military and civilian HGS displays.

Synthetic and Enhanced Vision Systems
Synthetic Vision System (SVS)

A synthetic vision system (SVS) is an electronic means to display a synthetic vision image of the external scene topography to the flight crew. [Figure 5-14] It is not a real-time image like that produced by an enhanced flight vision system (EFVS). Unlike EFVS, SVS requires a terrain and obstacle database, a precise navigation solution, and a display. The terrain image is based on the use of data

from a digital elevation model (DEM) that is stored within the SVS. With SVS, the synthetic terrain/vision image is intended to enhance pilot awareness of spatial position relative to important features in all visibility conditions. This is particularly useful during critical phases of flight, such as takeoff, approach, and landing where important features such as terrain, obstacles, runways, and landmarks may be depicted on the SVS display. [Figure 5-15] During approach operations, the obvious advantages of SVS are that the digital terrain image remains on the pilot's display regardless of how poor the visibility is outside. An SVS image can be displayed on either a head-down display or head-up display (HUD). Development efforts are currently underway that would combine SVS with a real-time sensor image produced by an EFVS. These systems will be known as Combined Vision Systems (CVS).

Enhanced Flight Vision System (EFVS)

For an in-depth discussion regarding Enhanced Flight Vision Systems, see Chapter 4.

Developing Combined Technology

The United States air transportation system is undergoing a transformation to accommodate a projected three-fold increase in air operations by 2025. Technological and systemic changes are being developed to significantly increase the capacity, safety, efficiency, and security for this Next Generation Air Transportation System (NextGen). One of the key capabilities envisioned to achieve these goals is the concept of equivalent visual operations (EVO), whereby visual flight rules (VFR) and operating procedures, such as separation assurance, are maintained independent of the actual weather conditions. One methodology by which the

Figure 5-18. Portable flight bag.

Figure 5-16. Enhanced and synthetic vision displayed on primary flight displays.

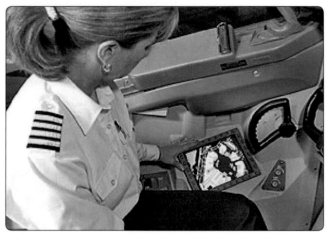

Figure 5-17. Installed flight bag.

goal of EVO might be attainable is to create a virtual visual flight environment for the flight crew, independent of the actual outside weather and visibility conditions, through application of EV and synthetic vision (SV) technologies. [Figure 5-16]

Electronic Flight Bag (EFB)

The electronic flight bag (EFB) is a system for pilots or crewmembers that provide a variety of electronic display, content manipulation, and calculation capabilities. Functions include, but are not limited to, aeronautical charts, documents, checklists, weight & balance, fuel calculations, moving maps, and logbooks.

EFB systems may manage information for use in the cockpit, cabin, and/or in support of ground operations and planning. The use of an EFB is unique to each aircraft operator and, depending on the type of operation, EFB use may require an authorization for use from the FAA issued as either an operations specification (OpSpec), maintenance specification (MSpec), or letter of authorization (LOA).

EFBs can be portable [Figure 5-18] or installed [Figure 5-19] in the aircraft. Portable EFBs may have a provision for securing in the cockpit for use during all phases of flight. The hardware device, whether it's an installed avionics display or portable commercial-off-the-shelf (COTS) device, commonly referred to as a portable electronic device (PED), is not considered to be an EFB unless the hardware device hosts and actively displays either Type A or B software application(s). A non-inclusive list of Type A and B software application examples can be found in appendix 1 and 2 of FAA Advisory Circular (AC) 120-76.

The purpose, technology, and functions for EFB use are rapidly evolving. New and advanced software applications and databases beyond traditional flight bag uses continue to be developed. The FAA has published and continues to update EFB policy and guidance to educate and assist aircraft operators interested in using or obtaining an EFB authorization as appropriate. The most current editions of the following FAA guidance and policy can be accessed from the FAA's website (http://www.faa.gov) or FAA's

Flight Standards Information Management System (FSIMS http://fsims.faa.gov).

- AC 120-76, Guidelines for the Certification, Airworthiness, and Operational Use of Electronic Flight Bags;

- AC 91-78, Use of Class 1 or Class 2 Electronic Flight Bag (EFB);

- AC 20-173, Installation of Electronic Flight Bag Components;

- FAA Order 8900.1 Volume 4, Chapter 15, Section 1, Electronic Flight Bag authorization for use; and

- FAA Order 8900.1 Volume 3, Chapter 18, Section 3, Part A Operations Specifications - General

Access to Special Use Airspace

Special use airspace consists of airspace of defined dimensions identified by an area on the surface of the earth wherein activities must be confined because of their nature, or wherein limitations are imposed upon aircraft operations that are not a part of those activities, or both. Special use airspace includes: restricted airspace, prohibited airspace,

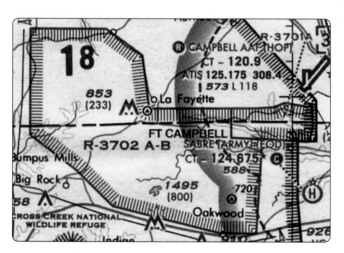

Figure 5-27. Restricted airspace.

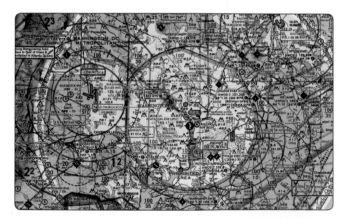

Figure 5-28. Prohibited airspace.

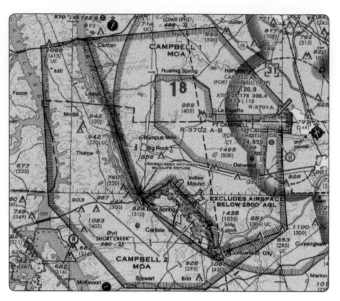

Figure 5-29. Military operations area (MOA).

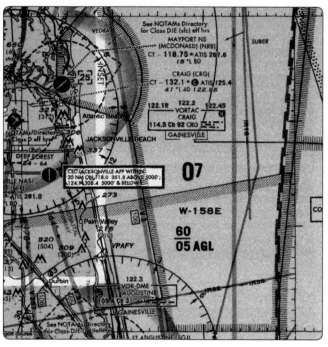

Figure 5-30. Warning area.

Military Operations Areas (MOA), warning areas, alert areas, temporary flight restriction (TFR), and controlled firing areas (CFAs). [Figures 5-27 through 5-32] Prohibited and restricted areas are regulatory special use airspace and are established in 14 CFR Part 73 through the rulemaking process. Warning areas, MOAs, alert areas, and CFAs are non-regulatory special use airspace. All special use airspace descriptions (except CFAs) are contained in FAA Joint Order 7400.8, Special Use Airspace, and are charted on IFR or visual charts and include the hours of operation, altitudes, and the controlling agency. [Figure 5-33]

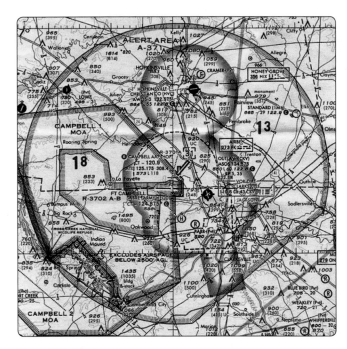

Figure 5-31. Alert area.

The vertical limits of special use airspace are measured by designated altitude floors and ceilings expressed as flight levels or as feet above mean sea level (MSL). Unless otherwise specified, the word "to" (an altitude or flight level) means "to and including" (that altitude or flight level). The horizontal limits of special use airspace are measured by boundaries described by geographic coordinates or other appropriate references that clearly define their perimeter. The period of time during which a designation of special use airspace is in effect is stated in the designation.

Civilians Using Special Use Airspace

The FAA and the Department of Defense (DOD) work together to maximize the use of special use airspace by opening such areas to civilian traffic when they are not being used by the military. The military airspace management system (MAMS) keeps an extensive database of information on the historical use of special use airspace, as well as schedules describing when each area is expected to be active. MAMS transmits the data to the special use airspace management system (SAMS), an FAA program that provides current and scheduled status information on special use airspace to civilian users. The two systems work together to ensure that the FAA and system users have current information on a daily basis. This information is available 24 hours a day at the following link: http://sua. faa.gov. The website merges information for both special use airspace and TFR making it a single comprehensive source to review airspace closure information.

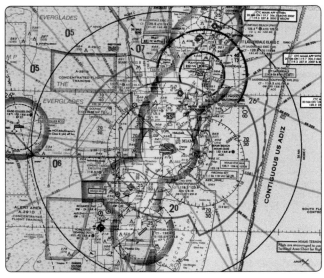

Figure 5-32. Temporary flight restriction (TFR).

The website contains two tabbed pages, List and Map, that display the scheduling and Notice to Airmen (NOTAM) data for SUAs, military training routes (MTRs), and TFRs. [Figure 5-34] By default, the List tabbed page displays all airspace types, and the Map tabbed page displays all airspace types apart from MTRs and ATC Assigned Airspaces (ATCAAs). Both the List and Map tabbed pages can be filtered to display specific data for an airspace name, type, or group. Groups include SUA, MTR, or TFR. The Map tabbed page provides a graphical depiction of scheduled airspaces that may be customized using a fly-out menu of map display options. This tabbed page also contains look-up functionality that allows a user to locate one or more airports within the map. [Figures 5-35 through 5-38]

Additional navigation features are included which allows the user to pan in any direction by dragging the cursor within the map. A permalink feature is also available that enables a user to bookmark a customized set of map layers that can easily be added to their Internet browser favorites list. Once a specific set of customized map layers has been bookmarked, a user may open that customized map display using the favorites option within their browser menu. The List tabbed page allows a user to view all SUA and MTR scheduling data and NOTAM text for a TFR. This text may be viewed for each NOTAM ID by expanding the NOTAM text section within the List grid or clicking the NOTAM ID to open a TFR Details page. The TFR Details page displays NOTAM text in a form layout for easy reading and includes a mapped image and sectional navigation map if available for the TFR.

SPECIAL USE AIRSPACE ON JACKSONVILLE SECTIONAL CHART

Unless otherwise noted are MSL and in feet.
Time is local. "TO" an altitude means "To and including."
FL – Flight Level
NO A/G – No air to ground communications.
Contact nearest FSS for information.

† Other times by NOTAM.
NOTAM – Use of this term in Restricted Areas
indicates FAA and DoD NOTAM systems.
Use of this term in all other Special Use areas indicates
the DoD NOTAM system.

U.S. P–PROHIBITED, R–RESTRICTED, W–WARNING, A–ALERT, MOA–MILITARY OPERATIONS AREA

NUMBER	ALTITUDE	TIME OF USE	CONTROLLING AGENCY/ CONTACT FACILITY	FREQUENCIES
P-50	TO BUT NOT INCL 3,000	CONTINUOUS	NO A/G	
R-2903 A	TO BUT NOT INCL 23,000	INTERMITTENT 0700-1900 TUE-SUN †24 HRS IN ADVANCE	JACKSONVILLE CNTR	
R-2903 C	TO 7,000	INTERMITTENT 0700-1900 TUE-SUN †24 HRS IN ADVANCE	JACKSONVILLE TRACON	
R-2903 D	TO 5,000	INTERMITTENT 0700-1900 TUE-SUN †24 HRS IN ADVANCE	JACKSONVILLE TRACON	
R-2904 A	TO BUT NOT INCL 1,800	0800-1700 (APR-AUG) 0800-1700 SAT-SUN (SEP-MAR) †24 HRS IN ADVANCE	JACKSONVILLE TRACON	
R-2906	TO 14,000	INTERMITTENT 0500-0100 †6 HRS IN ADVANCE	JACKSONVILLE TRACON	
R-2907 A	TO FL 230	INTERMITTENT 0500-0100 †6 HRS IN ADVANCE	JACKSONVILLE CNTR	

Figure 5-33. Special use airspace charted on an aeronautical chart.

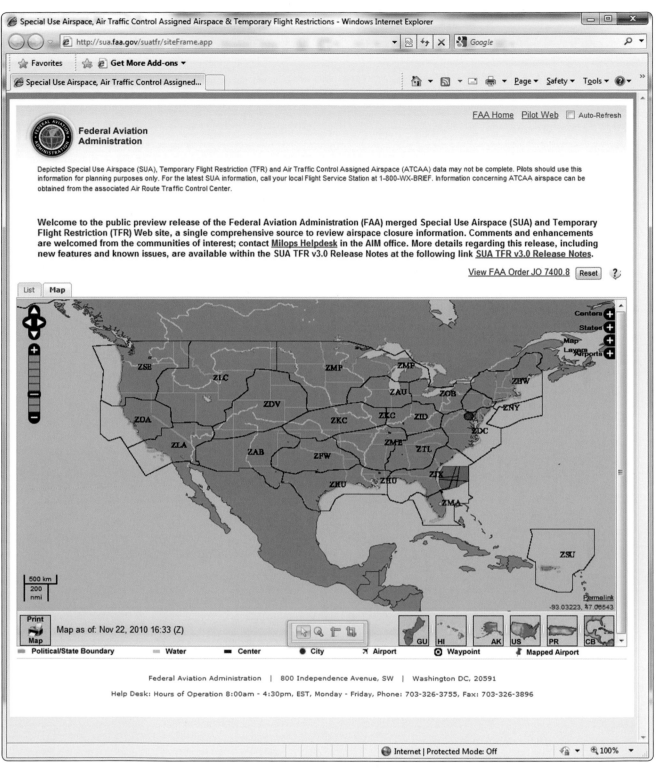

Figure 5-34. FAA website providing information for both special use airspace and temporary flight restrictions.

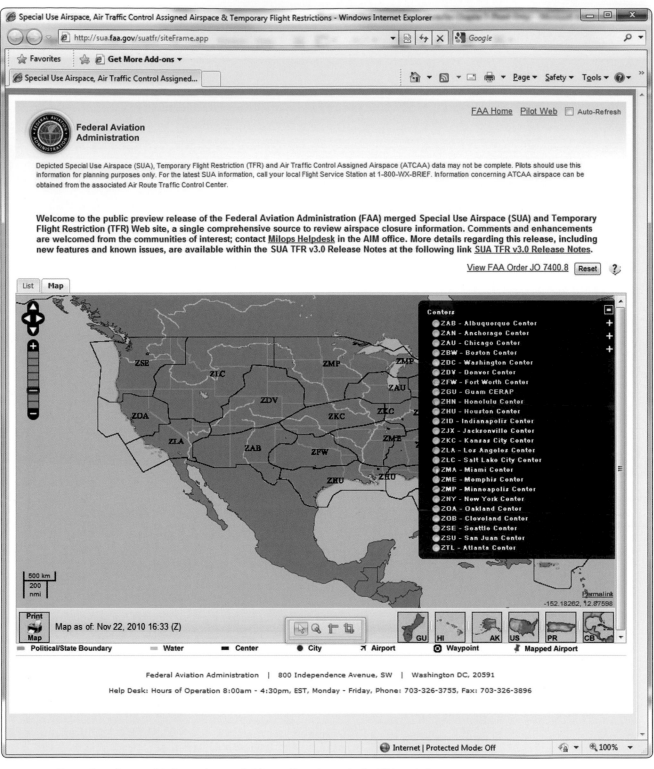

Figure 5-35. Center locations and information available to pilots through the FAA website.

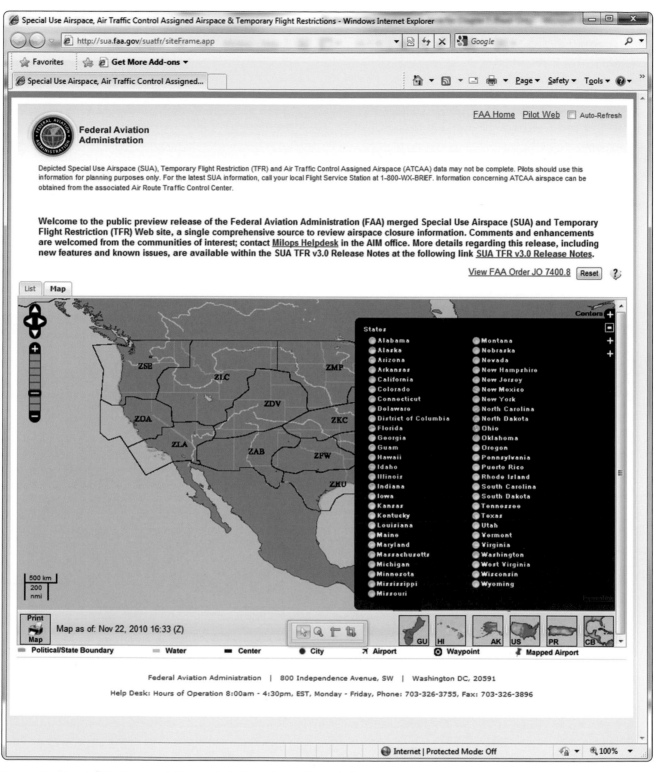

Figure 5-36. State information available to pilots through the FAA website.

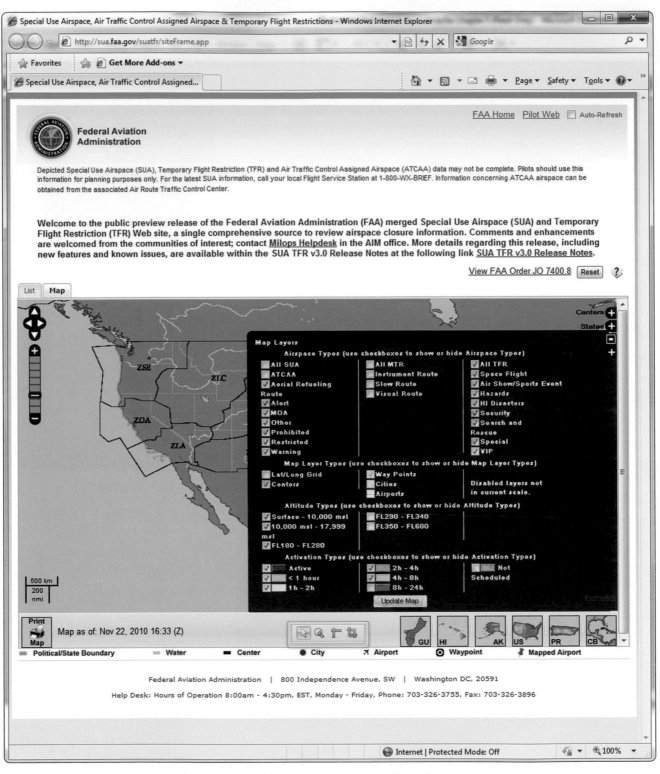

Figure 5-37. Map layer options and information available to pilots through the FAA website.

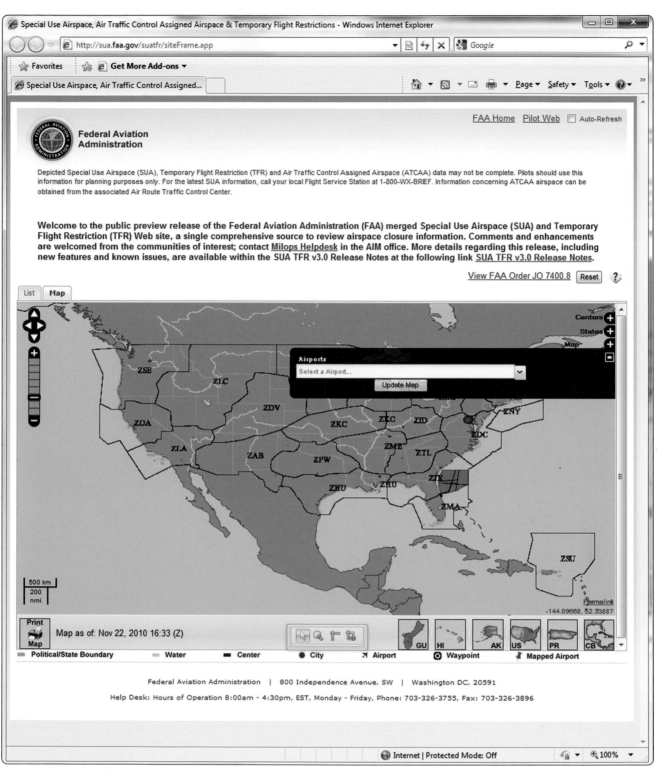

Figure 5-38. Airport information available to pilots through the FAA website.

Chapter 6
Airborne Navigation Databases

Introduction

Area Navigation (RNAV) systems, aeronautical applications, and functions that depend on databases are widespread. [Figure 6-1] Since the 1970s, installed flight systems have relied on airborne navigation databases to support their intended functions, such as navigation data used to facilitate the presentation of flight information to the flight crew and understanding and better visualization of the governing aeronautical flight charts. With the overwhelming upgrades to navigation systems and fully integrated flight management systems (FMS) that are now installed in almost all corporate and commercial aircraft, the need for reliable and consistent airborne navigation databases is more important than ever.

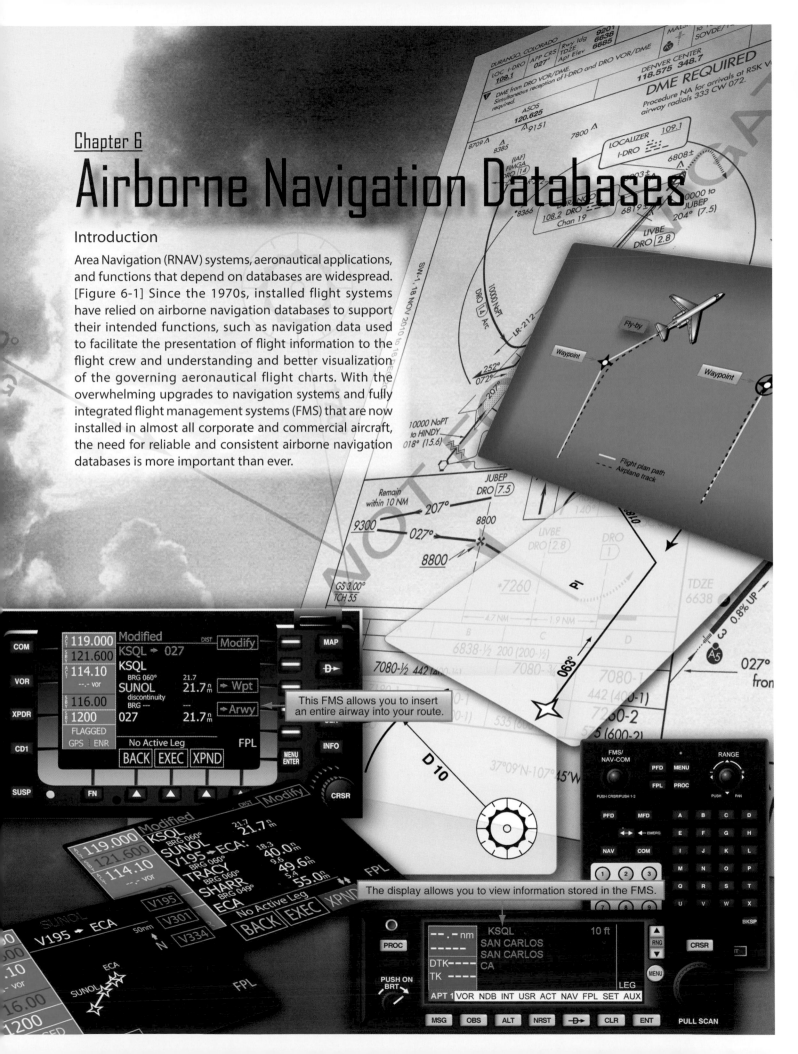

This FMS allows you to insert an entire airway into your route.

The display allows you to view information stored in the FMS.

Figure 6-1. Area navigation (RNAV) receivers.

The capabilities of airborne navigation databases depend largely on the way they are implemented by the avionics manufacturers. They can provide data about a large variety of locations, routes, and airspace segments for use by many different types of RNAV equipment. Databases can provide pilots with information regarding airports, air traffic control (ATC) frequencies, runways, and special use airspace. Without airborne navigation databases, RNAV would be extremely limited. In order to understand the capabilities and limitations of airborne navigation databases, pilots must understand the way databases are compiled and revised by the database provider and processed by the avionics manufacturer. Vital to this discussion is understanding of the regulations guiding database maintenance and use.

There are many different types of RNAV systems certified for instrument flight rules (IFR) use in the National Airspace System (NAS). The two most prevalent types are GPS and the multisensory FMS. [Figure 6-2] A modern GPS unit accurately provides the pilot with the aircraft's present position; however, it must use an airborne navigation database to determine its direction or distance from another location. The database provides the GPS with

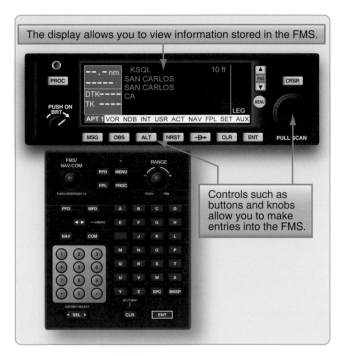

The display allows you to view information stored in the FMS.

Controls such as buttons and knobs allow you to make entries into the FMS.

Figure 6-2. GPS with a flight route on display.

position information for navigation fixes so it may perform the required geodetic calculations to determine the appropriate tracks, headings, and distances to be flown. [Figure 6-3]

Modern FMS are capable of a large number of functions including basic en route navigation, complex departure and arrival navigation, fuel planning, and precise vertical navigation. Unlike stand-alone navigation systems, most FMS use several navigation inputs. Typically, they formulate the aircraft's current position using a combination of conventional distance measuring equipment (DME) signals, inertial navigation systems (INS), GPS receivers, or other RNAV devices. Like stand-alone navigation avionics, they rely heavily on airborne navigation databases to provide the information needed to perform their numerous functions.

Airborne Navigation Database Standardization

Beginning in the 1970s, the requirement for airborne navigation databases became more critical. In 1973,

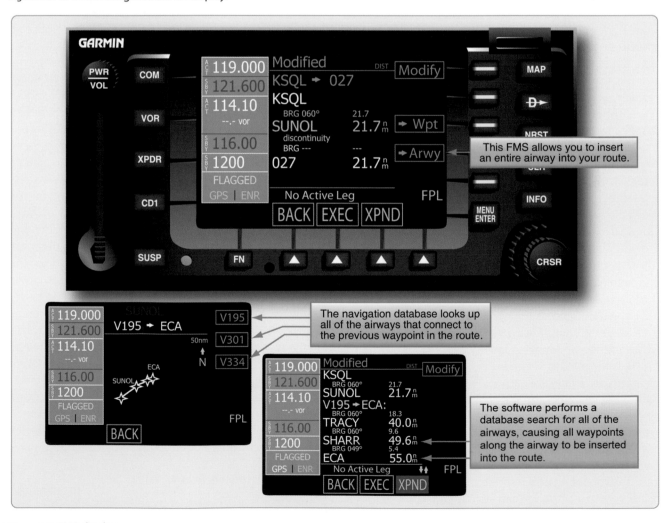

This FMS allows you to insert an entire airway into your route.

The navigation database looks up all of the airways that connect to the previous waypoint in the route.

The software performs a database search for all of the airways, causing all waypoints along the airway to be inserted into the route.

Figure 6-3. FMS display.

National Airlines installed the Collins ANS-70 and AINS-70 RNAV systems in their DC-10 fleet, which marked the first commercial use of avionics that required navigation databases. A short time later, Delta Air Lines implemented the use of an ARMA Corporation RNAV system that also used a navigation database. Although the type of data stored in the two systems was basically identical, the designers created the databases to solve the individual problems of each system, which meant that they were not interchangeable. As the implementation of RNAV systems expanded, a world standard for airborne navigation databases was needed.

In 1973, Aeronautical Radio, Inc. (ARINC) sponsored the formation of a committee to standardize aeronautical databases. In 1975, the committee published the first standard, ARINC Specification 424, which has remained the worldwide accepted format for transmission of navigation databases.

ARINC 424

ARINC 424 is the air transport industry's recommended standard for the preparation and transmission of data for the assembly of airborne system navigation databases. The data is intended for merging with the aircraft navigation system software to provide a source of navigation reference. Each subsequent version of ARINC 424 Specification provides additional capability for navigation systems to utilize. Merging of ARINC 424 data with each manufacturer's system software is unique and ARINC 424 leg types provide vertical guidance and ground track for a specific flight procedure. These leg types must provide repeatable flight tracks for the procedure design. The navigation database leg type is the path and terminator concept.

ARINC 424 Specification describes 23 leg types by their path and terminator. The path describes how the aircraft gets to the terminator by flying direct (a heading, a track, a course, etc.). The terminator is the event or condition that causes the navigation computer system to switch to the next leg (a fix, an altitude, an intercept, etc.). When a flight procedure instructs the pilot to fly runway heading to 2000 feet then direct to a fix, this is the path and terminator concept. The path is the heading and the terminator is 2000 feet. The next leg is then automatically sequenced. A series of leg types are coded into a navigation database to make a flight procedure. The navigation database allows an FMS or GPS navigator to create a continuous display of navigational data, thus enabling an aircraft to be flown along a specific route. Vertical navigation can also be coded.

The data included in an airborne navigation database is organized into ARINC 424 records. These records are strings of characters that make up complex descriptions of each navigation entity. ARINC records can be sorted into four general groups: fix records, simple route records, complex route records, and miscellaneous records. Although it is not important for pilots to have in-depth knowledge of all the fields contained in the ARINC 424 records, pilots should be aware of the types of records contained in the navigation database and their general content.

Fix Records

Database records that describe specific locations on the face of the earth can be considered fix records. Navigational aids (NAVAIDs), waypoints, intersections, and airports are all examples of this type of record. These records can be used directly by avionics systems and can be included as parts of more complex records like airways or approaches.

Another concept pilots should understand relates to how aircraft make turns over navigation fixes. Fixes can be designated as fly-over or fly-by, depending on how they are used in a specific route. [Figure 6-4] Under certain circumstances, a navigation fix is designated as fly-over. This simply means that the aircraft must actually pass directly over the fix before initiating a turn to a new course. Conversely, a fix may be designated fly-by, allowing an aircraft's navigation system to use its turn anticipation feature, which ensures that the proper radius of turn is commanded to avoid overshooting the new course. Some RNAV systems are not programmed to fully use this feature. It is important to remember a fix can be coded as fly-over and fly-by in the same procedure, depending on how the fix is used (i.e., holding at an initial approach fix). RNAV or GPS stand-alone IAPs are flown using data pertaining to the particular IAP obtained from an onboard database to include the sequence of all waypoints used for the approach and missed approach, except that step down waypoints may not be included in some TSO-C129 receiver databases. Included in the database, in most receivers, is coding that informs the navigation system of which WPs are fly-over or fly-by. The navigation system may provide guidance appropriately to include leading the turn prior to a fly-by waypoint; or causing over flight of a fly-over waypoint. Where the navigation system does not provide such guidance, the pilot must accomplish the turn lead or waypoint over flight manually. Chart symbology for the fly-by waypoint provides pilot awareness of expected actions.

Simple Route Records

Route records are those that describe a flightpath instead of a fixed position. Simple route records contain

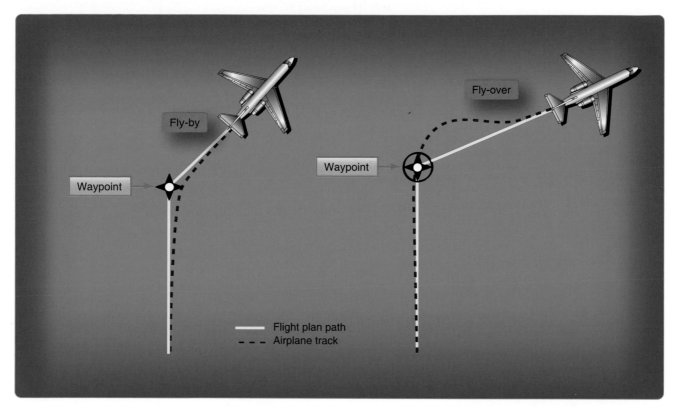

Figure 6-4. Fly-by-waypoints and fly-over-waypoints.

strings of fix records and information pertaining to how the fixes should be used by the navigation avionics.

A Victor Airway, for example, is described in the database by a series of en route airway records that contain the names of fixes in the airway and information about how those fixes make up the airway.

Complex Route Records

Complex route records include those strings of fixes that describe complex flightpaths like standard instrument departures (SIDs), standard terminal arrival routes (STARs), and instrument approach procedures (IAPs). Like simple routes, these records contain the names of fixes to be used in the route, as well as instructions on how the route is flown.

Miscellaneous Records

There are several other types of information that is coded into airborne navigation databases, most of which deal with airspace or communications. The receiver may contain additional information, such as restricted airspace, airport minimum safe altitudes, and grid minimum off route altitudes (MORAs).

Path and Terminator Concept

The path and terminator concept is a means to permit coding of terminal area procedures, SIDs, STARs, and approach procedures. Simply put, a textual description of a route or a terminal procedure is translated into a format that is useable in RNAV systems. One of the most important concepts for pilots to learn regarding the limitations of RNAV equipment has to do with the way these systems deal with the path and terminator field included in complex route records.

The first RNAV systems were capable of only one type of navigation; they could fly directly to a fix. This was not a problem when operating in the en route environment in which airways are mostly made up of direct routes between fixes. The early approaches for RNAV did not present problems for these systems and the databases they used because they consisted mainly of DME/DME overlay approaches flown only direct point-to-point navigation. The desire for RNAV equipment to have the ability to follow more complicated flightpaths necessitated the development of the path and terminator field that is included in complex route records.

Path and Terminator Legs

There are currently 23 different leg types, or path and terminators that have been created in the ARINC 424 standard that enable RNAV systems to follow the complex paths that make up instrument departures, arrivals, and approaches. They describe to navigation avionics a path to be followed and the criteria that must be met before the path concludes and the next path begins. Although there are 23 leg types available, none of the manufactured database equipment is capable of using all of the leg types. Pilots must continue to monitor procedures for accuracy and not rely solely on the information that the database is showing. If the RNAV system does not have the leg type

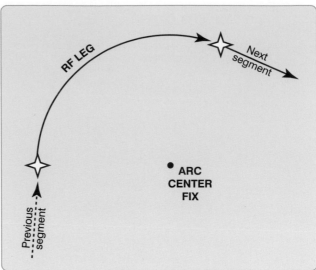

Figure 6-7. Constant radius arc or RF leg.

Figure 6-5. Initial fix.

demanded by procedures, data packers have to select one or a combination of available lleg types to give the best approximation, which can result in an incorrect execution of the procedure. Below is a list of the 23 leg types and their uses that may or may not be used by all databases.

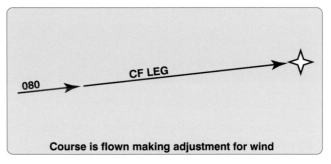

Figure 6-8. Course to a fix or CF leg.

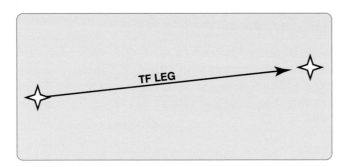

Figure 6-6. Track to a fix leg type.

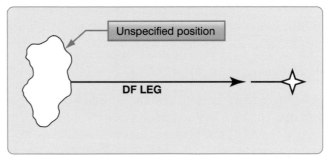

Figure 6-9. Direct to a fix or DF leg.

- Initial fix or IF leg—defines a database fix as a point in space and is only required to define the beginning of a route or procedure. [Figure 6-5]

- Track to a fix or TF leg—defines a great circle track over the ground between two known database fixes and the preferred method for specification of straight legs (course or heading can be mentioned on charts but designer should ensure TF leg is used for coding). [Figure 6-6]

- Constant radius arc or RF leg—defines a constant radius turn between two databases fixes, lines tangent to the arc, and a center fix. [Figure 6-7]

- Course to a fix or CF leg—defines a specified course to a specific database fix. Whenever possible, TF legs

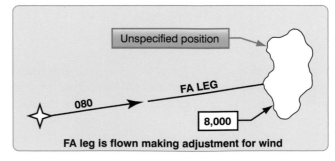

Figure 6-10. Fix to an altitude or FA leg.

should be used instead of CF legs to avoid magnetic variation issues. [Figure 6-8]

- Direct to a fix or DF leg—defines an unspecified track starting from an undefined position to a specified

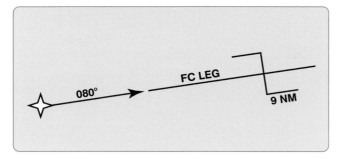

Figure 6-11. Track from a fix from a distance or FC leg.

fix. [Figure 6-9]

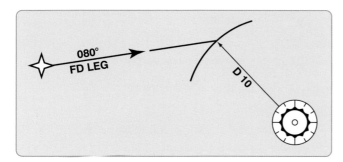

Figure 6-12. Track from a fix to a DME distance or FD leg.

- Fix to an altitude or FA leg—defines a specified track over the ground from a database fix to a specified altitude at an unspecified position. [Figure 6-10]

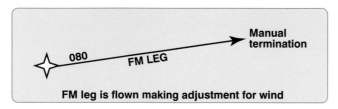

Figure 6-13. From a fix to a manual termination or FM leg.

- Track from a fix from a distance or FC leg—defines a specified track over the ground from a database fix for a specific distance. [Figure 6-11]
- Track from a fix to a distance measuring equipment (DME) distance or FD leg—defines a specified track over the ground from a database fix to a specific

DME distance that is from a specific database DME NAVAID. [Figure 6-12]

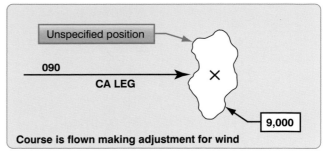

Figure 6-14. Course to an altitude or CA leg.

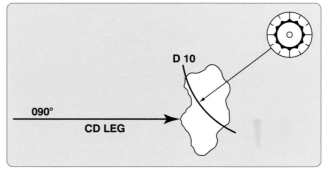

Figure 6-15. Course to a DME distance of CD leg.

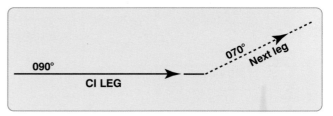

Figure 6-16. Course to an intercept or CI leg.

- From a fix to a manual termination or FM leg—defines a specified track over the ground from a database fix until manual termination of the leg. [Figure 6-13]

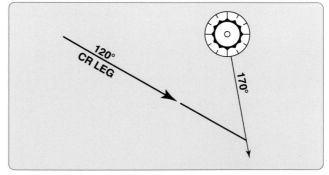

Figure 6-17. Course to a radial termination or CR leg.

- Course to an altitude or CA leg—defines a specified course to a specific altitude at an unspecified position. [Figure 6-14]

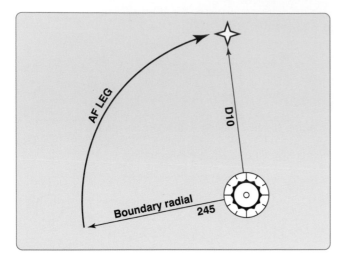

Figure 6-18. Arc to a fix or AF leg.

- Course to a DME distance or CD leg—defines a specified course to a specific DME distance that is from a specific database DME NAVAID. [Figure 6-15]

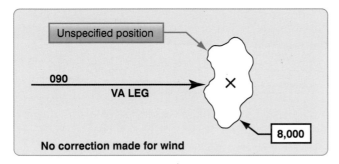

Figure 6-19. Heading to an altitude termination or VA leg.

- Course to an intercept or CI leg—defines a specified course to intercept a subsequent leg. [Figure 6-16]
- Course to a radial termination or CR leg—defines a course to a specified radial from a specific database VOR NAVAID. [Figure 6-17]

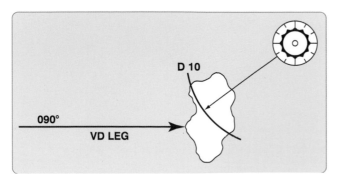

Figure 6-20. Heading to a DME distance termination or VD leg.

- Arc to a fix or AF leg—defines a track over the ground at a specified constant distance from a database DME NAVAID. [Figure 6-18]

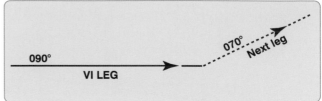

Figure 6-21. Heading to an intercept or VI leg.

- Heading to an altitude termination or VA leg—defines a specified heading to a specific altitude termination at an unspecified position. [Figure 6-19]

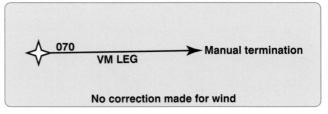

Figure 6-22. Heading to a manual termination or VM leg.

- Heading to a DME distance termination or VD leg—defines a specified heading terminating at a specified DME distance from a specific database DME NAVAID. [Figure 6-20]

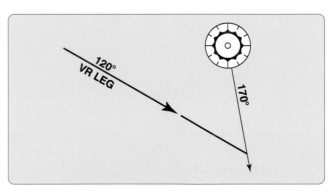

Figure 6-23. Heading to a radial termination or VR leg.

- Heading to an intercept or VI leg—defines a specified heading to intercept the subsequent leg at an unspecified position. [Figure 6-21]

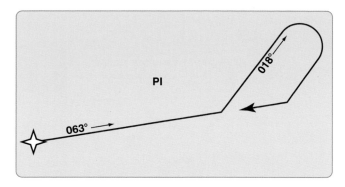

Figure 6-24. Procedure turn or PI leg.

- Heading to a manual termination or VM leg— defines a specified heading until a manual termination. [Figure 6-22]

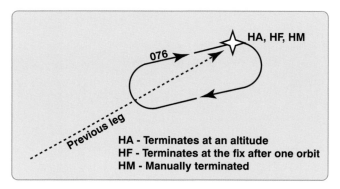

HA - Terminates at an altitude
HF - Terminates at the fix after one orbit
HM - Manually terminated

Figure 6-25. Racetrack course reversal or HA, HF, and HM leg.

- Heading to a radial termination or VR leg—defines a specified heading to a specified radial from a specific database VOR NAVAID. [Figure 6-23]

- Procedure turn or PI leg—defines a course reversal starting at a specific database fix and includes outbound leg followed by a left or right turn and 180° course reversal to intercept the next leg. [Figure 6-24]

- Racetrack course reversal or altitude termination (HA), single circuit terminating at the fix (base turn) (HF), or manual termination (HM) leg types—define racetrack pattern or course reversals at a specified database fix. [Figure 6-25]

The GRAND JUNCTION FIVE DEPARTURE for Grand Junction Regional in Grand Junction, Colorado, provides a good example of different types of path and terminator legs used. [Figure 6-26] When this procedure is coded into the navigation database, the person entering the data into the records must identify the individual legs of the flightpath and then determine which type of terminator should be used.

The first leg of the departure for Runway 11 is a climb via runway heading to 6,000 feet mean sea level (MSL) and then a climbing right turn direct to a fix. When this is entered into the database, a heading to an altitude (VA) value must be entered into the record's path and terminator field for the first leg of the departure route. This path and terminator tells the avionics to provide course guidance based on heading, until the aircraft reaches 6,000 feet, and then the system begins providing course guidance for the next leg. After reaching 6,000 feet, the procedure calls for a right turn direct to the Grand Junction (JNC) VORTAC. This leg is coded into the database using the path and terminator direct to a fix (DF) value, which defines an unspecified track starting from an undefined position to a specific database fix.

Another commonly used path and terminator value is heading to a radial (VR) which is shown in Figure 6-27 using the CHANNEL ONE DEPARTURE procedure for Santa Ana, California. The first leg of the runway 19L/R procedure requires a climb on runway heading until crossing the I-SNA 1 DME fix or the SLI R-118, this leg must be coded into the database using the VR value in the Path and Terminator field. After crossing the I-SNA 1 DME fix or the SLI R-118, the avionics should cycle to the next leg of the procedure that in this case, is a climb on a heading of 175° until crossing SLI R-132. This leg is also coded with a VR Path and Terminator. The next leg of the procedure consists of a heading of 200° until intercepting the SXC R-084. In order for the avionics to correctly process this leg, the database record must include the heading to an intercept (VI) value in the Path and Terminator field. This value directs the avionics to follow a specified heading to intercept the subsequent leg at an unspecified position.

The path and terminator concept is a very important part of airborne navigation database coding. In general, it is not necessary for pilots to have an in-depth knowledge of the ARINC coding standards; however, pilots should be familiar with the concepts related to coding in order to understand the limitations of specific RNAV systems that use databases.

Path and Terminator Limitations

How a specific RNAV system deals with Path and Terminators is of great importance to pilots operating with airborne navigation databases. Some early RNAV systems may ignore this field completely. The ILS or LOC/ DME RWY 3 approach at Durango, Colorado, provides an example of problems that may arise from the lack of path and terminator capability in RNAV systems. Although approaches of this type are authorized only for sufficiently equipped RNAV systems, it is possible that a pilot may elect to fly the approach with conventional navigation, and then

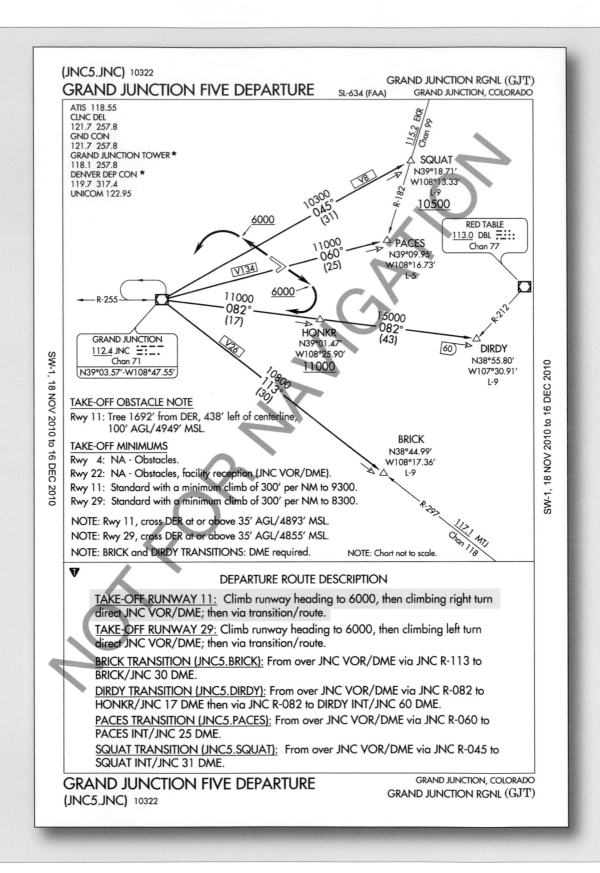

(JNC5.JNC) 10322
GRAND JUNCTION FIVE DEPARTURE
SL-634 (FAA)

GRAND JUNCTION RGNL (GJT)
GRAND JUNCTION, COLORADO

ATIS 118.55
CLNC DEL
121.7 257.8
GND CON
121.7 257.8
GRAND JUNCTION TOWER ★
118.1 257.8
DENVER DEP CON ★
119.7 317.4
UNICOM 122.95

115.2 EKR
Chan 99

△ SQUAT
N39°18.71'
W108°13.33'
L-9
10500

10300
045°
(31)

6000

V8

R-182

△ PACES
N39°09.95'
W108°16.73'
L-5

11000
060°
(25)

RED TABLE
113.0 DBL
Chan 77

V134

6000

R-255

11000
082°
(17)

GRAND JUNCTION
112.4 JNC
Chan 71
N39°03.57'-W108°47.55'

△ HONKR
N39°01.47'
W108°25.90'
11000

15000
082°
(43)

R-212

60 △ DIRDY
N38°55.80'
W107°30.91'
L-9

V26

10800
113°
(30)

TAKE-OFF OBSTACLE NOTE
Rwy 11: Tree 1692' from DER, 438' left of centerline,
100' AGL/4949' MSL.

TAKE-OFF MINIMUMS
Rwy 4: NA - Obstacles.
Rwy 22: NA - Obstacles, facility reception (JNC VOR/DME).
Rwy 11: Standard with a minimum climb of 300' per NM to 9300.
Rwy 29: Standard with a minimum climb of 300' per NM to 8300.

NOTE: Rwy 11, cross DER at or above 35' AGL/4893' MSL.
NOTE: Rwy 29, cross DER at or above 35' AGL/4855' MSL.
NOTE: BRICK and DIRDY TRANSITIONS: DME required.

BRICK
N38°44.99'
W108°17.36'
L-9

R-297

112.1 MTJ
Chan 118

NOTE: Chart not to scale.

SW-1, 18 NOV 2010 to 16 DEC 2010

SW-1, 18 NOV 2010 to 16 DEC 2010

DEPARTURE ROUTE DESCRIPTION

TAKE-OFF RUNWAY 11: Climb runway heading to 6000, then climbing right turn direct JNC VOR/DME; then via transition/route.

TAKE-OFF RUNWAY 29: Climb runway heading to 6000, then climbing left turn direct JNC VOR/DME; then via transition/route.

BRICK TRANSITION (JNC5.BRICK): From over JNC VOR/DME via JNC R-113 to BRICK/JNC 30 DME.

DIRDY TRANSITION (JNC5.DIRDY): From over JNC VOR/DME via JNC R-082 to HONKR/JNC 17 DME then via JNC R-082 to DIRDY INT/JNC 60 DME.

PACES TRANSITION (JNC5.PACES): From over JNC VOR/DME via JNC R-060 to PACES INT/JNC 25 DME.

SQUAT TRANSITION (JNC5.SQUAT): From over JNC VOR/DME via JNC R-045 to SQUAT INT/JNC 31 DME.

GRAND JUNCTION FIVE DEPARTURE
(JNC5.JNC) 10322

GRAND JUNCTION, COLORADO
GRAND JUNCTION RGNL (GJT)

Figure 6-26. Grand Junction Five Departure.

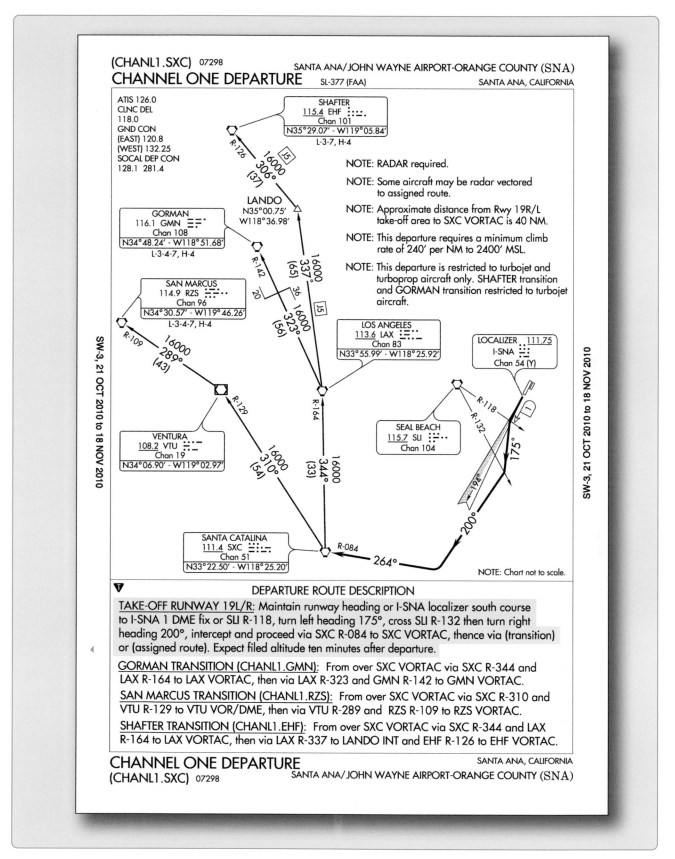

ATIS 126.0
CLNC DEL
118.0
GND CON
(EAST) 120.8
(WEST) 132.25
SOCAL DEP CON
128.1 281.4

SHAFTER
115.4 EHF
Chan 101
N35°29.07' - W119°05.84'
L-3-7, H-4

GORMAN
116.1 GMN
Chan 108
N34°48.24' - W118°51.68'
L-3-4-7, H-4

SAN MARCUS
114.9 RZS
Chan 96
N34°30.57' - W119°46.26'
L-3-4-7, H-4

VENTURA
108.2 VTU
Chan 19
N34°06.90' - W119°02.97'

SANTA CATALINA
111.4 SXC
Chan 51
N33°22.50' - W118°25.20'

LANDO
N35°00.75'
W118°36.98'

LOS ANGELES
113.6 LAX
Chan 83
N33°55.99' - W118°25.92'

SEAL BEACH
115.7 SLI
Chan 104

LOCALIZER 111.75
I-SNA
Chan 54 (Y)

NOTE: RADAR required.

NOTE: Some aircraft may be radar vectored to assigned route.

NOTE: Approximate distance from Rwy 19R/L take-off area to SXC VORTAC is 40 NM.

NOTE: This departure requires a minimum climb rate of 240' per NM to 2400' MSL.

NOTE: This departure is restricted to turbojet and turboprop aircraft only. SHAFTER transition and GORMAN transition restricted to turbojet aircraft.

NOTE: Chart not to scale.

SW-3, 21 OCT 2010 to 18 NOV 2010

SW-3, 21 OCT 2010 to 18 NOV 2010

DEPARTURE ROUTE DESCRIPTION

TAKE-OFF RUNWAY 19L/R: Maintain runway heading or I-SNA localizer south course to I-SNA 1 DME fix or SLI R-118, turn left heading 175°, cross SLI R-132 then turn right heading 200°, intercept and proceed via SXC R-084 to SXC VORTAC, thence via (transition) or (assigned route). Expect filed altitude ten minutes after departure.

GORMAN TRANSITION (CHANL1.GMN): From over SXC VORTAC via SXC R-344 and LAX R-164 to LAX VORTAC, then via LAX R-323 and GMN R-142 to GMN VORTAC.

SAN MARCUS TRANSITION (CHANL1.RZS): From over SXC VORTAC via SXC R-310 and VTU R-129 to VTU VOR/DME, then via VTU R-289 and RZS R-109 to RZS VORTAC.

SHAFTER TRANSITION (CHANL1.EHF): From over SXC VORTAC via SXC R-344 and LAX R-164 to LAX VORTAC, then via LAX R-337 to LANDO INT and EHF R-126 to EHF VORTAC.

Figure 6-27. Channel One Departure.

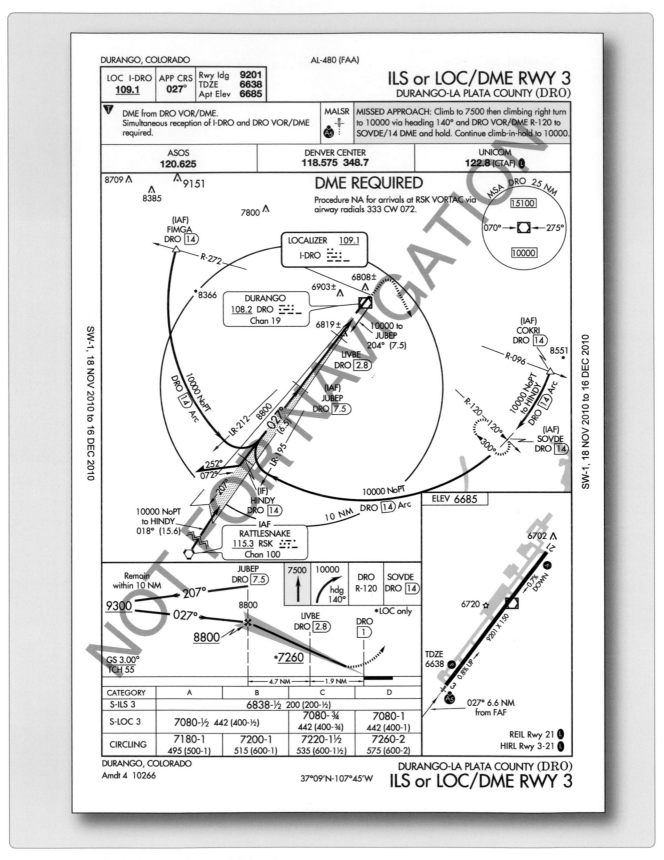

Figure 6-28. ILS or LOC/DME RWY 3 in Durango, Colorado.

reengage RNAV during a missed approach. If this missed approach is flown using an RNAV system that does not use Path and terminator values or the wrong leg types, then the system will most likely ignore the first two legs of the procedure. This will cause the RNAV equipment to direct the pilot to make an immediate turn toward the Durango VOR instead of flying the series of headings that terminate at specific altitudes as dictated by the approach procedure. [Figure 6-28] Pilots must be aware of their individual systems Path and Terminator handling characteristics and always review the manufacturer's documentation to familiarize themselves with the capabilities of the RNAV equipment they are operating. Pilots should be aware that some RNAV equipment was designed without the fly-over capability which can cause problems for pilots attempting to use this equipment to fly complex flightpaths in the departure, arrival, or approach environments.

Role of the Database Provider

Compiling and maintaining a worldwide airborne navigation database is a large and complex job. Within the United States, the FAA sources give the database providers information, in many different formats, which must be analyzed, edited, and processed before it can be coded into the database. In some cases, data from outside the United States must be translated into English so it may be analyzed and entered into the database. Once the data is coded, it must be continually updated and maintained.

Once the FAA notifies the database provider that a change is necessary, the update process begins. The change is incorporated into a 28-day airborne database revision cycle

based on its assigned priority. If the information does not reach the coding phase prior to its cutoff date (the date that new aeronautical information can no longer be included in the next update), it is held out of revision until the next cycle. The cutoff date for aeronautical databases is typically 21 days prior to the effective date of the revision.

The integrity of the data is ensured through a process called cyclic redundancy check (CRC). A CRC is an error detection algorithm capable of detecting small bit-level changes in a block of data. The CRC algorithm treats a data block as a single, large binary value. The data block is divided by a fixed binary number called a generator polynomial whose form and magnitude is determined based on the level of integrity desired. The remainder of the division is the CRC value for the data block. This value is stored and transmitted with the corresponding data block. The integrity of the data is checked by reapplying the CRC algorithm prior to distribution.

Role of the Avionics Manufacturer

When avionics manufacturers develop a piece of equipment that requires an airborne navigation database, they typically form an agreement with a database provider to supply the database for that new avionics platform. It is up to the manufacturer to determine what information to include in the database for their system. In some cases, the navigation data provider has to significantly reduce the number of records in the database to accommodate the storage capacity of the manufacturer's new product, which means that the database may not contain all procedures.

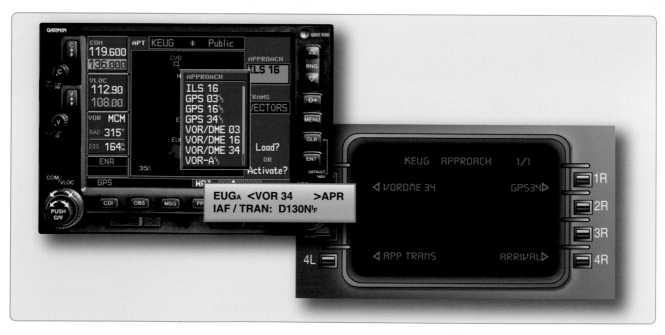

Figure 6-29. Naming conventions of three different systems for the VOR 34 Approach.

Another important fact to remember is that although there are standard naming conventions included in the ARINC 424 specification, each manufacturer determines how the names of fixes and procedures are displayed to the pilot. This means that although the database may specify the approach identifier field for the VOR/DME Runway 34 approach at Eugene Mahlon Sweet Airport (KEUG) in Eugene, Oregon, as "V34," different avionics platforms may display the identifier in any way the manufacturer deems appropriate. For example, a GPS produced by one manufacturer might display the approach as "VOR 34," whereas another might refer to the approach as "VOR/DME 34," and an FMS produced by another manufacturer may refer to it as "VOR34." [Figure 6-29]

These differences can cause visual inconsistencies between chart and GPS displays, as well as confusion with approach clearances and other ATC instructions for pilots unfamiliar with specific manufacturer's naming conventions. The manufacturer determines the capabilities and limitations of an RNAV system based on the decisions that it makes regarding that system's processing of the airborne navigation database.

Users Role

Like paper charts, airborne navigation databases are subject to revision. According to Title 14 of the Code of Federal Regulations (14 CFR) Part 91, section 91.503, the end user (operator) is ultimately responsible for ensuring that data meets the quality requirements for its intended application. Updating data in an aeronautical database is considered to be maintenance and all Part 91 operators

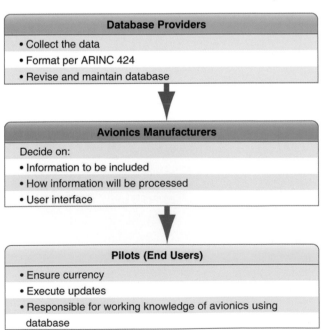

may update databases in accordance with 14 CFR Part 91, section 43.3(g). Parts 121, 125, and 135 operators must update databases in accordance with their approved maintenance program. For Part 135 helicopter operators, this includes maintenance by the pilot in accordance with 14 CFR Part 43, section 43.3(h).

Pilots using the databases are ultimately responsible for ensuring that the database they are operating with is current. This includes checking Notices to Airmen (NOTAM)-type information concerning errors that may be supplied by the avionics manufacturer or the database supplier. The database user is responsible for learning how the specific navigation equipment handles the navigation database. The manufacturer's documentation is the pilot's best source of information regarding the capabilities and limitations of a specific database. [Figure 6-30]

Operational Limitations of Airborne Navigation Databases

Understanding the capabilities and limitations of the navigation systems installed in an aircraft is one of the pilot's biggest concerns for IFR flight. Considering the vast number of RNAV systems and pilot interfaces available today, it is critical that pilots and flight crews be familiar with the manufacturer's operating manual for each RNAV system they operate and achieve and retain proficiency operating those systems in the IFR environment.

Most professional and general aviation pilots are familiar with the possible human factors issues related to flightdeck automation. It is particularly important to consider those issues when using airborne navigation databases. Although modern avionics can provide precise guidance throughout all phases of flight, including complex departures and arrivals, not all systems have the same capabilities.

RNAV equipment installed in some aircraft is limited to direct route point-to-point navigation. Therefore, it is very important for pilots to familiarize themselves with the capabilities of their systems through review of the manufacturer documentation. Most modern RNAV systems are contained within an integrated avionics system that receives input from several different navigation and aircraft system sensors. These integrated systems provide so much information that pilots may sometimes fail to recognize errors in navigation caused by database discrepancies or misuse. Pilots must constantly ensure that the data they enter into their avionics is accurate and current. Once the transition to RNAV is made during a flight, pilots and flight crews must always be capable and ready to revert to conventional means of navigation if problems arise.

Figure 6-30. Database rolls.

Closed Indefinitely Airports

Some U.S. airports have been closed for up to several years, with little or no chance that they will ever reopen; yet their "indefinite" closure status – as opposed to permanent or UFN closure, or abandonment – causes them to continue to appear on both VFR and IFR charts and in airborne navigation databases; and their instrument approach procedures, if any, continue to be included – and still appear to be valid – in the paper and electronic versions of the United States Terminal Procedures Publication (TPP) charts. Airpark South, 2K2, at Ozark, Missouri, is a case in point.

Even though this airport has been closed going on two years and, due to industrial and residential development surrounding it, likely will never be reopened, the airport is nonetheless still charted in a way that could easily lead a pilot to believe that it is still open and operating. Even the current U.S. Low Altitude En route chart displays a blue symbol for this airport, indicating that it still has a Department of Defense (DOD) approved instrument approach procedure available for use.

Aircrews need to use caution when selecting an airport in a cautionary or emergency situation, especially if the airport was not previously analyzed suitable for diversion during preflight. Aircrews could assume, based on charts and their FMS database, the airport is suitable and perhaps the only available diversionary or emergency option. The airport however, could be closed and hazardous even for emergency use. In these situations, Air Traffic Control may be queried for the airport's status.

Storage Limitations

As the data in a worldwide database grows, the required data storage space increases. Over the years that panel-mounted GPS and FMSs have developed, the size of the commercially available airborne navigation databases has grown exponentially.

Some manufacturer's systems have kept up with this growth and some have not. Many of the limitations of older RNAV systems are a direct result of limited data storage capacity. For this reason, avionics manufacturers must make decisions regarding which types of procedures will be included with their system. For instance, older GPS units rarely include all of the waypoints that are coded into master databases. Even some modern FMS equipment, which typically have much larger storage capacity, do not include all of the data that is available from the database producers. The manufacturers often choose not to include certain types of data that they think is of low importance

to the usability of the unit. For example, manufacturers of FMS used in large airplanes may elect not to include airports where the longest runway is less than 3,000 feet or to include all the procedures for an airport.

Manufacturers of RNAV equipment can reduce the size of the data storage required in their avionics by limiting the geographic area the database covers. Like paper charts, the amount of data that needs to be carried with the aircraft is directly related to the size of the coverage area. Depending on the data storage that is available, this means that the larger the required coverage area, the less detailed the database can be.

Again, due to the wide range of possible storage capacities, and the number of different manufacturers and product lines, the manufacturer's documentation is the pilot's best source of information regarding limitations caused by storage capacity of RNAV avionics.

Charting/Database Inconsistencies

It is important for pilots to remember that many inconsistencies may exist between aeronautical charts and airborne navigation databases. Since there are so many sources of information included in the production of these materials, and the data is manipulated by several different organizations before it is eventually displayed on RNAV equipment, the possibility is high that there will be noticeable differences between the charts and the databases. Because of this, pilots must be familiar with the capabilities of the database and have updated aeronautical charts while flying to ensure the proper course is being flown.

Naming Conventions

Obvious differences exist between the names of procedures shown on charts and those that appear on the displays of many RNAV systems. Most of these differences can be accounted for simply by the way the avionics manufacturers elect to display the information to the pilot. It is the avionics manufacturer that creates the interface between the pilot and the database. For example, the VOR 12R approach in San Jose, California, might be displayed several different ways depending on how the manufacturer designs the pilot interface. Some systems display procedure names exactly as they are charted, but many do not.

The naming of multiple approaches of the same type to the same runway is also changing. Multiple approaches with the same guidance will be annotated with an alphabetical suffix beginning at the end of the alphabet and working backwards for subsequent procedures (e.g., ILS Z RWY 28, ILS Y RWY 28, etc.). The existing annotations, such as

ILS 2 RWY 28 or Silver ILS RWY 28, will be phased out and replaced with the new designation.

NAVAIDs are also subject to naming discrepancies as well. This problem is complicated by the fact that multiple NAVAIDs can be designated with the same identifier. VOR XYZ may occur several times in a provider's database, so the avionics manufacturer must design a way to identify these fixes by a more specific means than the three-letter identifier. Selection of geographic region is used in most instances to narrow the pilot's selection of NAVAIDs with like identifiers.

Non-directional beacons (NDBs) and locator outer markers (LOMs) can be displayed differently than they are charted. When the first airborne navigation databases were being implemented, NDBs were included in the database as waypoints instead of NAVAIDs. This necessitated the use of five character identifiers for NDBs. Eventually, the NDBs were coded into the database as NAVAIDs, but many of the RNAV systems in use today continue to use the five-character identifier. These systems display the characters "NB" after the charted NDB identifier. Therefore, NDB ABC would be displayed as "ABCNB."

Other systems refer to NDB NAVAIDs using either the NDB's charted name if it is five or fewer letters, or the one to three character identifier. PENDY NDB located in North Carolina, for instance, is displayed on some systems as "PENDY," while other systems might only display the NDBs identifier "ACZ." [Figure 6-31]

Using the VOR/DME Runway 34 approach at Eugene Mahlon Sweet Airport (KEUG) in Eugene, Oregon, as another example, which is named V34, may be displayed differently by another avionics platform. For example, a GPS produced by one manufacturer might display the approach as VOR 34, whereas another might refer to the approach as VOR/DME 34, and an FMS produced by another manufacturer may refer to it as VOR34. These differences can cause visual inconsistencies between chart and GPS displays, as well as confusion with approach clearances and other ATC instructions for pilots unfamiliar with specific manufacturer's naming conventions.

For detailed operational guidance, refer to Advisory Circular (AC) 90-100, U.S. Terminal and En Route Area Navigation (RNAV) Operations; AC 90-101, Approval Guidance for RNP Procedures with Special Aircraft and Aircrew Authorization Required (SAAAR); AC 90-105, Approval Guidance for RNP Operations and Barometric Vertical Navigation in the U.S. National Airspace System; and AC 90-107, Guidance for Localizer Performance with Vertical Guidance and Localizer Performance without Vertical Guidance Approach Operations in the U.S. National Airspace System.

Issues Related To Magnetic Variation

Magnetic variations for locations coded into airborne navigation databases can be acquired in several ways. In many cases they are supplied by government agencies in the epoch year variation format. Theoretically, this value is determined by government sources and published for public use every five years. Providers of airborne navigation databases do not use annual drift values; instead the database uses the epoch year variation until it is updated by the appropriate source provider. In the United States, this is the National Oceanic and Atmospheric Administration (NOAA). In some cases the variation for a given location is a value that has been calculated by the avionics system. These dynamic magnetic variation values can be different than those used for locations during aeronautical charting and must not be used for conventional NAVAIDs or airports.

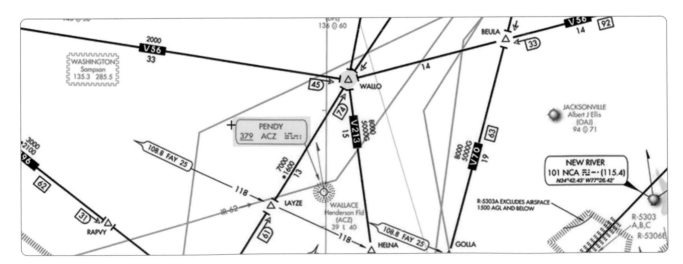

Figure 6-31. Manufacturer's naming conventions.

Discrepancies can occur for many reasons. Even when the variation values from the database are used, the resulting calculated course might be different from the course depicted on the charts. Using the magnetic variation for the region instead of the actual station declination can result in differences between charted and calculated courses and incorrect ground track. Station declination is only updated when a NAVAID is site checked by the governing authority that controls it, so it is often different than the current magnetic variation for that location. Using an onboard means of determining variation usually entails coding some sort of earth model into the avionics memory. Since magnetic variation for a given location changes predictably over time, this model may only be correct for one time in the lifecycle of the avionics. This means that if the intended lifecycle of a GPS unit were 20 years, the point at which the variation model might be correct would be when the GPS unit was 10 years old. The discrepancy would be greatest when the unit was new, and again near the end of its life span.

Another issue that can cause slight differences between charted course values and those in the database occurs when a terminal procedure is coded using magnetic variation of record. When approaches or other procedures are designed, the designers use specific rules to apply variation to a given procedure. Some controlling government agencies may elect to use the epoch year variation of an airport to define entire procedures at that airport. This may result in course discrepancies between the charted value and the value calculated using the actual variations from the database.

Issues Related To Revision Cycle

Pilots should be aware that the length of the airborne navigation database revision cycle could cause discrepancies between aeronautical charts and information derived from the database. One important difference between aeronautical charts and databases is the length of cutoff time. Cutoff refers to the length of time between the last day that changes can be made in the revision, and the date the information becomes effective. Aeronautical charts typically have a cutoff date of 10 days prior to the effective date of the charts.

Chapter 7
Helicopter Instrument Procedures

Introduction

This chapter presents information on IFR helicopter operations in the National Airspace System. Advances in avionics technology installed in helicopters such as Global Positioning System (GPS) and Wide Area Augmentation System (WAAS) are bringing approach procedures to heliports around the country.

The ability to operate helicopters under IFR increases their utility and safety. Helicopter IFR operators have an excellent safety record due to the investment in IFR-equipped helicopters, development of instrument approach procedures (IAPs), and IFR-trained flight crews. The safety record of IFR operations in the Gulf of Mexico is equivalent to the safety record of the best-rated airlines. Manufacturers are working to increase IFR all-weather capabilities of helicopters by providing slower minimum instrument airspeeds (V_{MINI}), faster cruising speeds, and better autopilots and flight management systems (FMS). As a result, in October 2005, the first civil helicopter in the United States was certified for flight into known icing conditions. [Figure 7-1]

Area	Non-Mountainous		Mountainous (14 CFR Part 95)	
Condition	Local	Cross Country	Local	Cross Country
	Ceiling-visibility			
Day	500-1	800-2	500-2	800-3
Night—High Lighting Conditions	500-2	1000-3	500-3	1000-3
Night—Low Lighting Conditions	800-3	1000-5	1000-3	1000-5

Figure 7-1. Icing tests. To safely provide an all-weather capability and flight into known icing conditions that would otherwise delay or cancel winter flight operations, the digital control of the S-92 Rotor Ice Protection System (RIPS) determines the temperature and moisture content of the air and removes any ice buildup by heating the main and tail rotor blades. The system is shown here during testing.

Helicopter Instrument Flight Rule (IFR) Certification

It is very important that pilots be familiar with the IFR requirements for their particular helicopter. Within the same make, model, and series of helicopter, variations in the installed avionics may change the required equipment or the level of augmentation for a particular operation. The Automatic Flight Control System/Autopilot/Flight Director (AFCS/AP/FD) equipment installed in IFR helicopters can be very complex. For some helicopters, the AFCS/AP/FD complexity requires formal training in order for the pilot(s) to obtain and maintain a high level of knowledge of system operation, limitations, failure indications, and reversionary modes. For a helicopter to be certified to conduct operations in instrument meteorological conditions (IMC), it must meet the design and installation requirements of Title 14 Code of Federal Regulations (14 CFR) Part 27, Appendix B (Normal Category) and Part 29, Appendix B (Transport Category), which is in addition to the visual flight rule (VFR) requirements.

These requirements are broken down into the following categories: flight and navigation equipment, miscellaneous requirements, stability, helicopter flight manual limitations, operations specifications, and minimum equipment list (MEL).

Flight and Navigation Equipment

The basic installed flight and navigation equipment for helicopter IFR operations is listed under 14 CFR Part 29, section 29.1303, with amendments and additions in Appendix B of 14 CFR Parts 27 and 29 under which they are certified. The list includes:

- Clock
- Airspeed indicator
- Sensitive altimeter (A "sensitive" altimeter relates to the instrument's displayed change in altitude over its range. For "Copter" Category (CAT) II operations, the scale must be in 20-foot intervals.) adjustable for barometric pressure
- Magnetic direction indicator
- Free-air temperature indicator
- Rate-of-climb (vertical speed) indicator
- Magnetic gyroscopic direction indicator
- Stand-by bank and pitch (attitude) indicator
- Non-tumbling gyroscopic bank and pitch (attitude) indicator
- Speed warning device (if required by 14 CFR Part 29)

Miscellaneous Requirements

- Overvoltage disconnect
- Instrument power source indicator
- Adequate ice protection of IFR systems
- Alternate static source (single-pilot configuration)
- Thunderstorm lights (transport category helicopters)

Stabilization and Automatic Flight Control System (AFCS)

Helicopter manufacturers normally use a combination of a stabilization and/or AFCS in order to meet the IFR stability requirements of 14 CFR Parts 27 and 29. These systems include:

- Aerodynamic surfaces, which impart some stability or control capability that generally is not found in the basic VFR configuration.
- Trim systems provide a cyclic centering effect. These systems typically involve a magnetic brake/spring device and may be controlled by a four-way switch on the cyclic. This system requires "hands on" flying of the helicopter.
- Stability Augmentation Systems (SAS) provide short-term rate damping control inputs to increase helicopter stability. Like trim systems, SAS requires "hands-on" flying.

- Attitude Retention Systems (ATT) return the helicopter to a selected attitude after a disturbance. Changes in attitude can be accomplished usually through a four- way "beep" switch or by actuating a "force trim" switch on the cyclic, which sets the desired attitude manually. Attitude retention may be a SAS function or may be the basic "hands off" autopilot function.

- Autopilot Systems (APs) provide for "hands off" flight along specified lateral and vertical paths. The functional modes may include heading, altitude, vertical speed, navigation tracking, and approach. APs typically have a control panel for mode selection and indication of mode status. APs may or may not be installed with an associated FD. APs typically control the helicopter about the roll and pitch axes (cyclic control) but may also include yaw axis (pedal control) and collective control servos.

- Flight Directors (FDs) provide visual guidance to the pilot to fly selected lateral and vertical modes of operation. The visual guidance is typically provided by a "single cue," commonly known as a "vee bar," which provides the indicated attitude to fly and is superimposed on the attitude indicator. Other FDs may use a "two cue" presentation known as a "cross pointer system." These two presentations only provide attitude information. A third system, known as a "three cue" system, provides information to position the collective as well as attitude (roll and pitch) cues. The collective control cue system identifies and cues the pilot what collective control inputs to use when path errors are produced or when airspeed errors exceed preset values. The three-cue system pitch command provides the required cues to control airspeed when flying an approach with vertical guidance at speeds slower than the best-rate-of-climb (BROC) speed. The pilot manipulates the helicopter's controls to satisfy these commands, yielding the desired flightpath or may couple the autopilot to the FD to fly along the desired flightpath. Typically, FD mode control and indication are shared with the autopilot. Pilots must be aware of the mode of operation of the augmentation systems and the control logic and functions in use. For example, on an instrument landing system (ILS) approach and using the three-cue mode (lateral, vertical, and collective cues), the FD collective cue responds to glideslope deviation, while the horizontal bar cue of the "cross-pointer" responds to airspeed deviations. However, the same system when operated in the two-cue mode on an ILS, the FD horizontal bar cue responds to glideslope deviations. The need to be aware of the FD mode of operation is particularly significant when operating using two pilots.

Pilots should have an established set of procedures and responsibilities for the control of FD/AP modes for the various phases of flight. Not only does a full understanding of the system modes provide for a higher degree of accuracy in control of the helicopter, it is the basis for crew identification of a faulty system.

Helicopter Flight Manual Limitations

Helicopters are certificated for IFR operations with either one or two pilots. Certain equipment is required to be installed and functional for two-pilot operations and additional equipment is required for single-pilot operation.

In addition, the Helicopter Flight Manual (HFM) defines systems and functions that are required to be in operation or engaged for IFR flight in either the single or two-pilot configurations. Often, in a two-pilot operation, this level of augmentation is less than the full capability of the installed systems. Likewise, a single-pilot operation may require a higher level of augmentation.

The HFM also identifies other specific limitations associated with IFR flight. Typically, these limitations include, but are not limited to:

- Minimum equipment required for IFR flight (in some cases, for both single-pilot and two-pilot operations)

- V_{MINI} (minimum speed—IFR) [Figure 7-2]

- V_{NEI} (never exceed speed—IFR)

- Maximum approach angle

- Weight and center of gravity (CG) limits

- Helicopter configuration limitations (such as door positions and external loads)

- Helicopter system limitations (generators, inverters, etc.)

- System testing requirements (many avionics and AFCS, AP, and FD systems incorporate a self-test feature)

- Pilot action requirements (for example, the pilot must have hands and feet on the controls during certain operations, such as an instrument approach below certain altitudes)

Final approach angles/descent gradient for public approach procedures can be as high as 7.5 degrees/795 FPNM. At 70 knots indicated airspeed (KIAS) (no wind), this equates to a descent rate of 925 feet per minute (fpm). With a 10-knot tailwind, the descent rate increases to 1,056 fpm. "Copter" Point-in-space (PinS) approach procedures are restricted to helicopters with a maximum V_{MINI} of 70 KIAS and an IFR approach angle that enables them to meet the final approach angle/descent gradient. Pilots of helicopters with

Manufacturer	V$_{MINI}$ Limitations	MAX IFR Approach Angle	G/A Mode Speed
Augusta			
A-109	60 (80 coupled)		
A-109C	40	9.0	
Bell			
BH 212	40		
BH 214ST	70		
BH 222	50		
BH 222B	50		
BH 412	60	5.0	
BH 430	50 (65 coupled)	4.0	
Eurocopter			
AS-355	55	4.5	
AS-365	75	4.5	
BK-117	45 (70 coupled)	6.0	
EC-135	60	4.6	
EC-155	70	4.0	
Sikorsky			
S-76A	60 (AFCS Phase II)	3.5	75 KIAS
S-76A	50 (AFCS Phase III)	7.5	75 KIAS
S-76B	60	7.5	75 KIAS
S-76C	60		
SK-76C++	50 (60 coupled)	6.5	

NOTE: The V$_{MINI}$, MAX IFR Approach Angle and G/A Mode Speed for a specific helicopter may vary with avionics/autopilot installation. Pilots are therefore cautioned to refer only to the Rotorcraft Flight Manual limitations for their specific helicopter. The maximum rate of descent for many autopilots is 1,000 FPM.

LEGEND

In some helicopters with the autopilot engaged, the V$_{MINI}$ may increase to a speed greater than 70 KIAS, or in the "go around" mode requires a speed faster than 70 KIAS.

Figure 7-2. V$_{MINI}$ limitations, maximum IFR approach angles and G/A mode speeds for selected IFR certified helicopters.

a V$_{MINI}$ of 70 KIAS may have inadequate control margins to fly an approach that is designed with the maximum allowable angle/descent gradient or minimum allowable deceleration distance from the missed approach point (MAP) to the heliport. The "Copter" PinS final approach segment is limited to 70 KIAS since turn containment and the deceleration distance from the MAP to the heliport may not be adequate at faster speeds. For some helicopters, engaging the autopilot may increase the V$_{MINI}$ to a speed greater than 70 KIAS, or in the "go around" (G/A) mode, require a speed faster than 70 KIAS. [Figure 7-2] It may be possible for these helicopters to be flown manually on the approach or on the missed approach in a mode other than the G/A mode.

Since slower IFR approach speeds enable the helicopter to fly steeper approaches and reduces the distance from the heliport that is required to decelerate the helicopter, you may want to operate your helicopter at speeds slower than its established V$_{MINI}$. The provision to apply for a determination of equivalent safety for instrument flight below V$_{MINI}$ and the minimum helicopter requirements are specified in Advisory Circulars (AC) 27-1, Certification of Normal Category Rotorcraft and AC 29-2, Certification of Transport Category Rotorcraft. Application guidance is available from the Rotorcraft Directorate Standards Staff, ASW-110, 2601 Meacham Blvd., Fort Worth, Texas, 76137-4298, (817) 222-5111.

Performance data may not be available in the HFM for speeds other than the best rate of climb speed. To meet missed approach climb gradients, pilots may use observed performance for similar weight, altitude, temperature, and speed conditions to determine equivalent performance. When missed approaches utilizing a climbing turn are flown with an autopilot, set the heading bug on the missed approach heading, and then at the MAP, engage the indicated airspeed mode, followed immediately by applying climb power and selecting the heading mode. This is important since the autopilot roll rate and maximum bank angle in the Heading Select mode are significantly more robust than in the NAV mode. Figure 7-3 represents the bank angle and roll limits of the S76 used by the Federal Aviation Administration (FAA) for flight testing. It has a roll rate in the Heading Select mode of 5 degrees per second with only 1 degree per second in the NAV mode. The bank angle in the Heading Select mode is 20 degrees, with only 17 degrees in the NAV Change Over mode. Furthermore, if the Airspeed Hold mode is not selected on some autopilots when commencing the missed approach, the helicopter accelerates in level flight until the best rate of climb is

Autopilot Mode	Bank Angle Limit (Degrees)	Roll Rate Limit (Degrees/Sec)
Heading hold	<6	None specified
VOR/RNAV (Capture)	+/-22	5
VOR/RNAV (On Course)	+/-13	1
		5 VOR/RNAV Approach
Heading Select	+/-20	5
VOR/RNAV (Course Change Over Station/Fix)	+/-17	1

Figure 7-3. Autopilot bank angle and roll rate limits for the S-76 used by the William J. Hughes Technical Center for Flight Tests.

attained, and only then will a climb begin.

WAAS localizer performance (LP) lateral-only PinS testing conducted in 2005 by the FAA at the William J. Hughes Technical Center in New Jersey for helicopter PinS also captured the flight tracks for turning missed approaches. [Figure 7-4] The large flight tracks that resulted during the turning missed approach were attributed in part to operating the autopilot in the NAV mode and exceeding the 70 KIAS limit.

Operations Specifications

A flight operated under 14 CFR Part 135 has minimums and procedures more restrictive than a flight operated under 14 CFR Part 91. These Part 135 requirements are detailed in their operations specifications (OpSpecs). Helicopter Emergency Medical Service (HEMS) operators have even more restrictive OpSpecs. Shown in Figure 7-5 is an excerpt from an OpSpecs detailing the minimums for precision approaches. The inlay in Figure 7-5 shows the minimums for the ILS Runway 3R approach at Detroit Metro Airport. With all lighting operative, the minimums for helicopter Part 91 operations are a 200-foot ceiling, and 1,200-feet runway visual range (RVR) - one-half airplane Category A visibility but no less than 1/4 SM/1,200 RVR. However, as shown in the OpSpecs, the minimum visibility this Part 135 operator

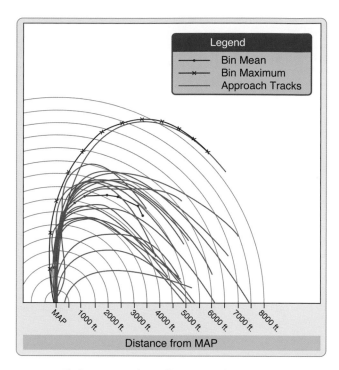

Figure 7-4. Flight tests at the William J. Hughes Technical Center point out the importance of airspeed control and using the correct technique to make a turning missed approach.

must adhere to is 1,600 RVR. Pilots operating under 14 CFR Part 91 are encouraged to develop their own personal

OpSpecs based on their own equipment, training, and experience.

Minimum Equipment List (MEL)

A helicopter operating under 14 CFR Part 135 with certain installed equipment inoperative is prohibited from taking off unless the operation is authorized in the approved MEL. The MEL provides for some equipment to be inoperative if certain conditions are met. [Figure 7-6] In many cases, a helicopter configured for single-pilot IFR may depart IFR with certain equipment inoperative provided a crew of two pilots is used. Under 14 CFR Part 91, a pilot may defer certain items without an MEL if those items are not required by the type certificate, CFRs, or airworthiness directives (ADs), and the flight can be performed safely without them. If the item is disabled, removed, or marked inoperative, a logbook entry is made.

Pilot Proficiency

Helicopters of the same make and model may have variations in installed avionics that change the required equipment or the level of augmentation for a particular operation. The complexity of modern AFCS, AP, and FD systems requires a high degree of understanding to safely

U.S. Department
of Transportation
Federal Aviation
Administration

Operations Specifications

H117. **Straight-in Category I Precision Instrument Approach**
Procedures - All Airports

HQ Control: 11/22/00
HQ Revision: 000

a. Except as provided in this paragraph, the certificate holder shall not use any Category I IFR landing minimum lower than that prescribed by any applicable published instrument approach procedure. The IFR landing minimums prescribed in this paragraph are the lowest authorized (other than Airborne Radar approaches) for use at any airport. Provided that the fastest approach speed used in the final approach segment is less than 91 knots, the certificate holder is authorized to conduct straight-in precision instrument approach procedures using the following:

    (1) The published Category A minimum descent altitude (MDA) or decision height (DH), as appropriate.

    (2) One-half of the published Category A visibility/RVR minimum or the visibility/RVR minimums prescribed by this paragraph, whichever is higher.

b. <u>Straight-In Category I Precision Approach Procedures.</u> The certificate holder shall not use an IFR landing minimum for straight-in precision approach procedures lower than that specified in the following table. Touchdown zone RVR reports, when available for a particular runway, are controlling for all approaches to and landings on that runway. (See NOTE 2.)

Precision Approaches	Full ILS (See NOTE 1), MLS, or PAR			
Approach Light Configuration	HAT	Helicopters Operated at Speeds of 90 Knots or Less		Helicopters Operated at Speeds More Than 90 Knots	
		Visibility In SM.	TDZ RVR In Feet	Visibility In SM.	TDZ RVR In Feet
No Lights or ODALS or MALS or SSALS	200	3/4	3500	3/4	4000
MALSR or SSALR or ALSF-1 or ALSF-2	200	1/4	1600	1/2	2400
MALSR with TDZ and CL or SSALR with TDZ and CL or ALSF-1/ALSF-2 with TDZ and CL	200	1/4	1600	1/2	1800

NOTE 1: A full ILS requires an operative LOC, GS, and OM ... fix, an NDB, VOR, DME fix, or a published minimum GSIA f...
NOTE 2: The Mid RVR and Rollout RVR reports (if available) provide advisory information to pilots ...
The Mid RVR report may be substituted for the TDZ RVR report if the TDZ RVR report is not available.

c. <u>Special Limitations and Provisions for Instrument Approac...</u>
certificate holder is authorized operations at foreign airpor...

    (1) Foreign approach lighting systems equivalent to U.S. ...
and nonprecision approaches. Sequenced flashing lig...
equivalence of a foreign lighting system to U.S. stand...

    (2) For straight-in landing minimums at foreign airports w...
lowest authorized MDA or DH shall be obtained as fo...

Print Date: 4/16/2008

H117-1

Figure 7-5. Operations Specifications.

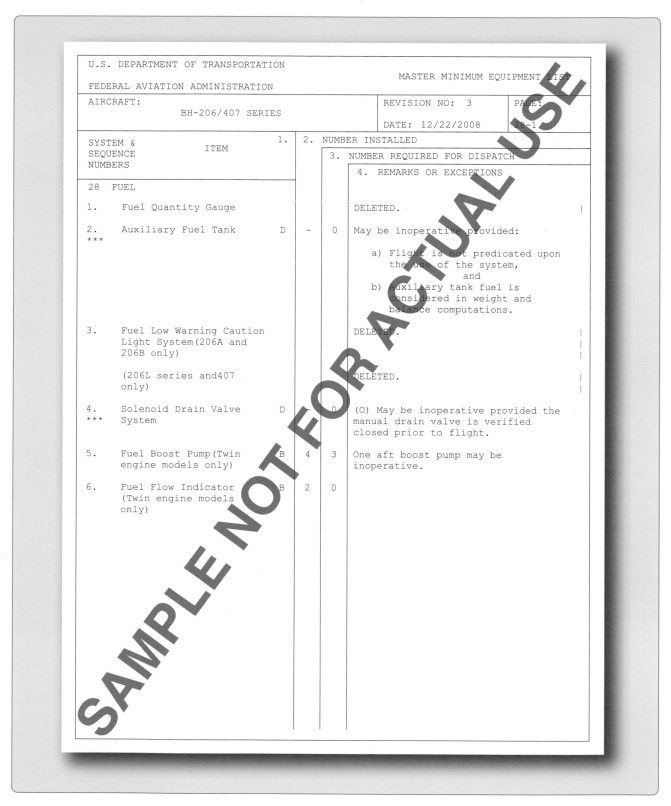

U.S. DEPARTMENT OF TRANSPORTATION

FEDERAL AVIATION ADMINISTRATION

MASTER MINIMUM EQUIPMENT LIST

AIRCRAFT: BH-206/407 SERIES	REVISION NO: 3	PAGE:
	DATE: 12/22/2008	28-1

SYSTEM & SEQUENCE NUMBERS	ITEM	1.	2. NUMBER INSTALLED

2. NUMBER INSTALLED

3. NUMBER REQUIRED FOR DISPATCH

4. REMARKS OR EXCEPTIONS

28 FUEL

1.	Fuel Quantity Gauge				DELETED.

2.
***
	Auxiliary Fuel Tank	D	-	0	May be inoperative provided:
					
					   a) Flight is not predicated upon the use of the system,
					          and
					   b) Auxiliary tank fuel is considered in weight and balance computations.

3.	Fuel Low Warning Caution Light System(206A and 206B only)				DELETED.
	(206L series and407 only)				DELETED.
4. ***	Solenoid Drain Valve System	D	-	0	(O) May be inoperative provided the manual drain valve is verified closed prior to flight.
5.	Fuel Boost Pump(Twin engine models only)	B	4	3	One aft boost pump may be inoperative.
6.	Fuel Flow Indicator (Twin engine models only)	B	2	0	

Figure 7-6. Example of a Minimum Equipment List (MEL).

and efficiently control the helicopter in IFR operations. Formal training in the use of these systems is highly recommended for all pilots.

During flight operations, you must be aware of the mode of operation of the augmentation system and the control logic and functions employed. [Figure 7-2]

Helicopter VFR Minimums

Helicopters have the same VFR minimums as airplanes with two exceptions. In Class G airspace or under a special visual flight rule (SVFR) clearance, helicopters have no minimum visibility requirement but must remain clear of clouds and operate at a speed that is slow enough to give the pilot an adequate opportunity to see other aircraft or an obstruction in time to avoid a collision. Helicopters are also authorized (14 CFR Part 91, appendix D, section 3) to obtain SVFR clearances at airports with the designation NO SVFR in the Airport Facility Directory (A/FD) or on the sectional chart. Figure 7-7 shows the visibility and cloud clearance requirements for VFR and SVFR. However, lower minimums associated with Class G airspace and SVFR do not take the place of the VFR minimum requirements of either Part 135 regulations or respective OpSpecs.

Knowledge of all VFR minimums is required in order to determine if a PinS approach can be conducted or if a SVFR clearance is required to continue past the (MAP). These approaches and procedures are discussed in detail later.

Helicopter IFR Takeoff Minimums

A pilot operating under 14 CFR Part 91 has no takeoff minimums to comply with other than the requirement to attain V_{MINI} before entering IMC. For most helicopters, this requires a distance of approximately 1/2 mile and an altitude of 100 feet. If departing with a steeper climb gradient, some helicopters may require additional altitude to accelerate to V_{MINI}. To maximize safety, always consider using the Part 135 operator standard takeoff visibility minimum of 1/2 statute mile (SM) or the charted departure minima, whichever is higher. A charted departure that provides protection from obstacles has either a higher visibility requirement, climb gradient, and/or departure path. Part 135 operators are required to adhere to the takeoff minimums prescribed in the instrument approach procedures (IAPs) for the airport

Helicopter IFR Alternates

The pilot must file for an alternate if weather reports and forecasts at the proposed destination do not meet certain minimums. These minimums differ for Part 91 and Part 135 operators.

Helicopter VFR Minimums		
Airspace	Flight visibility	Distance from clouds
Class A	Not applicable	Not applicable
Class B	3 SM	Clear of clouds
Class C	3 SM	500 feet below 1,000 feet above 2,000 feet horizontal
Class D	3 SM	500 feet below 1,000 feet above 2,000 feet horizontal
Class E Less than 10,000 feet MSL	3 SM	500 feet below 1,000 feet above 2,000 feet horizontal
At or above 10,000 feet MSL	5 SM	1,000 feet below 1,000 feet above 1 statue mile horizontal
Class G 1,200 feet or less above the surface (regardless of MSL altitude)		
Day, except as provided in §91.155(b)	None	Clear of clouds
Night, except as provided in §91.155(b)	None	Clear of clouds
More than 1,200 feet above the surface but less than 10,000 feet MSL		
Day	1 SM	500 feet below 1,000 feet above 2,000 feet horizontal
Night	3 SM	500 feet below 1,000 feet above 2,000 feet horizontal
More than 1,200 feet above the surface and at or above 10,000 feet MSL	5 SM	1,000 feet below 1,000 feet above 1 statute mile horizontal
B, C, D, E, Surface Area Airspace SVFR Minimums		
Day	None	Clear of clouds
Night	None	Clear of clouds

Figure 7-7. Helicopter VFR minimums.

Part 91 Operators

Part 91 operators are not required to file an alternate if, at the estimated time of arrival (ETA) and for 1 hour after, the ceiling is at least 1,000 feet above the airport elevation or 400 feet above the lowest applicable approach minima, whichever is higher, and the visibility is at least 2 SM. If an alternate is required, an airport can be used if the ceiling is at least 200 feet above the minimum for the approach to be flown and visibility is at least 1 SM, but never less than the minimum required for the approach to be flown. If no instrument approach procedure has been published for the alternate airport, the ceiling and visibility minima are those allowing descent from the MEA, approach, and landing under basic VFR.

Part 135 Operators

Part 135 operators are not required to file an alternate if, for at least 1 hour before and 1 hour after the ETA, the ceiling is at least 1,500 feet above the lowest circling approach minimum descent altitude (MDA). If a circling instrument approach is not authorized for the airport, the ceiling must be at least 1,500 feet above the lowest published minimum or 2,000 feet above the airport elevation, whichever is higher. For the IAP to be used at the destination airport, the forecasted visibility for that airport must be at least 3 SM or 2 SM more than the lowest applicable visibility minimums, whichever is greater.

Alternate landing minimums for flights conducted under 14 CFR Part 135 are described in the OpSpecs for that operation. All helicopters operated under IFR must carry enough fuel to fly to the intended destination, fly from that airport to the filed alternate, if required, and continue for 30 minutes at normal cruising speed.

Helicopter Instrument Approaches

Many new helicopter IAPs have been developed to take advantage of advances in both avionics and helicopter technology.

Standard Instrument Approach Procedures to an Airport

Helicopters flying standard instrument approach procedures (SIAP) must adhere to the MDA or decision altitude for Category A airplanes and may apply the 14 CFR Part 97.3 (d-1) rule to reduce the airplane Category A

97.3 Symbols and Terms Used in Procedures

b) (1)   "Copter procedures" means helicopter procedures with applicable minimums as prescribed in §97.35 of this part. Helicopters may also use other procedures prescribed in Subpart C of this part and may use the Category A minimum descent altitude (MDA) or decision height (DH). The required visibility minimum may be reduced to ½ the published visibility minimum, but in no case may it be reduced to less than one-quarter mile or 1,200 feet RVR.

Figure 7-8. Part 97 excerpt.

visibility by half but in no case less than ¼ SM or 1,200 RVR. [Figure 7-8] The approach can be initiated at any speed up to the highest approach category authorized; however, the speed on the final approach segment must be reduced to the Category A speed of less than 90 KIAS before the MAP in order to apply the visibility reduction. A constant airspeed is recommended on the final approach segment to comply with the stabilized approach concept since a decelerating approach may make early detection of wind shear on the approach path more difficult. [Figure 7-9]

When visibility minimums must be increased for inoperative components or visual aids, use the Inoperative Components and Visual Aids Table (provided in the front cover of the U.S. Terminal Procedures) to derive the Category A minima before applying any visibility reduction. The published visibility may be increased above the standard visibility minima due to penetrations of the 20:1 and 34:1 final approach obstacle identification surfaces (OIS). The minimum visibility required for 34:1 penetrations is ¾ SM and for 20:1 penetrations 1 SM, which is discussed in chapter 5 of this handbook. When there are penetrations of the final approach OIS, a visibility credit for approach lighting systems is not allowed for either airplane or helicopter procedures that would result in values less than the appropriate ¾ SM or 1 SM visibility

Helicopter Use of Standard Instrument Approach Procedures			
Procedure	Helicopter Visibility Minima	Helicopter MDA/DA	Maximum Speed Limitations
Standard	The greater of: one half the Category A visibility minima, ¼ statute mile visibility or 1,200 RVR unless annotated (Visibility Reduction by Helicopters NA.)	As published for Category A	The helicopter may initiate the final approach segment at speeds up to the upper limit of the highest Approach Category authorized by the procedure, but must be slowed to no more than 90 KIAS at the MAP in order to apply the visibility reduction.
Copter Procedure	As published	As published	90 KIAS when on a published route/track.
GPS Copter Procedure	As published	As published	90 KIAS when on a published route, track, or holding, 70 KIAS when on the final approach or missed approach segment. Military procedures are limited to 90 KIAS for all segments.

Figure 7-9. Helicopter use of standard instrument approach procedures.

requirement. The 14 CFR Part 97.3 visibility reduction rule does not apply, and you must take precautions to avoid any obstacles in the visual segment. Procedures with penetrations of the final approach OIS are annotated at the next amendment with "Visibility Reduction by Helicopters NA."

Until all the affected SIAPs have been annotated, an understanding of how the standard visibilities are established is the best aid in determining if penetrations of the final approach OIS exists. Some of the variables in determining visibilities are: density altitude (DA)/MDA height above touchdown (HAT), height above airport (HAA), distance of the facility to the MAP (or the runway threshold for non- precision approaches), and approach lighting configurations.

The standard visibility requirement, without any credit for lights, is 1 SM for non-precision approaches and 3⁄4 SM for precision approaches. This is based on a Category A airplane 250–320 feet HAT/HAA; for non-precision approaches a distance of 10,000 feet or less from the facility to the MAP (or runway threshold). For precision approaches, credit for any approach light configuration; for non-precision approaches (with a 250 HAT) configured with a Medium Intensity Approach Lighting System (MALSR), Simplified Short Approach Lighting System (SSALR), or Approach Lighting System With Sequenced Flashing Lights (ALSF)-1 normally results in a published visibility of 1⁄2 SM.

Consequently, if an ILS is configured with approach lights or a non-precision approach is configured with MALSR, SSALR, or ALSF-1 lighting configurations and the procedure has a published visibility of 3⁄4 SM or greater, a penetration of the final approach OIS may exist. Also, pilots are unable to determine whether there are penetrations of the final approach OIS if a non-precision procedure does not have approach lights or is configured with ODALS, MALS, or SSALS/ SALS lighting since the minimum published visibility is 3⁄4 SM or greater.

As a rule of thumb, approaches with published visibilities of 3⁄4 SM or more should be regarded as having final approach OIS penetrations and care must be taken to avoid any obstacles in the visual segment.

Approaches with published visibilities of 1⁄2 SM or less are free of OIS penetrations and the visibility reduction in Part 97.3 is authorized.

Copter Only Approaches to An Airport or Heliport

Pilots flying Copter SIAPs, other than GPS, may use the published minima with no reductions in visibility allowed. The maximum airspeed is 90 KIAS on any segment of the approach or missed approach. Figure 7-10 illustrates the COPTER ILS or LOC RWY 13 approach at New York/La Guardia (LGA) airport.

Copter ILS approaches to Category (CAT) I facilities with DAs no lower than a 200-foot HAT provide an advantage over a conventional ILS of shorter final segments and lower minimums (based on the 20:1 missed approach surface). There are also Copter approaches with minimums as low as 100-foot HAT and 1⁄4 SM visibility. Approaches with a HAT below 200 feet are annotated with the note: "Special Aircrew & Aircraft Certification Required" since the FAA must approve the helicopter and its avionics, and the flight crew must have the required experience, training, and checking.

The ground facilities (approach lighting, signal in space, hold lines, maintenance, etc.) and air traffic infrastructure for CAT II ILS approaches are required to support these procedures. The helicopter must be equipped with an AP, FD, or head up guidance system, alternate static source (or heated static source), and radio altimeter. The pilot must have at least a private pilot helicopter certificate, an instrument helicopter rating, and a type rating if the helicopter requires a type rating. Pilot experience requires the following flight times: 250 pilot in command (PIC), 100 helicopter PIC, 50 night PIC, 75 hours of actual or simulated instrument flight time, including at least 25 hours of actual or simulated instrument flight time in a helicopter or a helicopter flight simulator, and the appropriate recent experience, training and check. For Copter CAT II ILS operations below 200 feet HAT, approach deviations are limited to 1⁄4 scale of the localizer or glide slope needle. Deviations beyond that require an immediate missed approach unless the pilot has at least one of the visual references in sight and otherwise meets the requirements of 14 CFR Part 91.175(c). The reward for this effort is the ability to fly Copter ILS approaches with minima that are sometimes below the airplane CAT II minima. The procedure to apply for this certification is available from your local Flight Standards District Office (FSDO).

Copter GPS Approaches to an Airport or Heliport

Pilots flying Copter GPS or WAAS SIAPs must limit the speed to 90 KIAS on the initial and intermediate segment of the approach and to no more than 70 KIAS on the final and missed approach segments. If annotated, holding may also be limited to 90 KIAS to contain the helicopter within the small airspace provided for helicopter holding patterns. During testing for helicopter holding, the optimum airspeed and leg length combination was determined to

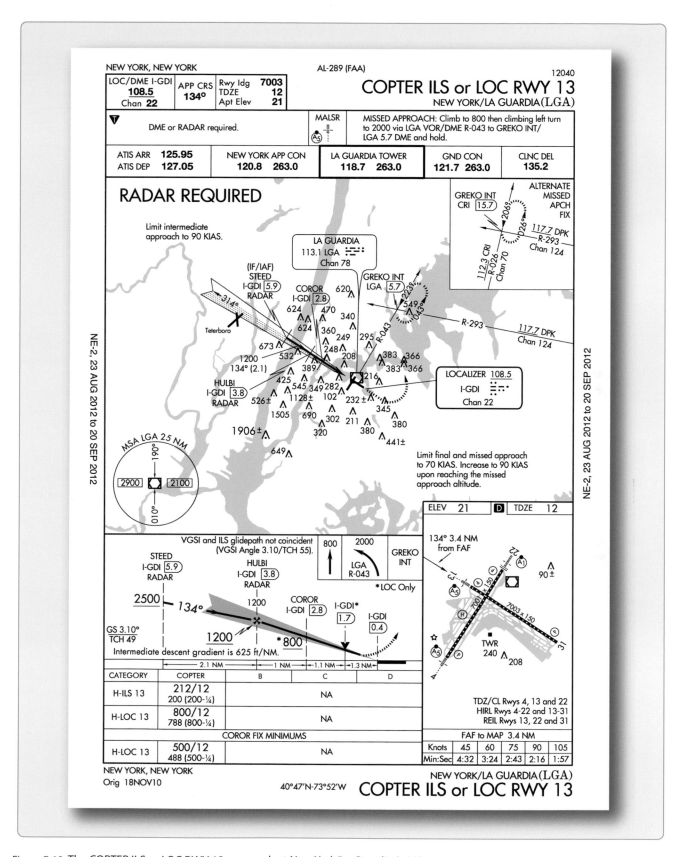

Figure 7-10. The COPTER ILS or LOC RWY 13 approach at New York/La Guardia (LGA) airport.

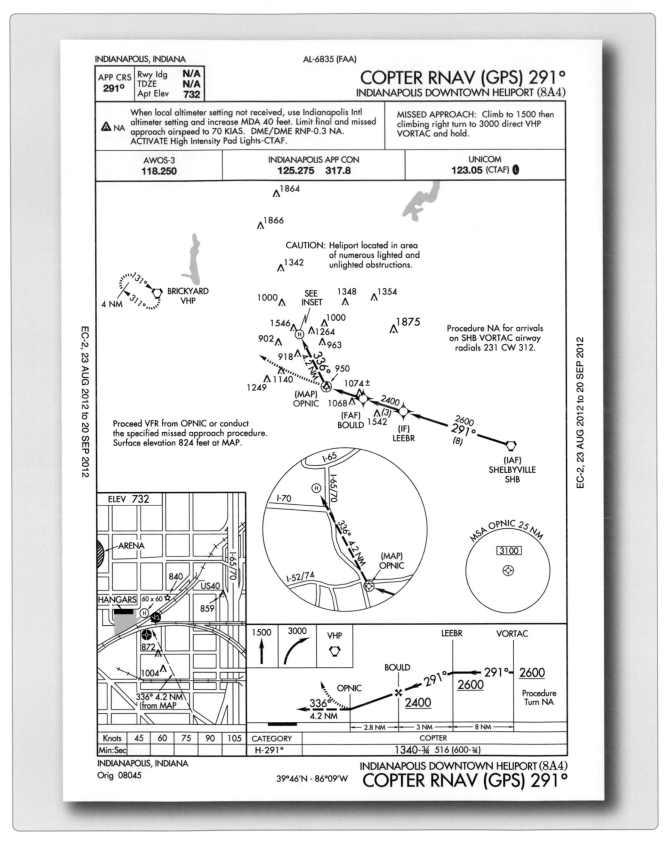

Figure 7-11. COPTER RNAV (GPS) 291° at Indianapolis Downtown Heliport.

be 90 KIAS with a 3 NM outbound leg length. Consideration was given to the wind drift on the dead reckoning entry leg at slower speeds, the turn radius at faster airspeeds, and the ability of the helicopter in strong wind conditions to intercept the inbound course prior to the holding fix. The published minimums are to be used with no visibility reductions allowed. Figure 7-11 is an example of a Copter GPS PinS approach that allows the helicopter to fly VFR from the MAP to the heliport.

The final and missed approach protected airspace providing obstacle and terrain avoidance is based on 70 KIAS, with a maximum 10-knot tailwind component. It is absolutely essential that pilots adhere to the 70 KIAS limitation in procedures that include an immediate climbing and turning missed approach. Exceeding the airspeed restriction increases the turning radius significantly and can cause the helicopter to leave the missed approach protected airspace. This may result in controlled flight into terrain (CFIT) or obstacles.

If a helicopter has a V_{MINI} greater than 70 knots, then it is not capable of conducting this type of approach. Similarly, if the autopilot in "go-around" mode climbs at a V_{YI} greater than 70 knots, then that mode cannot be used. It is the responsibility of the pilot to determine compliance with missed approach climb gradient requirements when operating at speeds other than V_Y or V_{YI}. Missed approaches that specify an "IMMEDIATE CLIMBING TURN" have no provision for a straight ahead climbing segment before turning. A straight segment results in exceeding the protected airspace limits.

Protected obstacle clearance areas and surfaces for the missed approach are established on the assumption that the missed approach is initiated at the DA point and for non-precision approaches no lower than the MDA at the MAP (normally at the threshold of the approach end of the runway). The pilot must begin the missed approach at those points. Flying beyond either point before beginning the missed approach results in flying below the protected obstacle clearance surface (OCS) and can result in a collision with an obstacle.

The missed approach segment U.S. Standard for Terminal Instrument Procedures (TERPS) criteria for all Copter approaches take advantage of the helicopter's climb capabilities at slow airspeeds, resulting in high climb gradients. [Figure 7-12] The OCS used to evaluate the missed approach is a 20:1 inclined plane. This surface is twice as steep for the helicopter as the OCS used to evaluate the airplane missed approach segment. The helicopter climb gradient is therefore required to be double that of the airplane's required missed approach climb gradient.

A minimum climb gradient of at least 400 FPNM is required unless a higher gradient is published on the approach chart (e.g., a helicopter with a ground speed of 70 knots is required to climb at a rate of 467 fpm (467 fpm = 70 KIAS × 400 feet per NM/60 seconds)). The advantage of using the 20:1 OCS for the helicopter missed approach segment instead of the 40:1 OCS used for the airplane is that obstacles that penetrate the 40:1 missed approach segment may not have to be considered. The result is

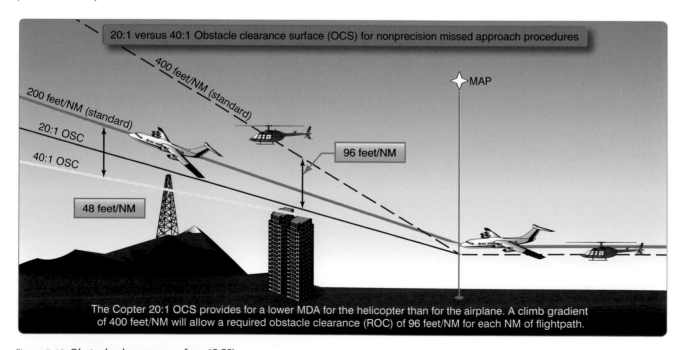

Figure 7-12. Obstacle clearance surface (OCS).

7-13

the DA/MDA may be lower for helicopters than for other aircraft. The minimum required climb gradient of 400 FPNM for the helicopter in a missed approach provides 96 feet of required obstacle clearance (ROC) for each NM of flightpath.

Helicopter Approaches to VFR Heliports

Helicopter approaches to VFR heliports are normally developed either as public procedures to a PinS that may serve more than one heliport or as a special procedure to a specific VFR heliport that requires pilot training due to its unique characteristics. These approaches can be developed using very high frequency omni-directional range (VOR) or automatic direction finder (ADF), but area navigation (RNAV) using GPS is the most common system used today. RNAV using the WAAS offers the most advantages because it can provide lower approach minimums, narrower route widths to support a network of approaches, and may allow the heliport to be used as an alternate. A majority of the special procedures to a specific VFR heliport are developed in support of HEMS operators and have a "Proceed Visually" segment between the MAP and the heliport. Public procedures are developed as a PinS approach with a "Proceed VFR" segment between the MAP and the landing area. These PinS "Proceed VFR" procedures specify a course and distance from the MAP to the available heliports in the area.

Approach to a PinS

The note associated with these procedures is: "PROCEED VFR FROM (NAMED MAP) OR CONDUCT THE SPECIFIED MISSED APPROACH." They may be developed as a special or public procedure where the MAP is located more than 2 SM from the landing site, the turn from the final approach to the visual segment is greater than 30 degrees, or the VFR segment from the MAP to the landing site has obstructions that require pilot actions to avoid them. Figure 7-13 is an example of a public PinS approach that allows the pilot to fly to one of four heliports after reaching the MAP.

For Part 135 operations, pilots may not begin the instrument approach unless the latest weather report indicates that the weather conditions are at or above the authorized IFR or VFR minimums as required by the class of airspace, operating rule and/or OpSpecs, whichever is higher. Visual contact with the landing site is not required; however, prior to the MAP, for either Part 91 or 135 operators, the pilot must determine if the flight visibility meets the basic VFR minimums required by the class of airspace, operating rule and/or OpSpecs (whichever is higher). The visibility is limited to no lower than that published in the procedure until canceling IFR. If VFR minimums do not exist, then the published MAP must

be executed. The pilot must contact air traffic control (ATC) upon reaching the MAP, or as soon as practical after that, and advise whether executing the missed approach or canceling IFR and proceeding VFR. Figure 7-14 provides examples of the procedures used during a PinS approach for Part 91 and Part 135 operations.

To proceed VFR in uncontrolled airspace, Part 135 operators are required to have at least 1/2 SM visibility and a 300-foot ceiling. Part 135 HEMS operators must have at least 1 SM day or 2 SM night visibility and a 500-foot ceiling provided the heliport is located within 3 NM of the MAP. These minimums apply regardless of whether the approach is located on the plains of Oklahoma or in the Colorado mountains. However, for heliports located farther than 3 NM from the heliport, Part 135 HEMS operators are held to an even higher standard and the minimums and lighting conditions contained in Figure 7-15 apply to the entire route. Mountainous terrain at night with low light conditions requires a ceiling of 1,000 feet and either 3 SM or 5 SM visibility depending on whether it has been determined as part of the operator's local flying area.

In Class B, C, D, and E surface area airspace, a SVFR clearance may be obtained if SVFR minimums exist. On your flight plan, give ATC a heads up about your intentions by entering the following in the remarks section: "Request SVFR clearance after the MAP."

Approach to a Specific VFR Heliport

The note associated with these procedures is: "PROCEED VISUALLY FROM (NAMED MAP) OR CONDUCT THE SPECIFIED MISSED APPROACH." Due to their unique characteristics, these approaches require training. They are developed for hospitals, oil rigs, private heliports, etc. As Specials, they require Flight Standards approval by a Letter of Authorization (LOA) for Part 91 operators or by OpSpecs for Part 135 operators. The heliport associated with these procedures must be located within 10,560 feet of the MAP, the visual segment between the MAP and the heliport evaluated for obstacle hazards, and the heliport must meet the appropriate VFR heliport recommendations of AC 150/5390-2, "Heliport Design."

The PinS optimum location is 0.65 NM from the heliport. This provides an adequate distance to decelerate and land from an approach speed of 70 KIAS. Certain airframes may be certified to fly at reduced V_{MINI} or below V_{MINI} speeds as a result of flight control design or adherence to AC 29-2, Certification of Transport Category Rotorcraft. In these cases, an approach procedure stating the minimum certified airspeed or flight below V_{MINI} should be annotated on the approach procedure. The distance also permits

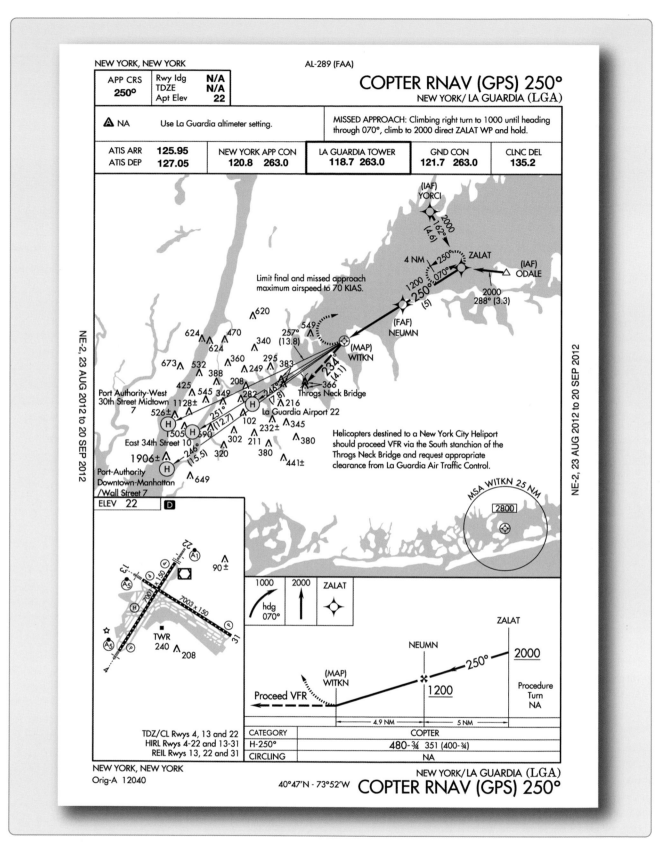

Figure 7-13. COPTER RNAV (GPS) 250° at New York/La Guardia Airport.

		Non-Mountainous		Mountainous (14 CFR Part 95)	
	Area	Local	Cross Country	Local	Cross Country
Condition		Ceiling-visibility			
Day		500-1	800-2	500-2	800-3
Night—High Lighting Conditions		500-2	1000-3	500-3	1000-3
Night—Low Lighting Conditions		800-3	1000-5	1000-3	1000-5

Figure 7-14. Weather minimums and lighting conditions for HEMS operators.

Point-in-Space Approach Examples

EXAMPLE 1

Under Part 91, the operator flies the published IFR PinS approach procedure that has a charted MDA of 340 mean sea level (MSL) and visibility of ¾ SM. When approaching the MAP at an altitude of 340 feet MSL, the pilot transitions from Instrument Meteorological Conditions (IMC) to Visual Meteorological Conditions (VMC) and determines that the flight visibility is ½ SM. The pilot must determine prior to the MAP whether the applicable basic VFR weather minimums can be maintained from the MAP to the heliport or execute a missed approach. If the pilot determines that the applicable basic VFR weather minimums can be maintained to the heliport, the pilot may proceed VFR. If the visual segment is in Class B, C, D, or the surface area of Class E airspace, it may require the pilot to obtain a Special VFR clearance.

EXAMPLE 2

For an operator to proceed VFR under Part 135, a minimum visibility of ½ SM during the day and 1 SM at night with a minimum ceiling of 300 feet. If prior to commencing the approach, the pilot determines the reported visibility is ¾ SM during the day, the pilot descends IMC to an altitude no lower than the MDA and transitions to VMC. If the pilot determines prior to the MAP that the flight visibility is less than ½ SM in the visual segment, a missed approach must be executed at the MAP.

Figure 7-15. Point-in-space (PinS) approach examples for Part 91 and Part 135 operations.

optimal blending of obstacle clearance criteria with non-instrument heliport approach areas.

The visibility minimum is based on the distance from the MAP to the heliport, among other factors (e.g., height above the heliport elevation when at the MAP MDA). The pilot is required to acquire and maintain visual contact with the heliport final approach and takeoff (FATO) area at or prior to the MAP. Obstacle or terrain avoidance from the MAP to the heliport is the responsibility of the pilot. Pilots need to level off when reaching the MDA, which may occur before arriving at the MAP, until reaching the visual approach angle on the approach path to clear the obstacles. If the required weather minimums do not exist, then the published MAP must be executed at the MAP because IFR obstruction clearance areas are not applied to the visual segment of the approach and a missed approach segment protection is not provided between the MAP and the heliport. As soon as practicable after reaching the MAP, the pilot advises ATC whether cancelling IFR and proceeding visually or executing the missed approach.

Inadvertent IMC

Whether it is a corporate or HEMS operation, helicopter pilots sometimes operate in challenging weather conditions. An encounter with weather that does not permit continued flight under VFR might occur when conditions do not allow for the visual determination of a usable horizon (e.g., fog, snow showers, or night operations over unlit surfaces such as water). Flight in conditions of limited visual contrast should be avoided since this can result in a loss of horizontal or surface reference, and obstacles such as wires become perceptually invisible. To prevent spatial disorientation, loss of control (LOC) or CFIT, pilots should slow the helicopter to a speed that provides a controlled deceleration in the distance equal to the forward visibility. The pilot should look for terrain that provides sufficient contrast to either continue the flight or to make a precautionary landing. If spatial disorientation occurs and a climb into IMC is not feasible due to fuel state, icing conditions, equipment, etc., make every effort to land the helicopter with a slight forward descent to prevent any sideward or rearward motion.

All helicopter pilots should receive training on avoidance and recovery from inadvertent IMC with emphasis on avoidance. An unplanned transition from VMC to IMC flight is an emergency that involves a different set of pilot actions. It requires the use of different navigation and operational procedures, interaction with ATC, and crewmember resource management (CRM). Consideration should be given to the local flying area's terrain, airspace, air traffic facilities, weather (including seasonal affects such as icing and thunderstorms), and available airfield/heliport approaches.

Training should emphasize the identification of circumstances conducive to inadvertent IMC and a strategy to abandon continued VFR flight in deteriorating conditions. This strategy should include a minimum altitude/airspeed combination that provides for an off-airport/heliport landing, diverting to better conditions, or initiating an emergency transition to IFR. Pilots should be able to readily identify the minimum initial altitude and course in order to avoid CFIT. Current IFR en route and approach charts for the route of flight are essential. A GPS navigation receiver with a moving map provides exceptional situational awareness for terrain and obstacle avoidance.

Training for an emergency transition to IFR should include full and partial panel instrument flight, unusual attitude recovery, ATC communications, and instrument approaches. If an ILS is available and the helicopter is equipped, an ILS approach should be made. Otherwise, if the helicopter is equipped with an IFR approach-capable GPS receiver with a current database, a GPS approach should be made. If neither, an ILS or GPS procedure is available use another instrument approach.

Upon entering inadvertent IMC, priority must be given to control of the helicopter. Keep it simple and take one action at a time.

- Control. First use the wings on the attitude indicator to level the helicopter. Maintain heading and increase to climb power. Establish climb airspeed at the best angle of climb but no slower than V_{MINI}.

- Climb. Climb straight ahead until your crosscheck is established. Then make a turn only to avoid terrain or objects. If an altitude has not been previously established with ATC to climb to for inadvertent IMC, then you should climb to an altitude that is at least 1,000 feet above the highest known object and that allows for contacting ATC.

- Communicate. Attempt to contact ATC as soon as the helicopter is stabilized in the climb and headed away from danger. If the appropriate frequency is not known, you should attempt to contact ATC on either very high frequency (VHF) 121.5 or ultra high frequency (UHF) 243.0. Initial information provided to ATC should be your approximate location, that inadvertent IMC has been encountered and an emergency climb has been made, your altitude, amount of flight time remaining (fuel state), and number of persons on board. You should then request a vector to either VFR weather conditions or to the nearest suitable airport/heliport that conditions will support a successful approach. If unable to contact ATC and a transponder code has not been previously established with ATC for inadvertent IMC, change the transponder code to 7700.

A radio altimeter is a necessity for alerting the pilot when inadvertently going below the minimum altitude. Barometric altimeters are subject to inaccuracies that become important in helicopter IFR operations, especially in cold temperatures.

IFR Heliports

AC 150/5390-2, Heliport Design, provides recommendations for heliport design to support non-precision, approach with vertical guidance (APV), and precision approaches to a heliport. When a heliport does not meet the criteria of this AC, FAA Order 8260.42, United States Standard for Helicopter Area Navigation (RNAV), requires that an instrument approach be published as a SPECIAL procedure with annotations that special aircrew qualifications are required to fly the procedure. Currently, there are no operational civil IFR heliports in the U.S., although the U.S. military has some non-precision and precision approach procedures to IFR heliports.

Appendix A

Emergency Procedures

Introduction

Changing weather conditions, air traffic control (ATC), aircraft, and pilots are variables that make instrument flying an unpredictable and challenging operation. Safety of the flight depends on the pilot's ability to manage these variables while maintaining positive aircraft control and adequate situational awareness (SA). This appendix discusses recognition and suggested remedies for emergency events related to un-forecasted, adverse weather, aircraft system malfunctions, communication/navigation system malfunctions, loss of SA, and inadvertent instrument meteorological conditions (IIMCs).

Emergencies

An emergency can be either a distress or urgency condition as defined in the pilot/controller glossary. Distress is defined as a condition of being threatened by serious and/or imminent danger and requiring immediate assistance. Urgency is defined as a condition of being concerned about safety and requiring timely but not immediate assistance; a potential distress condition.

Pilots do not hesitate to declare an emergency when faced with distress conditions, such as fire, mechanical failure, or structural damage. However, some are reluctant to report an urgency condition when encountering situations that may not be immediately perilous but are potentially catastrophic. An aircraft is in an urgency condition the moment that the pilot becomes doubtful about position, fuel endurance, weather, or any other condition that could adversely affect flight safety. The time for a pilot to request assistance is when an urgent situation may, or has just occurred, not after it has developed into a distress situation.

The pilot in command (PIC) is responsible for crew, passengers, and operation of the aircraft at all times. Title 14 of the Code of Federal Regulations (14 CFR) part 91, section 91.3 allows deviations from regulations during emergencies that allow the PIC to make the best decision to ensure safety of all personnel during these contingencies. Also, by declaring an emergency during flight, that aircraft becomes a priority to land safely. Pilots who become apprehensive for their safety for any reason should request assistance immediately. Assistance is available in the form of radio, radar, direction finding (DF) stations, and other aircraft.

Inadvertent Thunderstorm Encounter

A pilot should always avoid intentionally flying through a thunderstorm of any intensity; however, certain conditions may be present that could lead to an inadvertent thunderstorm encounter. For example, flying in areas where thunderstorms are embedded in large cloud masses may make thunderstorm avoidance difficult, even when the aircraft is equipped with thunderstorm detection equipment. Pilots must be prepared to deal with inadvertent thunderstorm penetration. At the very least, a thunderstorm encounter subjects the aircraft to turbulence that could be severe. The pilot, as well as the crew and any passengers, should tighten seat belts and shoulder harnesses and secure any loose items in the cabin or flight deck.

As with any emergency, the first order of business is to fly the aircraft. The pilot workload is high; therefore, increased concentration is necessary to maintain an instrument scan. Once in a thunderstorm, it is better to maintain a course straight through the thunderstorm rather than turning around. A straight course most likely gets the pilot out of the hazard in the least amount of time, and turning maneuvers only increase structural stress on the aircraft.

Reduce power to a setting that maintains a recommended turbulence penetration speed as described in the appropriate aircraft operator's manual, and try to minimize additional power adjustments. Concentrate on keeping the aircraft in a level attitude while allowing airspeed and altitude to fluctuate. Similarly, if using autopilot, disengage altitude and speed hold modes because they only increase the aircraft's maneuvering, which increases structural stress.

During a thunderstorm encounter, the potential for icing also exists. As soon as possible, if the aircraft is so equipped, turn on anti-icing/deicing equipment. Icing can be rapid at any altitude, and may lead to power failure and/or loss of airspeed indication. Lightning is also present in a thunderstorm and can temporarily blind the pilot. To reduce risk, turn up flight deck lights to the highest intensity, concentrate on flight instruments, and resist the urge to look outside.

Inadvertent Icing Encounter

Because icing is unpredictable, pilots may find themselves in icing conditions although they have done everything to avoid the condition. To stay alert to this possibility while operating in visible moisture, pilots should monitor the outside air temperature (OAT).

Anti-icing/deicing equipment is critical to safety of the flight. If anti-icing/deicing equipment is not used before sufficient ice has accumulated, it may not be able to remove all ice accumulation. Use of anti-icing/deicing reduces power availability; therefore, pilots should be familiar with the aircraft operator's manual for use of anti-icing/deicing equipment.

Before entering visible moisture with temperatures at five degrees above freezing or cooler, activate appropriate anti-icing/ deicing equipment in anticipation of ice accumulation; early ice detection is critical. Detecting ice may be particularly difficult during night flight. The pilot may need to use a flashlight to check for ice accumulation on the wings, fuselage, landing gear, and horizontal stabilizer. At the first indication of ice accumulation, the pilot must act to circumvent icing conditions. Options for action once ice has begun to accumulate on the aircraft are the following:

- Move to an altitude with significantly colder temperatures.
- Move to an altitude with temperatures above freezing.
- Fly to an area clear of visible moisture.
- Change the heading, and fly to an area of known non-icing conditions.

If these options are not available, consider an immediate landing at the nearest suitable airport. Anti-icing/deicing equipment does not allow aircraft to operate in icing conditions indefinitely; it only provides more time to evade icing conditions. If icing is encountered, an aircraft controllability check should be considered in the landing configuration. Give careful consideration to configuration changes that might produce unanticipated aircraft flight dynamics.

Precipitation Static

Precipitation static occurs when accumulated static electricity discharges from extremities of the aircraft. This discharge has the potential to create problems with the aircraft's instruments. These problems range from serious, such as complete loss of VHF communications and erroneous magnetic compass readings, to the annoyance of high-pitched audio squealing.

Precipitation static is caused when an aircraft encounters airborne particles during flight (rain or snow) and develops a negative charge. It can also result from atmospheric electric fields in thunderstorm clouds. When a significant negative voltage level is reached, the aircraft discharges it, creating electrical disturbances. To reduce problems associated with precipitation static, the pilot ensures that the aircraft's static wicks are maintained and accounted for. All broken or missing static wicks should be replaced before an instrument flight.

Aircraft System Malfunction

Preventing aircraft system malfunctions that might lead to an in-flight emergency begins with a thorough preflight inspection. In addition to items normally checked before visual flight rules (VFR) flight, pilots intending to fly instrument flight rules (IFR) should pay particular attention to antennas, static wicks, anti-icing/deicing equipment, pitot tube, and static ports. During taxi, verify operation and accuracy of all flight instruments. The pilots must ensure that all systems are operational before departing into IFR conditions.

Generator Failure

Depending on aircraft being flown, a generator failure is indicated in different ways. Some aircraft use an ammeter that indicates the state of charge or discharge of the battery. A positive indication on the ammeter indicates a charge condition; a negative indication reveals a discharge condition. Other aircraft use a load meter to indicate the load being carried by the generator. If the generator fails, a zero load indication is shown on the load meter. Review the appropriate aircraft operator's manual for information on the type of systems installed in the aircraft.

Once a generator failure is detected, the pilot must reduce electrical load on the battery and land as soon as practical. Depending on electrical load and condition of the battery, sufficient power may be available for an hour or more of flight or for only a matter of minutes. The pilot must be familiar with systems requiring electricity to run and which continue to operate without power. In aircraft with multiple generators, care should be taken to reduce electrical load to avoid overloading the operating generator(s). The pilot can attempt to troubleshoot generator failure by following established procedures published in the appropriate aircraft operator's manual. If the generator cannot be reset, inform ATC of an impending electrical failure.

Instrument Failure

System or instrument failure is usually identified by a warning indicator or an inconsistency between indications on the attitude indicator, supporting performance instruments, and instruments at the other pilot station, if so equipped. Aircraft control must be maintained while the pilot identifies the failed components and expedite cross-check including all flight instruments. The problem may be individual instrument failure or a system failure affecting several instruments.

One method of identification involves an immediate comparison of the attitude indicator with rate-of-turn indicator and vertical speed indicator (VSI). Along with providing pitch-and-bank information, this technique compares the static system with the pressure system and electrical system. Identify the failed components and use remaining functional instruments to maintain aircraft control. Attempt to restore inoperative components by checking the appropriate power source, changing to a backup or alternate system, and resetting the instrument if possible. Covering failed instruments may enhance the ability to maintain aircraft control and navigate the aircraft. ATC should be notified of the problem and, if necessary, declare an emergency before the situation deteriorates beyond the ability to recover.

Pitot/Static System Failure

A pitot or static system failure can also cause erratic and unreliable instrument indications. When a static system problem occurs, it affects the airspeed indicator, altimeter, and VSI. In the absence of an alternate static source in an unpressurized aircraft, the pilot could break the glass on the VSI because it is not required for instrument flight. Breaking the glass provides both the altimeter and airspeed indicator a source of static pressure, but pilots should be cautious because breaking the glass can cause additional instrument errors. Before considering, pilots should be familiar with their aircraft's specific procedures for static problems.

Loss of Situational Awareness (SA)

SA is an overall assessment of environmental elements and how they affect flight. SA permits the pilot to make decisions ahead of time and allows evaluation of several different options. Conversely, a pilot who is missing important information about the flight is apt to make reactive decisions. Poor SA means that the pilot lacks vision regarding future events that can force him or her to make decisions quickly often with limited options. During an IFR flight, pilots operate at varying levels of SA. For example, a pilot may be en route to a destination with a high level of SA when ATC issues an unexpected standard terminal arrival route (STAR). Because the STAR is unexpected and the pilot is unfamiliar with the procedure, SA is reduced. However, after becoming familiar with the STAR and resuming normal navigation, the pilot returns to a higher level of SA.

Factors reducing SA include distractions, unusual or unexpected events, complacency, high workload, unfamiliar situations, and inoperative equipment. In some situations, a loss of SA may be beyond a pilot's control. With an electrical system failure and associated loss of an attitude indication, a pilot may find the aircraft in an unusual attitude. In this situation, established procedures are used to regain SA and aircraft control. Pilots must be alert to loss of SA especially when hampered by a reactive mindset. To regain SA, reassess the situation and work toward understanding what the problem is. The pilot may need to seek additional information from other sources, such as navigation instruments, other crewmembers, or ATC.

Inadvertent Instrument Meteorological Condition (IIMC)

Some pilots have the misconception that inadvertent instrument meteorological condition (IIMC) does not apply to an IFR flight. The following examples could cause a pilot to inadvertently encounter IMC.

1. The aircraft has entered visual meteorological conditions (VMC) during an instrument approach procedure (IAP) and while circling to land encounters IMC.

2. During a non-precision IAP, the aircraft, in VMC, levels at the MDA just below the overcast. Suddenly, the aircraft re-enters the overcast because either the pilot was unable to correctly hold his or her altitude and climbed back into the overcast, or the overcast sloped downward ahead of the aircraft and, while maintaining the correct MDA, the aircraft re-entered the clouds.

3. After inadvertently re-entering the clouds, the pilot maintains aircraft control, and then maneuvers to the published holding fix, while contacting ATC. If navigational guidance or pilot SA were lost, the pilot would then climb to the published MSA (see AIM 5-4-7c).

In order to survive an encounter with IIMC, a pilot must recognize and accept the seriousness of the situation. The pilot will need to immediately commit to the instruments and perform the proper recovery procedures.

Maintaining Aircraft Control

Once the crewmembers recognize the situation, they commit to controlling the aircraft by using and trusting flight instruments. Attempting to search outside the flight deck for visual confirmation can result in spatial disorientation and complete loss of control. The crew must rely on instruments and depend on crew coordination to facilitate that transition. The pilot or flight crew must abandon their efforts to establish visual references and fly the aircraft by their flight instruments.

The most important concern, along with maintaining aircraft control, is to initiate a climb immediately. An immediate climb provides a greater separation from natural and manmade obstacles, as well as improve radar reception of the aircraft by ATC. An immediate climb should be appropriate for the current conditions, environment, and known or perceived obstacles. Listed below are procedures that can assist in maintaining aircraft control after encountering IIMC with the most critical action being to immediately announce IIMC and begin a substantial climb while procedures are being performed. These procedures are performed nearly simultaneously:

- Attitude—level wings on the attitude indicator.
- Heading—maintain heading; turn only to avoid known obstacles.
- Power—adjust power as necessary for desired climb rate.
- Airspeed—adjust airspeed as necessary. Complete the IIMC recovery according to local and published regulations and policies.

In situations where the pilot encounters IIMC while conducting an instrument maneuver, the best remedy is immediate execution of the published missed approach.

The pilot must trust the flight instruments concerning the aircraft's attitude regardless of intuition or visual interpretation. The vestibular sense (motion sensing by the inner ear) can confuse the pilot. Because of inertia, sensory areas of the inner ear cannot detect slight changes in aircraft attitude nor can they accurately sense attitude changes that occur at a uniform rate over time. Conversely, false sensations often push the pilot to believe that the attitude of the aircraft has changed when in fact it has not, resulting in spatial disorientation.

ATC Requirements During an In-Flight Emergency

ATC personnel can help pilots during in-flight emergency situations. Pilots should understand the services provided by ATC and the resources and options available. These services enable pilots to focus on aircraft control and help them make better decisions in a time of stress.

Provide Information

During emergency situations, pilots should provide as much information as possible to ATC. ATC uses the information to determine what kind of assistance it can provide with available assets and capabilities. Information requirements vary depending on the existing situation. ATC requires at a minimum, the following information for in-flight emergencies:

- Aircraft identification and type
- Nature of the emergency
- Pilot's desires

If time and the situation permits, the pilot should provide ATC with more information. Listed below is additional information that would help ATC in further assisting the pilot during an emergency situation.

- Aircraft altitude
- Point of departure and destination
- Airspeed
- Fuel remaining in time
- Heading since last known position
- Visible landmarks
- Navigational aids (NAVAID) signals received
- Time and place of last known position
- Aircraft color
- Pilot reported weather
- Emergency equipment on board
- Number of people on board
- Pilot capability for IFR flight
- Navigation equipment capability

When the pilot requests, or when deemed necessary, ATC can enlist services of available radar facilities and DF facilities operated by the FAA. ATC can also coordinate with other agencies, such as the U.S. Coast Guard (USCG) and other local authorities and request their emergency services.

Radar Assistance

Radar is an invaluable asset that can be used by pilots during emergencies. With radar, ATC can provide navigation assistance to aircraft and provide last-known location during catastrophic emergencies. If a VFR aircraft encounters or is about to encounter IMC weather conditions, the pilot can request radar vectors to VFR airports or VFR conditions. If the pilot determines that he or she is qualified and the aircraft is capable of conducting IFR flight, the pilot should file an IFR flight plan and request a clearance from ATC to the destination airport as appropriate. If the aircraft has already encountered IFR conditions, ATC can inform the pilot of appropriate terrain/obstacle clearance minimum altitude. If the aircraft is below appropriate terrain/obstacle clearance minimum altitude and sufficiently accurate position information has been received or radar identification is established, ATC can furnish a heading or radial on which to climb to reach appropriate terrain/ obstacle clearance minimum altitude.

Emergency Airports

ATC personnel consider how much remaining fuel in relation to the distance to the airport and weather conditions when recommending an emergency airport to aircraft requiring assistance. Depending on the nature of the emergency, certain weather phenomena may deserve weighted consideration. A pilot may elect to fly further to land at an airport with VFR conditions instead of closer airfield with IFR conditions. Other considerations are airport conditions, NAVAID status, aircraft type, pilot's qualifications, and vectoring or homing capability to the emergency airport. In addition, ATC and pilots should determine which guidance can be used to fly to the emergency airport. The following options may be available:

- Radar
- DF
- Following another aircraft
- NAVAIDs
- Pilotage by landmarks
- Compass headings

Emergency Obstruction Video Map (EOVM)

The emergency obstruction video map (EOVM) is intended to facilitate advisory service in an emergency situation when appropriate terrain/obstacle clearance minimum altitude cannot be maintained. The EOVM, and the service provided, are used only under the following conditions:

1. The pilot has declared an emergency.

2. The controller has determined an emergency condition exists or is imminent because of the pilots inability to maintain an appropriate terrain/obstacle clearance minimum altitude.

NOTE: Appropriate terrain/obstacle clearance minimum altitudes may be defined as minimum IFR altitude (MIA), minimum en route altitude (MEA), minimum obstacle clearance altitude (MOCA), or minimum vectoring altitude (MVA).

When providing emergency vectoring service, the controller advises the pilot that any headings issued are emergency advisories intended only to direct the aircraft toward and over an area of lower terrain/obstacle elevation. Altitudes and obstructions depicted on the EOVM are actual altitudes and locations of the obstacle/terrain and contain no lateral or vertical buffers for obstruction clearance.

Responsibility

ATC, in communication with an aircraft in distress, should handle the emergency and coordinate and direct the activities of assisting facilities. ATC will not transfer this responsibility to another facility unless that facility can better handle the situation. When an ATC facility receives information about an aircraft in distress, they forward detailed data to the center in the area of the emergency. Centers serve as central points for collecting information, coordinating with search and rescue (SAR) and distributing information to appropriate agencies.

Although 121.5 megahertz and 243.0 megahertz are emergency frequencies, the pilot should keep the aircraft on the initial contact frequency. The pilot should change frequencies only when a valid reason exists. When necessary, and if weather and circumstances permit, ATC should recommend that aircraft maintain or increase altitude to improve communications, radar, or DF reception.

Escort

An escort aircraft, if available, should consider and evaluate an appropriate formation. Special consideration must be given if maneuvers take the aircraft through clouds. Aircraft should not execute an in-flight join up during emergency conditions unless both crews involved are familiar with and capable of formation flight and can communicate and have visual contact with each other.

Appendix B

Acronyms

A

AAC—Aircraft Administration Communications

AAUP—Attention All Users Page

AC—Advisory Circular

ACARS—Aircraft Communications Addressing and Reporting System

ADCs—Air Data Computers

ADDS—Aviation Digital Data Services

ADF—Automatic Direction Finder

ADS—Automatic Dependent Surveillance

ADS-B—Automatic Dependent Surveillance-Broadcast

AEG—Aircraft Evaluation Group

A/FD—Airport/Facility Directory

AFM—Airplane Flight Manual or Aircraft Flight Manual

AFMS—Airplane Flight Manual Supplements

AFS—Aircraft Flight Safety

AFS—Flight Standards Service

AFSS—Automated Flight Service Station

AGL—Above Ground Level

AIM—Aeronautical Information Manual

AIP—Aeronautical Information Publication

AIR—Aircraft Certification Service

AIRMETs—Airman's Meteorological Information

AIS—Aeronautical Information Services

AJV-5—Aeronautical Information Services

AJW-3—Flight Inspection Services

ALT ACQ—Altitude Acquire

ALT Hold—Altitude Hold

ANP—Actual Navigation Performance

AOC—Aircraft Operational Communications

APT WP—Airport Waypoint

APV—Approach with Vertical Guidance

ARINC—Aeronautical Radio Incorporated

A-RNAV—Advanced Area Navigation

ARTCC—Air Route Traffic Control Center

ARTS—Automated Radar Terminal System

ASDA—Accelerate-Stop Distance Available

ASDE-X—Airport Surface Detection Equipment-Model X

A-SMGCS—Advanced Surface Movement Guidance and Control System

ASOS—Automated Surface Observing System

ASR—Airport Surveillance Radar

ASRS—Aviation Safety Reporting System

ATA—Air Transport Association

ATC—Air Traffic Control

ATCAA—Air Traffic Control Assigned Airspace

ATCRBS—Air Traffic Control Radar Beacon System
ATCS—Air Traffic Control Specialist
ATD—Along-Track Distance

ATIS—Automatic Terminal Information Service
ATS—Air Traffic Service
AWC—Aviation Weather Center
AWOS—Automated Weather Observing System
AWSS—Automated Weather Sensor System

B

Baro-VNAV—Barometric Vertical Navigation
B-RNAV—European Basic RNAV

C

CAT—Category
CDI—Course Deviation Indicator
CDL—Configuration Deviation List
CENRAP—Center Radar ARTS Processing
CFA—Controlled Firing Areas
CFIT—Controlled Flight Into Terrain
CFR—Code of Federal Regulations
CNF—Computer Navigation Fix
COP—Changeover Point
COTS—Commercial Off-The-Shelf
CRC—Cyclic Redundancy Check
CTAF—Common Traffic Advisory Frequency
CVFP—Charted Visual Flight Procedure
CWAS—Center Weather Advisories

D

DA—Density Altitude; Decision Altitude
D-ATIS—Digital Automatic Terminal Information Service
DDA—Derived Decision Altitude
DEM—Digital Elevation Model
DER—Departure End of the Runway
DF—Direction Finding
DH—Decision Height
DME—Distance Measuring Equipment
DOD—Department of Defense
DOT—Department of Transportation
DPs—Departure Procedures
DRVSM—Domestic Reduced Vertical Separation Minimums
DUATS—Direct User Access Terminal System
DVA—Diverse Vector Area

E

E/D—End of Decent
EDCT—Expect Departure Clearance Time
EFAS—En-Route Flight Advisory Service
EFB—Electronic Flight Bag
EFC—Expect Further Clearance

EFV—Enhanced Flight Visibility
EFVS—Enhanced Flight Vision System
EGPWS—Enhanced Ground Proximity Warning Systems
EOPs—Engine Out Procedures
EOVM—Emergency Obstruction Video Map
ER-OPS—Extended Range Operations
ETA—Estimated Time of Arrival
EV—Enhanced Vision
EVO—Equivalent Visual Operations
EWINS—Enhanced Weather Information System

F

FAA—Federal Aviation Administration
FAF—Final Approach Fix
FAP—Final Approach Point
FB—Fly-By
FBWP—Fly-By Waypoint
FCC—Federal Communications Commission
FD—Flight Director
FDP—Flight Data Processing
FE—Flight Engineer
FIR—Flight Information Region
FIS—Flight Information System
FIS-B—Flight Information Services–Broadcast
FISDL—Flight Information Services Data Link
FL—Flight Level
FL CHG—Flight Level Change
FLIR—Forward Looking Infra-Red
FMC—Flight Management Computer
FMS—Flight Management System
FO—Fly-Over
FOM—Flight Operations Manual
FOWP—Fly-Over Waypoint
FPA—Flight Path Angle
FPM—Feet Per Minute
FPNM—Feet Per Nautical Mile
FPV—Flight Path Vector
FSB—Flight Standardization Board
FSDO—Flight Standards District Office
FSL—NOAA Forecast Systems Laboratory
FSS—Flight Service Station

G

GBAS—Ground-Based Augmentation System
GCA—Ground Controlled Approach
GCO—Ground Communication Outlet
GDP—Ground Delay Programs
GDPE—Ground Delay Program Enhancements
GLS—Ground Based Augmentation System Landing System
GNSS—Global Navigation Satellite System GPS—Global Positioning System
GPWS—Ground Proximity Warning System G/S—Glide Slope

GS—Ground Speed
GWS—Graphical Weather Service

H

HAA—Height Above Airport
HAR—High Altitude Redesign
HAT—Height Above Touchdown
HAZMAT—Hazardous Materials
HDD—Head-Down Display
HF—High Frequency
HGS—Head-up Guidance System
HIWAS—Hazardous In-flight Weather Advisory Service
HSI—Horizontal Situation Indicator
HUD—Head-Up Display

I

IAF—Initial Approach Fix
IAP—Instrument Approach Procedure
IAS—Indicated Air Speed
ICAO—International Civil Aviation Organization
IF—Intermediate Fix
IFR—Instrument Flight Rules
IIMC—Inadvertent Instrument Meteorological Condition
ILS—Instrument Landing System
IMC—Instrument Meteorological Conditions
INS—Inertial Navigation System
IOC—Initial Operational Capability
IRU—Inertial Reference Unit

K

KIAS—knots indicated airspeed
KLAX—Los Angeles International Airport
KLMU—Monroe Regional Airport

L

LAAS—Local Area Augmentation System
LAHSO—Land And Hold Short Operations
LDA—Localizer-type Directional Aid; Landing Distance Available
LF—Low Frequency
LNAV—Lateral Navigation
LOC—Localizer
LOM—Locator Outer Marker
LPV—Localizer Performance with Vertical Guidance
LTP—Landing Threshold Point
LVL CHG—Level Change

M

MAA—Maximum Authorized Altitude
MAHWP—Missed Approach Holding Waypoint
MAMS—Military Airspace Management System

MAP—Missed Approach Point
MAWP—Missed Approach Waypoint
MCA—Minimum Crossing Altitude

MDA—Minimum Descent Altitude
MDH—Minimum Descent Height
MEA—Minimum En-route Altitude
MEL—Minimum Equipment List
METAR—Aviation Routine Weather Report
MF—Medium Frequency
MIA—Minimum IFR Altitude
MLS—Microwave Landing System
MMWR—Millimeter Wave Radar
MNPS—Minimum Navigation Performance Specifications
MOA—Military Operations Area
MOCA—Minimum Obstruction Clearance Altitude
MOPS—Minimum Operational Performance Standards
MORA—Minimum Off-Route Altitude
MRA—Minimum Reception Altitude
MSA—Minimum Safe Altitude
MSAW—Minimum Safe Altitude Warning
MSL—Mean Sea Level
MTA—Minimum Turning Altitude
MTR—Military Training Route
MVA—Minimum Vectoring Altitude

N

NA—Not Authorized
NACO—National Aeronautical Charting Office
NAS—National Airspace System
NASA—National Aeronautics and Space Administration
NAT—North Atlantic
NAVAID—Navigational Aid
NCAR—National Center for Atmospheric Research
NDB—Non-Directional Beacon
NextGen—Next Generation Air Transportation System
NFDC—National Flight Data Center
NFPO—National Flight Procedures Office
NGA—National Geospatial- Intelligence Agency
NM—Nautical Mile
NMAC—Near Mid-Air Collision
NNEW—Next Generation Network-Enabled
NOAA—National Oceanic and Atmospheric Administration
NO A/G—No Air-to-Ground Communication
NOPAC—North Pacific
NOTAM—Notice to Airmen
NOZ—Normal Operating Zone
NPA—Non-Precision Approach
NRR—Non-Restrictive Routing
NRS—National Reference System
NTAP—Notice to Airmen Publication
NTSB—National Transportation Safety Board

NTZ—No Transgression Zone
NVS—NAS voice switch
NWS—National Weather Service

O

OAT—Outside Air Temperature
OBS—Omni Bearing Selector
OCS—Obstacle Clearance Surface
ODP—Obstacle Departure Procedure
OpSpecs—Operations Specifications
OROCA—Off-Route Obstruction Clearance Altitude
OSV—Operational Service Volume

P

PA—Precision Approach
PAR—Precision Approach Radar
PCG—Positive Course Guidance
PDC—Pre-Departure Clearance
PDR—Preferential Departure Route
PF—Pilot Flying
PFD—Primary Flight Display
PI—Principal Inspector
PIC—Pilot In Command
PIREP—Pilot Weather Report
PKI—Public/Private Key Technology
PM—Pilot Monitoring
POH—Pilot's Operating Handbook
POI—Principle Operations Inspector
PRM—Precision Runway Monitor
P-RNAV—European Precision RNAV
PT—Procedure Turn
PTP—Point-To-Point

Q

QFE—Transition Height
QNE—Transition Level
QNH—Transition Altitude

R

RA—Resolution Advisory; Radio Altitude
RAIM—Receiver Autonomous Integrity Monitoring
RAL—Research Applications Laboratory
RCO—Remote Communications Outlet
RJ—Regional Jet
RNAV—Area Navigation
RNP—Required Navigation Performance
RNP AR—Required Navigation Performance Authorization Required
ROC—Required Obstacle Clearance
RSP—Runway Safety Program
RVR—Runway Visual Range
RVSM—Reduced Vertical Separation Minimums
RVV—Runway Visibility Value

RWY—Runway
RWY WP—Runway Waypoint

S

SA—Situational Awareness
SAAAR—Special Aircraft and Aircrew Authorization Required
SAAR—Special Aircraft and Aircrew Requirements
SAMS—Special Use Airspace Management System
SAR—Search and Rescue
SAS—Stability Augmentation System
SATNAV—Satellite Navigation
SDF—Simplified Directional Facility
SDR—Service Difficulty Reports
SER—Start End of Runway
SIAP—Standard Instrument Approach Procedure
SID—Standard Instrument Departure
SIGMET—Significant Meteorological Information
SM—Statute Mile
SMA—Surface Movement Advisor
SMGCS—Surface Movement Guidance and Control System
SOIA—Simultaneous Offset Instrument Approaches
SOP—Standard Operating Procedure
SPECI—Non-routine (Special) Aviation Weather Report
SSV—Standard Service Volume
STAR—Standard Terminal Arrival
STARS—Standard Terminal Automation Replacement System
STC—Supplemental Type Certificate
SUA—Special Use Airspace
SUA/ISE—Special Use Airspace/In-flight Service Enhancement
SVFR—Special Visual Flight Rules
SVS—Synthetic Vision System
SWAP—Severe Weather Avoidance Plan
SWIM—System Wide Information Management

T

TA—Traffic Advisory
TAA—Terminal Arrival Area
TACAN—Tactical Air Navigation
TAF—Terminal Aerodrome Forecast
TAS—True Air Speed
TAWS—Terrain Awareness and Warning Systems
TB—Track Bar
TCAS—Traffic Alert and Collision Avoidance System
TCH—Threshold Crossing Height
TDLS—Terminal Data Link System
TDZ—Touchdown Zone
TDZE—Touchdown Zone Elevation
TEC—Tower En-route Control
TERPS—U.S. Standard for Terminal Instrument Procedures
TFR—Temporary Flight Restriction
TIS-B—Traffic Information Service-Broadcast
TIBS—Telephone Information Briefing Service

TIS-B—Traffic Information Service-Broadcast
TIBS—Telephone Information Briefing Service
TOC—Top Of Climb
TOD—Top Of Descent
TODA—Takeoff Distance Available
TOGA—Takeoff/Go Around
TORA—Takeoff Runway Available
TPP—Terminal Procedures Publication
TRACAB—Terminal Radar Approach Control in Tower Cab
TSO—Technical Standard Order
TSOA—Technical Standing Order Authorization

U

UHF—Ultra High Frequency
UNICOM—Universal Communications
U.S.—United States
USAF—United States Air Force
USDA—United States Department of Agriculture

V

VCOA—Visual Climb Over Airport / Airfield
VDA—Vertical Descent Angle
VDP—Visual Descent Point
VFR—Visual Flight Rules
VGSI—Visual Glide Slope Indicator
VHF—Very High Frequency
VLJ—Very Light Jet
VMC—Visual Meteorological Conditions
VMINI—Minimum Speed–IFR
VNAV—Vertical Navigation
VNEI—Never-Exceed Speed-IFR
VSI—Vertical Speed Indicator
VOR—Very High-Frequency Omnidirectional Range
VORTAC—Very High Frequency Omnidirectional Range/Tactical Air Navigation
VOT—VOR Test Facility Signal
VPA—Vertical Path Angle
VREF—Reference Landing Speed
V/S—Vertical Speed
VSO—Stalling Speed or the Minimum Steady Flight Speed in the Landing Configuration

W

WAAS—Wide Area Augmentation System
WAC—World Aeronautical Chart
WAM—Wide Area Multi-lateration
WP—Waypoint

Z

ZFW—Fort Worth Air Route Traffic Control Center

Glossary

Abeam Fix. A fix, NAVAID, point, or object positioned approximately 90 degrees to the right or left of the aircraft track along a route of flight. Abeam indicates a general position rather than a precise point.

Accelerate-Stop Distance Available (ASDA). The runway plus stopway length declared available and suitable for the acceleration and deceleration of an airplane aborting a takeoff.

Advanced Surface Movement Guidance and Control System (A-SMGCS). A system providing routing, guidance and surveillance for the control of aircraft and vehicles, in order to maintain the declared surface movement rate under all weather conditions within the aerodrome visibility operational level (AVOL) while maintaining the required level of safety.

Aircraft Approach Category. A grouping of aircraft based on reference landing speed (V_{REF}), if specified, or if V_{REF} is not specified, $1.3 V_{SO}$ (the stalling speed or minimum steady flight speed in the landing configuration) at the maximum certificated landing weight.

Airport Diagram. A full-page depiction of the airport that includes the same features of the airport sketch plus additional details, such as taxiway identifiers, airport latitude and longitude, and building identification. Airport diagrams are located in the U.S. Terminal Procedures booklet following the instrument approach charts for a particular airport.

Airport/Facility Directory (A/FD). Regional booklets published by the Aeronautical Information Services branch (AJV-5) that provide textual information about all airports, both VFR and IFR. The A/FD includes runway length and width, runway surface, load bearing capacity, runway slope, airport services, and hazards, such as birds and reduced visibility.

Airport Sketch. Depicts the runways and their length, width, and slope, the touchdown zone elevation, the lighting system installed on the end of the runway, and taxiways. Airport sketches are located on the lower left or right portion of the instrument approach chart.

Airport Surface Detection Equipment-Model X (ASDE-X). Enables air traffic controllers to detect potential runway conflicts by providing detailed coverage of movement on runways and taxiways. By collecting data from a variety of sources, ASDE-X is able to track vehicles and aircraft on the airport movement area and obtain identification information from aircraft transponders.

Air Route Traffic Control Center (ARTCC). A facility established to provide air traffic control service to aircraft operating on IFR flight plans within controlled airspace and principally during the en route phase of flight

Air Traffic Service (ATS). Air traffic service is an ICAO generic term meaning variously, flight information service, alerting service, air traffic advisory service, air traffic control service (area control service, approach control service, or aerodrome control service).

Approach End of Runway (AER). The first portion of the runway available for landing. If the runway threshold is displaced, use the displaced threshold latitude/longitude as the AER.

Approach Fix. From a database coding standpoint, an approach fix is considered to be an identifiable point in space from the intermediate fix (IF) inbound. A fix located between the initial approach fix (IAF) and the IF is considered to be associated with the approach transition or feeder route.

Approach Gate. An imaginary point used by ATC to vector aircraft to the final approach course. The approach gate is established along the final approach course 1 NM from the final approach fix (FAF) on the side away from the airport and is located no closer than 5 NM from the landing threshold.

Area Navigation (RNAV). A method of navigation that permits aircraft operations on any desired course within the coverage of station referenced navigation signals or within the limits of self contained system capability.

Automated Surface Observing System (ASOS)/Automated Weather Sensor System (AWSS). The ASOS/AWSS is the primary surface weather observing system of the U.S.

Automated Surface Observing System (ASOS). A weather observing system that provides minute-by minute weather observations, such as temperature, dew point, wind, altimeter setting, visibility, sky condition, and precipitation. Some ASOS stations include a precipitation discriminator that can differentiate between liquid and frozen precipitation.

Automated Weather Observing System (AWOS). A suite of sensors that measure, collect, and disseminate weather data. AWOS stations provide a minute-by-minute update of weather parameters, such as wind speed and direction, temperature and dew point, visibility, cloud heights and types, precipitation, and barometric pressure. A variety of AWOS system types are available (from AWOS 1 to AWOS 3), each of which includes a different sensor array.

Automated Weather Sensor System (AWSS). The AWSS is part of the Aviation Surface Weather Observation Network suite of programs and provides pilots and other users with weather information through the Automated Surface Observing System. The AWSS sensor suite automatically collects, measures, processes, and broadcasts surface weather data.

Automated Weather System. Any of the automated weather sensor platforms that collect weather data at airports and disseminate the weather information via radio and/or landline. The systems currently consist of the Automated Surface Observing System (ASOS), Automated Weather Sensor System (AWSS), and Automated Weather Observation System (AWOS).

Automatic Dependent Surveillance-Broadcast (ADS-B). A surveillance system that continuously broadcasts GPS position information, aircraft identification, altitude, velocity vector, and direction to all other aircraft and air traffic control facilities within a specific area. Automatic dependent surveillance-broadcast (ADS-B) information is displayed in the flight deck via a flight deck display of traffic information (CDTI) unit, providing the pilot with greater situational awareness. ADS-B transmissions also provides controllers with a more complete picture of traffic and updates that information more frequently than other surveillance equipment.

Automatic Terminal Information Service (ATIS). A recorded broadcast available at most airports with an operating control tower that includes crucial information about runways and instrument approaches in use, specific outages, and current weather conditions, including visibility.

Center Radar ARTS Presentation/Processing (CENRAP). CENRAP was developed to provide an alternative to a non-radar environment at terminal facilities should an ASR fail or malfunction. CENRAP sends aircraft radar beacon target information to the ASR terminal facility equipped with ARTS.

Changeover Point (COP). A COP indicates the point where a frequency change is necessary between navigation aids, when other than the midpoint on an airway, to receive course guidance from the facility ahead of the aircraft instead of the one behind. These COPs divide an airway or route segment and ensure continuous reception of navigational signals at the prescribed minimum en route IFR altitude.

Charted Visual Flight Procedure (CVFP). A CVFP may be established at some towered airports for environmental or noise considerations, as well as when necessary for the safety and efficiency of air traffic operations. Designed primarily for turbojet aircraft, CVFPs depict prominent landmarks, courses, and recommended altitudes to specific runways.

Cockpit Display of Traffic Information (CDTI). The display and user interface for information about air traffic within approximately 80 miles. It typically combines and shows traffic data from TCAS, TIS-B, and ADS-B. Depending on features, the display may also show terrain, weather, and navigation information.

Collision Hazard. A condition, event, or circumstance that could induce an occurrence of a collision or surface accident or incident.

Columns. See database columns.

Contact Approach. An approach where an aircraft on an IFR flight plan, having an air traffic control authorization, operating clear of clouds with at least one mile flight visibility, and a reasonable expectation of continuing to the destination airport in those conditions, may deviate from the instrument approach procedure and proceed to the destination airport by visual reference to the surface. This approach is only authorized when requested by the pilot and the reported ground visibility at the destination airport is at least one statute mile.

Controlled Flight Into Terrain (CFIT). A situation where a mechanically normally functioning airplane is inadvertently flown into the ground, water, or an obstacle. There are two basic causes of CFIT accidents; both involve flight crew situational awareness. One definition of situational awareness is an accurate perception by pilots of the factors and conditions currently affecting the safe operation of the aircraft and the crew. The causes of CFIT are the flight crews' lack of vertical position awareness or their lack of horizontal position awareness in relation to terrain and obstacles.

Database Columns. The spaces for data entry on each record. One column can accommodate one character.

Database Field. The collection of characters needed to define one item of information.

Database Identifier. A specific geographic point in space identified on an aeronautical chart and in a naviation database, officially designated by the controlling state authority or derived by Jeppesen. It has no ATC function and should not be used in filing flight plans nor used when communicating with ATC.

Database Record. A single line of computer data made up of the fields necessary to define fully a single useful piece of data.

Decision Altitude (DA). A specified altitude in the precision approach at which a missed approach must be initiated if the required visual reference to continue the approach has not been established. The term "Decision Altitude (DA)" is referenced to mean sea level and the term "Decision Height (DH)" is referenced to the threshold elevation. Even though DH is charted as an altitude above MSL, the U.S. has adopted the term "DA" as a step toward harmonization of the United States and international terminology. At some point, DA will be published for all future instrument approach procedures with vertical guidance.

Decision Height (DH). See Decision Altitude.

Departure End of Runway (DER). The end of runway available for the ground run of an aircraft departure. The end of the runway that is opposite the landing threshold, sometimes referred to as the stop end of the runway. Altitude, velocity vector, and direction to all other aircraft and air traffic control facilities within a specific area. Automatic dependent surveillance-broadcast (ADS-B) information is displayed in the flight deck via a cockpit display of traffic information (CDTI) unit, providing the pilot with greater situational awareness. ADS-B transmissions also provide controllers with a more complete picture of traffic and update that information more frequently than other surveillance equipment.

Descend Via. A descend via clearance instructs you to follow the altitudes published on a STAR. You are not authorized to leave your last assigned altitude unless specifically cleared to do so. If ATC amends the altitude or route to one that is different from the published procedure, the rest of the charted descent procedure is canceled. ATC will assign you any further route, altitude, or airspeed clearances, as necessary.

Digital ATIS (D-ATIS). An alternative method of receiving ATIS reports by aircraft equipped with datalink services capable of receiving information in the flight deck over their Aircraft Communications Addressing and Reporting System (ACARS) unit.

Digital elevation model (DEM). A digital representation of ground surface topography or terrain.

Diverse Vector Area (DVA). An airport may establish a diverse vector area if it is necessary to vector aircraft below the minimum vectoring altitude to assist in the efficient flow of departing traffic. DVA design requirements are outlined in TERPS and allow for the vectoring of aircraft immediately off the departure end of the runway below the MVA.

Dynamic Magnetic Variation. A field that is simply a computer model calculated value instead of a measured value contained in the record for a waypoint.

Electronic Flight Bag (EFB). An electronic display system intended primarily for flight deck or cabin use. EFB devices can display a variety of aviation data or perform basic calculations (e.g., performance data, fuel calculations, etc.). In the past, some of these functions were traditionally accomplished using paper references or were based on data provided to the flight crew by an airline's flight dispatch function. The scope of the EFB system functionality may also include various other hosted databases and applications. Physical EFB displays may use various technologies, formats, and forms of communication. These devices are sometimes referred to as auxiliary performance computers (APC) or laptop auxiliary performance computers (LAPC).

Ellipsoid of Revolution. The surface that results when an ellipse is rotated about one of its axes.

En Route Obstacle Clearance Areas. Obstacle clearance areas for en route planning are identified as primary, secondary, and turning areas, and they are designed to provide obstacle clearance route protection width for airways and routes.

Expanded Service Volume. When ATC or a procedures specialist requires the use of a NAVAID beyond the limitations specified for standard service volume, an expanded service volume (ESV) may be established. See standard service volume.

Feeder Route. A feeder route is a route depicted on IAP charts to designate courses for aircraft to proceed from the en route structure to the IAF. Feeder routes, also referred to as approach transitions, technically are not considered approach segments but are an integral part of many IAPs.

Field. See database field.

Final Approach and Takeoff Area (FATO). The FATO is a defined heliport area over which the final approach to a

hover or a departure is made. The touchdown and lift-off area (TLOF) where the helicopter is permitted to land is normally centered in the FATO. A safety area is provided around the FATO.

Fix. A geographical position determined by visual reference to the surface, by reference to one or more radio NAVAIDs, by celestial plotting, or by another navigational device.

NOTE: Fix is a generic name for a geographical position and is referred to as a fix, waypoint, intersection, reporting point, etc.

Flight Information Region (FIR). A FIR is an airspace of defined dimensions within which Flight Information Service and Alerting Service are provided. Flight Information Service (FIS) is a service provided for the purpose of giving advice and information useful for the safe and efficient conduct of flights. Alerting Service is a service provided to notify appropriate organizations regarding aircraft in need of search and rescue aid, and assist such organizations as required.

Flight Level (FL). A flight level is a level of constant atmospheric pressure related to a reference datum of 29.92 "Hg. Each flight level is stated in three digits that represents hundreds of feet. For example, FL 250 represents an altimeter indication of 25,000 feet.

Floating Waypoints. Floating waypoints represent airspace fixes at a point in space not directly associated with a conventional airway. In many cases, they may be established for such purposes as ATC metering fixes, holding points, RNAV-direct routing, gateway waypoints, STAR origination points leaving the en route structure, and SID terminating points joining the en route structure.

Fly-by (FB) Waypoint. A waypoint that requires the use of turn anticipation to avoid overshooting the next flight segment.

Fly-over (FO) Waypoint. A waypoint that precludes any turn until the waypoint is overflown, and is followed by either an intercept maneuver of the next flight segment or direct flight to the next waypoint.

Four Corner Post Configuration. An arrangement of air traffic pathways in a terminal area that brings incoming flights over fixes at four corners of the traffic area, while outbound flights depart between the fixes, thus minimizing conflicts between arriving and departing traffic.

Gateway Fix. A navigational aid or fix where an aircraft transitions between the domestic route structure and the oceanic route airspace.

Geodetic Datum. The reference plane from which geodetic calculations are made. Or, according to ICAO Annex 15, the numerical or geometrical quantity or set of such quantities (mathematical model) that serves as a reference for computing other quantities in a specific geographic region, such as the latitude and longitude of a point.

Glidepath Angle (GPA). The angular displacement of the vertical guidance path from a horizontal plane that passes through the reference datum point (RDP). This angle is published on approach charts (e.g., 3.00°, 3.20°, etc.). GPA is sometimes referred to as vertical path angle (VPA).

Global Navigation Satellite System (GNSS). An umbrella term adopted by the International Civil Aviation Organization (ICAO) to encompass any independent satellite navigation system used by a pilot to perform onboard position determinations from the satellite data.

Gross Navigation Error (GNE). In the North Atlantic area of operations, a gross navigation error is a lateral separation of more than 25 NM from the centerline of an aircraft's cleared route, which generates an Oceanic Navigation Error Report. This report is also generated by a vertical separation if you are more than 300 feet off your assigned flight level.

Ground Communication Outlet (GCO). An unstaffed, remotely controlled ground/ground communications facility. Pilots at uncontrolled airports may contact ATC and AFSS via very high frequency (VHF) radio to a telephone connection. This lets pilots obtain an instrument clearance or close a VFR/IFR flight plan.

Head-Up Display (HUD). See head-up guidance system (HGS).

Head-up Guidance System (HGS). A system that projects critical flight data on a display positioned between the pilot and the windscreen. In addition to showing primary flight information, the HUD computes an extremely accurate instrument approach and landing guidance solution, and displays the result as a guidance cue for head-up viewing by the pilot.

Height Above Touchdown (HAT). The height of the DA above touchdown zone elevation (TDZE).

Highway in the Sky (HITS). A graphically intuitive pilot interface system that provides an aircraft operator with all of the attitude and guidance inputs required to safely fly an aircraft in close conformance to air traffic procedures.

Initial Climb Area (ICA). An area beginning at the departure end of runway (DER) to provide unrestricted climb to at least 400 feet above DER elevation.

Instrument Approach Waypoint. Fixes used in defining RNAV IAPs, including the feeder waypoint (FWP), the initial approach waypoint (IAWP), the intermediate waypoint (IWP), the final approach waypoint (FAWP), the RWY WP, and the APT WP, when required.

Instrument Landing System (ILS). A precision instrument approach system that normally consists of the following electronic components and visual aids: localizer, glide slope, outer marker, middle marker, and approach lights.

Instrument Procedure with Vertical Guidance (IPV). Satellite or flight management system (FMS) lateral navigation (LNAV) with computed positive vertical guidance based on barometric or satellite elevation. This term has been renamed APV.

International Civil Aviation Organization (ICAO). ICAO is a specialized agency of the United Nations whose objective is to develop standard principles and techniques of international air navigation and to promote development of civil aviation.

Intersection. Typically, the point at which two VOR radial position lines cross on a route, usually intersecting at a good angle for positive indication of position, resulting in a VOR/VOR fix.

Landing Distance Available (LDA). ICAO defines LDA as the length of runway, that is declared available and suitable for the ground run of an aeroplane landing.

Lateral Navigation (LNAV). Azimuth navigation without positive vertical guidance. This type of navigation is associated with nonprecision approach procedures or en route.

Local Area Augmentation System (LAAS). LAAS further increases the accuracy of GPS and improves signal integrity warnings.

Localizer Performance with Vertical Guidance (LPV). LPV is one of the four lines of approach minimums found on an RNAV (GPS) approach chart. Lateral guidance accuracy is equivalent to a localizer. The HAT is published as a DA since it uses an electronic glide path that is not dependent on any ground equipment or barometric aiding and may be as low as 200 feet and ½ SM visibility depending on the airport terrain and infrastructure. WAAS avionics approved for LPV is required. Baro-VNAV is not authorized to fly the LPV line of minimums on a RNAV (GPS) procedure since it uses an internally generated descent path that is subject to cold temperature effects and incorrect altimeter settings.

Loss of Separation. An occurrence or operation that results in less than prescribed separation between aircraft, or between an aircraft and a vehicle, pedestrian, or object.

LPV. See Localizer Performance with Vertical Guidance.

Magnetic Variation. The difference in degrees between the measured values of true north and magnetic north at that location.

Maximum Authorized Altitude (MAA). An MAA is a published altitude representing the maximum usable altitude or flight level for an airspace structure or route segment. It is the highest altitude on a Federal airway, jet route, RNAV low or high route, or other direct route for which an MEA is designated at which adequate reception of navigation signals is assured.

Metering Fix. A fix along an established route over which aircraft are metered prior to entering terminal airspace. Normally, this fix should be established at a distance from the airport which facilitates a profile descent 10,000 feet above airport elevation (AAE) or above.

Mid-RVR. The RVR readout values obtained from sensors located midfield of the runway.

Mileage Break. A point on a route where the leg segment mileage ends, and a new leg segment mileage begins, often at a route turning point.

Military Airspace Management System (MAMS). A Department of Defense system to collect and disseminate information on the current status of special use airspace. This information is provided to the Special Use Airspace

Management System (SAMS). The electronic interface also provides SUA schedules and historical activation and utilization data.

Minimum Crossing Altitude (MCA). An MCA is the lowest altitude at certain fixes at which the aircraft must cross when proceeding in the direction of a higher minimum en route IFR altitude. MCAs are established in all cases where obstacles intervene to prevent pilots from maintaining obstacle clearance during a normal climb to a higher MEA after passing a point beyond which the higher MEA applies.

Minimum Descent Altitude (MDA). The lowest altitude, expressed in feet above mean sea level, to which descent is authorized on final approach or during circle-to-land maneuvering in execution of a standard instrument approach procedure where no electronic glide slope is provided.

Minimum En Route Altitude (MEA). The MEA is the lowest published altitude between radio fixes that assures acceptable navigational signal coverage and meets obstacle clearance requirements between those fixes. The MEA prescribed for a Federal Airway or segment, RNAV low or high route, or other direct route applies to the entire width of the airway, segment, or route between the radio fixes defining the airway, segment, or route.

Minimum IFR Altitude (MIA). Minimum altitudes for IFR operations are prescribed in 14 CFR Part 91. These MIAs are published on IFR charts and prescribed in 14 CFR Part 95 for airways and routes, and in 14 CFR Part 97 for standard instrument approach procedures.

Minimum Navigation Performance Specifications (MNPS). A set of standards that require aircraft to have a minimum navigation performance capability in order to operate in MNPS designated airspace. In addition, aircraft must be certified by their State of Registry for MNPS operation. Under certain conditions, non-MNPS aircraft can operate in MNPS airspace, however, standard oceanic separation minima is provided between the non-MNPS aircraft and other traffic.

Minimum Obstruction Clearance Altitude (MOCA). The MOCA is the lowest published altitude in effect between radio fixes on VOR airways, off-airway routes, or route segments that meets obstacle clearance requirements for the entire route segment. This altitude also assures acceptable navigational signal coverage only within 22 NM of a VOR.

Minimum Reception Altitude (MRA). An MRA is determined by FAA flight inspection traversing an entire route of flight to establish the minimum altitude the navigation signal can be received for the route and for off-course NAVAID facilities that determine a fix. When the MRA at the fix is higher than the MEA, an MRA is established for the fix, and is the lowest altitude at which an intersection can be determined.

Minimum Safe Altitude (MSA). MSAs are published for emergency use on IAP charts. For conventional navigation systems, the MSA is normally based on the primary omnidirectional facility on which the IAP is predicated. For RNAV approaches, the MSA is based on the runway waypoint (RWY WP) for straight-in approaches, or the airport waypoint (APT WP) for circling approaches. For GPS approaches, the MSA center is the Missed Approach Waypoint (MAWP).

Minimum Vectoring Altitude (MVA). Minimum vectoring altitude charts are developed for areas where there are numerous minimum vectoring altitudes due to variable terrain features or man-made obstacles. MVAs are established for use by ATC when radar ATC is exercised.

Missed Approach Holding Waypoint (MAHWP). An approach waypoint sequenced during the holding portion of the missed approach procedure that is a fly-over waypoint.

Missed Approach Waypoint (MAWP). An approach waypoint sequenced during the missed approach procedure that is a fly-over waypoint.

National Airspace System (NAS). Consists of a complex collection of facilities, systems, equipment, procedures, and airports operated by thousands of people to provide a safe and efficient flying environment.

Navigational Gap. A navigational course guidance gap, referred to as an MEA gap, describes a distance along an airway or route segment where a gap in navigational signal coverage exists. The navigational gap may not exceed a specific distance that varies directly with altitude.

Next Generation Air Transportation System (NextGen). Ongoing, wide-ranging transformation of the National Airspace System (NAS). NextGen represents an evolution from a ground-based system of air traffic control to a satellite-based system of air traffic management.

Non-Directional Radio Beacon (NDB). An L/MF or UHF radio beacon transmitting nondirectional signals whereby the pilot of an aircraft equipped with direction finding equipment can determine bearing to or from the radio beacon and "home" on or track to or from the station. When the radio beacon is installed in conjunction with the ILS marker, it is normally called a compass locator.

Non-RNAV DP. A DP whose ground track is based on ground-based NAVAIDS and/or dead reckoning navigation.

Obstacle Clearance Surface (OCS). An inclined or level surface associated with a defined area for obstruction evaluation.

Obstacle Departure Procedure (ODP). A procedure that provides obstacle clearance. ODPs do not include ATC related climb requirements. In fact, the primary emphasis of ODP design is to use the least erroneous route of flight to the en route structure while attempting to accommodate typical departure routes.

Obstacle Identification Surface (OIS). The design of a departure procedure is based on TERPS, a living document that is updated frequently. Departure design criteria assumes an initial climb of 200 FPNM after crossing the departure end of the runway (DER) at a height of at least 35 feet above the ground. Assuming a 200 FPNM climb, the departure is structured to provide at least 48 FPNM of clearance above objects that do not penetrate the obstacle slope. The slope, known as the obstacle identification slope (OIS), is based on a 40 to 1 ratio, which is the equivalent of a 152 FPNM slope.

Off-Airway Routes. The FAA prescribes altitudes governing the operation of aircraft under IFR for fairway routes in a similar manner to those on federal airways, jet routes, area navigation low or high altitude routes, and other direct routes for which an MEA is designated.

Off-Route Obstruction Clearance Altitude (OROCA). An off-route altitude that provides obstruction clearance with a 1,000-foot buffer in nonmountainous terrain areas and a 2,000-foot buffer in designated mountainous areas within the U.S. This altitude may not provide signal coverage from groundbased navigational aids, air traffic control radar, or communications coverage.

Operations Specifications (OpSpecs). A published document providing the conditions under which an air carrier and operator for compensation or hire must operate in order to retain approval from the FAA.

Pilot Briefing Information. The current format for charted

IAPs issued by AJV-5. The information is presented in a logical order facilitating pilot briefing of the procedures. Charts include formatted information required for quick pilot or flight crew reference located at the top of the chart.

Point-in-Space (PinS) Approach. An approach normally developed to heliports that do not meet the IFR heliport design standards but meet the standards for a VFR heliport. A helicopter PinS approach can be developed using conventional NAVAIDs or RNAV systems. These procedures have either a VFR or visual segment between the MAP and the landing area. The procedure specifies a course and distance from the MAP to the heliport(s) and includes a note to proceed VFR or visually from the MAP to the heliport, or conduct the missed approach.

Positive Course Guidance (PCG). A continuous display of navigational data that enables an aircraft to be flown along a specific course line (e.g., radar vector, RNAV, ground-based NAVAID).

Precision Runway Monitor (PRM). Provides air traffic controllers with high precision secondary surveillance data for aircraft on final approach to parallel runways that have extended centerlines separated by less than 4,300 feet. High resolution color monitoring displays (FMA) are required to present surveillance track data to controllers along with detailed maps depicting approaches and a no transgression zone.

Preferential Departure Route (PDR). A specific departure route from an airport or terminal area to an en route point where there is no further need for flow control. It may be included in an instrument Departure Procedure (DP) or a Preferred IFR Route.

Preferred IFR Routes. A system of preferred IFR routes guides you in planning your route of flight to minimize route changes during the operational phase of flight, and to aid in the efficient orderly management of air traffic using federal airways.

Principal Operations Inspector (POI). Scheduled air carriers and operators for compensation or hire are assigned a principal operations inspector (POI) who works directly with the company and coordinates FAA operating approval.

Record. See Database Record.

Reduced Vertical Separation Minimums (RVSM). RVSM airspace is where air traffic control separates aircraft by a minimum of 1,000 feet vertically between flight level (FL) 290 and FL 410 inclusive. RVSM airspace is special

qualification airspace; the operator and the aircraft used by the operator must be approved by the Administrator. Air traffic control notifies operators of RVSM by providing route planing information.

Reference Landing Speed (V_{REF}). The speed of the airplane, in a specified landing configuration, at the point where it descends through the 50-foot height in the determination of the landing distance.

Remote Communications Outlet (RCO). An unmanned communications facility remotely controlled by air traffic personnel. RCOs serve FSSs and may be UHF or VHF. RCOs extend the communication range of the air traffic facility. RCOs were established to provide ground-to-ground communications between air traffic control specialists and pilots located at a satellite airport for delivering en route clearances, issuing departure authorizations, and acknowledging IFR cancellations or departure/landing times.

Reporting Point. A geographical location in relation to which the position of an aircraft is reported. (See Compulsory Reporting Points).

Required Navigation Performance (RNP). RNP is a statement of the navigation performance necessary for operation within a defined airspace. On-board monitoring and alerting is required.

RNAV DP. A Departure Procedure developed for RNAV-equipped aircraft whose ground track is based on satellite or DME/DME navigation systems.

Roll-Out RVR. The RVR readout values obtained from sensors located nearest the rollout end of the runway.

Runway Heading. The magnetic direction that corresponds with the runway centerline extended, not the painted runway numbers on the runway. Pilots cleared to "fly or maintain runway heading" are expected to fly or maintain the published heading that corresponds with the extended centerline of the departure runway (until otherwise instructed by ATC), and are not to apply drift correction (e.g., RWY 4, actual magnetic heading of the runway centerline 044.22°, fly 044°).

Runway Hotspots. Locations on a particular airport that historically have hazardous intersections. Hot spots alert pilots to the fact that there may be a lack of visibility at certain points or the tower may be unable to see that particular intersection. Whatever the reason, pilots need to be aware that these hazardous intersections exist and they should be increasingly vigilant when approaching and taxiing through these intersections. Pilots are typically notified of these areas by a Letter to Airmen or by accessing the FAA Office of Runway Safety.

Runway Incursion. An occurrence at an airport involving an aircraft, vehicle, person, or object on the ground that creates a collision hazard or results in a loss of separation with an aircraft that is taking off, intending to take off, landing, or intending to land.

Runway Safety Program (RSP). Designed to create and execute a plan of action that reduces the number of runway incursions at the nation's airports.

Runway Visual Range (RVR). An estimate of the maximum distance at which the runway, or the specified lights or markers delineating it, can be seen from a position above a specific point on the runway centerline. RVR is normally determined by visibility sensors or transmissometers located alongside and higher than the centerline of the runway. RVR is reported in hundreds of feet.

Runway Visibility Value (RVV). The visibility determined for a particular runway by a transmissometer. A meter provides a continuous indication of the visibility (reported in miles or fractions of miles) for the runway. RVV is used in lieu of prevailing visibility in determining minimums for a particular runway.

Significant Point. [ICAO Annex 11] A specified geographical location used in defining an ATS route or the flightpath of an aircraft and for other navigation and ATS purposes.

Special Instrument Approach Procedure. A procedure approved by the FAA for individual operators, but not published in FAR 97 for public use.

Special Use Airspace Management System (SAMS). A joint FAA and military program designed to improve civilian access to special use airspace by providing information on whether the airspace is active or scheduled to be active. The information is available to authorized users via an Internet website.

Standard Instrument Departure (SID). An ATC requested and developed departure route designed to increase capacity of terminal airspace, effectively control the flow of traffic with minimal communication, and reduce environmental impact through noise abatement procedures.

Standard Service Volume. Most air navigation radio aids that provide positive course guidance have a designated standard service volume (SSV). The SSV defines the reception limits of unrestricted NAVAIDS that are usable for random/unpublished route navigation. Standard service volume limitations do not apply to published IFR routes or procedures. See the AIM (Chapter 1) for the SSV for specific NAVAID types.

Standard Terminal Arrival (STAR). Provides a common method for departing the en route structure and navigating to your destination. A STAR is a preplanned instrument flight rule ATC arrival procedure published for pilot use in graphic and textual form to simplify clearance delivery procedures. STARs provide you with a transition from the en route structure to an outer fix or an instrument approach fix or arrival waypoint in the terminal area, and they usually terminate with an instrument or visual approach procedure.

Standardized Taxi Routes. Coded taxi routes that follow typical taxiway traffic patterns to move aircraft between gates and runways. ATC issues clearances using these coded routes to reduce radio communication and eliminate taxi instruction misinterpretation.

STAR Transition. A published segment used to connect one or more en route airways, jet routes, or RNAV routes to the basic STAR procedure. It is one of several routes that bring traffic from different directions into one STAR. STARs are published for airports with procedures authorized by the FAA, and these STARs are included at the front of each Terminal Procedures Publication regional booklet.

Start End of Runway (SER). The beginning of the takeoff runway available.

Station Declination. The angular difference between true north and the zero radial of a VOR at the time the VOR was last site-checked.

Surface Incident. An event during which authorized or unauthorized/unapproved movement occurs in the movement area or an occurrence in the movement area associated with the operation of an aircraft that affects or could affect the safety of flight.

Surface Movement Guidance Control System (SMGCS). Facilitates the safe movement of aircraft and vehicles at airports where scheduled air carriers are conducting authorized operations. The SMGCS low visibility taxi plan includes the improvement of taxiway and runway signs, markings, and lighting, as well as the creation of SMGCS low visibility taxi route charts.

Synthetic Vision System (SVS). A visual display of terrain,

obstructions, runways, and other surface features that creates a virtual view of what the pilot would see out the window. This tool could be used to supplement normal vision in low visibility conditions, as well as to increase situational awareness in IMC.

System Wide Information Management (SWIM). An advanced technology program designed to facilitate greater sharing of Air Traffic Management (ATM) system information, such as airport operational status, weather information, flight data, status of special use airspace, and National Airspace System (NAS) restrictions. SWIM supports current and future NAS programs by providing flexible and secure information management architecture for sharing NAS information.

Takeoff Distance Available (TODA). ICAO defines TODA as the length of the takeoff runway available plus the length of the clearway, if provided.

Takeoff Runway Available (TORA). ICAO defines TORA as the length of runway declared available and suitable for the ground run of an aeroplane takeoff.

Tangent Point (TP). The point on the VOR/DME RNAV route centerline from which a line perpendicular to the route centerline would pass through the reference facility.

Terminal Arrival Area (TAA). TAAs are the method by which aircraft are transitioned from the RNAV en route structure to the terminal area with minimal ATC interaction. The TAA consists of a designated volume of airspace designed to allow aircraft to enter a protected area, offering guaranteed obstacle clearance where the initial approach course is intercepted based on the location of the aircraft relative to the airport.

Threshold. The beginning of the part of the runway usable for landing.

Top Of Climb (TOC). An identifiable waypoint representing the point at which cruise altitude is first reached. TOC is calculated based on your current aircraft altitude, climb speed, and cruise altitude. There can only be one TOC waypoint at a time.

Top Of Descent (TOD). Generally utilized in flight management systems, top of descent is an identifiable waypoint representing the point at which descent is first initiated from cruise altitude. TOD is generally calculated using the destination elevation (if available) and the descent speed schedule.

Touchdown and Lift-Off Area (TLOF). The TLOF is a load bearing, usually paved, area at a heliport where the helicopter is permitted to land. The TLOF can be located at ground or rooftop level, or on an elevated structure. The TLOF is normally centered in the FATO.

Touchdown RVR. The RVR visibility readout values obtained from sensors serving the runway touchdown zone.

Touchdown Zone Elevation (TDZE). The highest elevation in the first 3,000 feet of the landing surface.

Tower En Route Control (TEC). The control of IFR en route traffic within delegated airspace between two or more adjacent approach control facilities. This service is designed to expedite air traffic and reduces air traffic control and pilot communication requirements.

TRACAB. A new type of air traffic facility that consists of a radar approach control facility located in the tower cab of the primary airport, as opposed to a separate room.

Traffic Information Service-Broadcast (TIS-B). An air traffic surveillance system that combines all available traffic information on a single display.

Traffic Management Advisor (TMA). A software suite that helps air traffic controllers to sequence arriving air traffic.

Transition Altitude (QNH). The altitude in the vicinity of an airport at or below which the vertical position of an aircraft is controlled by reference to altitudes (MSL).

Transition Height (QFE). Transition height is the height in the vicinity of an airport at or below which the vertical position of an aircraft is expressed in height above the airport reference datum.

Transition Layer. Transition layer is the airspace between the transition altitude and the transition level. Aircraft descending through the transition layer will set altimeters to local station pressure, while departing aircraft climbing through the transition layer will be using standard altimeter setting (QNE) of 29.92 inches of Mercury, 1013.2 millibars, or 1013.2 hectopascals.

Transition Level (QNE). The lowest flight level available for use above the transition altitude.

Turn Anticipation. The capability of RNAV systems to determine the point along a course, prior to a turn WP, where a turn should be initiated to provide a smooth path to intercept the succeeding course and to enunciate the information to the pilot.

Turn WP (Turning Point). A WP that identifies a change from one course to another.

Universal Communications (UNICOM). An air-ground communication facility operated by a private agency to provide advisory service at uncontrolled aerodromes and airports.

User-Defined Waypoint. User-defined waypoints typically are created by pilots for use in their own random RNAV direct navigation. They are newly established, unpublished airspace fixes that are designated geographic locations/ positions that help provide positive course guidance for navigation and a means of checking progress on a flight. They may or may not be actually plotted by the pilot on en route charts, but would normally be communicated to ATC in terms of bearing and distance or latitude/longitude. An example of user-defined waypoints typically includes those derived from database-driven area navigation (RNAV) systems whereby latitude/longitude coordinate-based waypoints are generated by various means including keyboard input, and even electronic map mode functions used to establish waypoints with a cursor on the display. Another example is an offset phantom waypoint, which is a point in space formed by a bearing and distance from NAVAIDs, such as VORs, VORTACs, and TACANs, using a variety of navigation systems.

User Request Evaluation Tool (URET). The URET helps provide enhanced, automated flight data management. URET is an automated tool provided at each radar position in selected en route facilities. It uses flight and radar data to determine present and future trajectories for all active and proposed aircraft flights. A graphic plan display depicts aircraft, traffic, and notification of predicted conflicts. Graphic routes for current plans and trial plans are displayed upon controller request. URET can generate a predicted conflict of two aircraft, or between aircraft and airspace.

Vertical Navigation (VNAV). Traditionally, the only way to get glidepath information during an approach was to use a ground-based NAVAID, but modern area navigation systems allow flight crews to display an internally generated descent path that allows a constant rate descent to minimums during approaches that would otherwise include multiple level-offs.

Vertical Navigation Planning. Included within certain STARs is information provided to help you reduce the amount of low altitude flying time for high performance aircraft, like jets and turboprops. An expected altitude is given for a key fix along the route. By knowing an intermediate altitude in advance when flying a high performance aircraft, you can plan the power or thrust settings and aircraft configurations that result in the most efficient descent, in terms of time, fuel requirements, and engine wear.

Visual Approach. A visual approach is an ATC authorization for an aircraft on an IFR flight plan to proceed visually to the airport of intended landing; it is not an IAP. Also, there is no missed approach segment. When it is operationally beneficial, ATC may authorize pilots to conduct a visual approach to the airport in lieu of the published IAP. A visual approach can be initiated by a pilot or the controller.

Visual Climb Over the Airport (VCOA). An option to allow an aircraft to climb over the airport with visual reference to obstacles to attain a suitable altitude from which to proceed with an IFR departure.

Waypoints. Area navigation waypoints are specified geographical locations, or fixes, used to define an area navigation route or the flightpath of an aircraft employing area navigation. Waypoints may be any of the following types: predefined, published, floating, user-defined, fly-by, or fly-over.

Waypoint (WP). A predetermined geographical position used for route/instrument approach definition, progress reports, published VFR routes, visual reporting points or points for transitioning and/or circumnavigating controlled and/or special use airspace, that is defined relative to a VORTAC station or in terms of latitude/longitude coordinates.

Wide Area Augmentation System (WAAS). A method of navigation based on GPS. Ground correction stations transmit position corrections that enhance system accuracy and add vertical navigation (VNAV) features.

Index